2022 台达杯国际太阳能建筑设计竞赛获奖作品集

Awarded Works from International Solar Building Design Competition 2022

阳光·山水驿

SUNSHINE & STATION IN THE SCENERY

中国建筑设计研究院有限公司 编
Edited by China Architecture Design & Research Group
执行主编：张 磊 鞠晓磊 张星儿
Chief Editor: Zhang Lei Ju Xiaolei Zhang Xinger

中国建筑工业出版社
CHINA ARCHITECTURE & BUILDING PRESS

图书在版编目（CIP）数据

阳光·山水驿：2022台达杯国际太阳能建筑设计竞赛获奖作品集 = SUNSHINE & STATION IN THE SCENERY Awarded Works from International Solar Building Design Competition 2022/中国建筑设计研究院有限公司编；张磊，鞠晓磊，张星儿执行主编. —北京：中国建筑工业出版社，2023.9
　　ISBN 978-7-112-29141-0

　　Ⅰ.①阳… Ⅱ.①中… ②张… ③鞠… ④张… Ⅲ.①太阳能建筑－建筑设计－作品集－世界－现代 Ⅳ.①TU18

中国国家版本馆CIP数据核字（2023）第173195号

　　生态文明是人类文明发展的历史趋势，生态文明建设是关系中华民族永续发展的根本大计。本次作品集以"阳光·山水驿"为主题，聚焦生态环境保护，以尊重自然、顺应自然、保护自然为主旨，发掘具有原创性、前瞻性、兼具深度与广度的创新设计。此次竞赛四川省平武县关坝沟流域自然保护小区科学考察站设计为赛题，进一步探寻人与自然的和谐发展理念，将科学技术与理论实践相结合，推动可再生能源在建筑上的应用，探索生态保护新途径与新思路。本书适用于高校建筑设计相关专业本科生、研究生以及从事相关专业的设计公司、建筑师等人员阅读参考。

责任编辑：唐　旭　吴　绫　张　华
责任校对：王　烨

阳光·山水驿
SUNSHINE & STATION IN THE SCENERY
2022台达杯国际太阳能建筑设计竞赛获奖作品集
Awarded Works from International Solar Building Design Competition 2022
中国建筑设计研究院有限公司　编
Edited by China Architecture Design & Research Group
执行主编：张　磊　鞠晓磊　张星儿
Chief Editor: Zhang Lei　Ju Xiaolei　Zhang Xinger
＊
中国建筑工业出版社出版、发行（北京海淀三里河路9号）
各地新华书店、建筑书店经销
北京雅盈中佳图文设计公司制版
临西县阅读时光印刷有限公司印刷
＊
开本：787毫米×1092毫米　1/12　印张：32$\frac{2}{3}$　字数：983千字
2023年10月第一版　2023年10月第一次印刷
定价：218.00元
ISBN 978-7-112-29141-0
　　　（41868）
版权所有　翻印必究
如有内容及印装质量问题，请联系本社读者服务中心退换
电话：（010）58337283　QQ：2885381756
（地址：北京海淀三里河路9号中国建筑工业出版社604室　邮政编码：100037）

建设生态文明，关系人民福祉，关乎民族未来。习近平总书记深刻指出："建设生态文明、推动绿色低碳循环发展，不仅可以满足人民日益增长的优美生态环境需要，而且可以推动实现更高质量、更有效率、更加公平、更可持续、更为安全的发展，走出一条生产发展、生活富裕、生态良好的文明发展道路。"本次竞赛以建设保护区科学考察站为赛题，参赛团队施展才华，呈现出精彩纷呈的竞赛作品，探索人与自然的和谐共处方式，为维持生态系统稳定探索新的途径。

感谢台达集团资助举办2022台达杯国际太阳能建筑设计竞赛！

谨以本书献给致力于生态文明建设与绿色低碳发展的同仁们！

Building an ecological civilization is related to the well-being of the people and the future of the nation. General Secretary Xi Jinping has profoundly pointed out that "building an ecological civilization and promoting green, low-carbon and circular development can not only meet the people's growing expectation for a beautiful eco-environment, but also promote the realization of higher-quality, more efficient, fairer, more sustainable and safer development, and lead out of a civilized path of development with development in production, a rich life and a sound ecology. " The competition was based on the theme of building a scientific research station in a protected area, and the participating teams showed their talents and presented exciting competition entries, exploring ways to harmonize the coexistence of man and nature and exploring new ways to maintain the stability of the ecosystem.

Thank Delta Electronics for sponsoring the International Solar Building Design Competition 2022!

This publication is dedicated to our fellow researchers and professionals who are committed to the advancement of ecological civilization and green, low-carbon development.

目 录
CONTENTS

阳光·山水驿　SUNSHINE & STATION IN THE SCENERY

2022台达杯国际太阳能建筑设计竞赛回顾
General Background of International Solar Building Design Competition 2022

2022 台达杯国际太阳能建筑设计竞赛评审专家介绍
Introduction to Jury Members of International Solar Building Design Competition 2022

获奖作品　Prize Awarded Works　　001

综合奖·一等奖　General Prize Awarded · First Prize

光·影·巷　Light · Shadow · Lane　　002

综合奖·二等奖　General Prize Awarded · Second Prize

山间·山现　Roof Hiding in the Mountain　　006

光·构　Light and Structure　　010

风之谷　Valley of the Winds　　016

综合奖 · 三等奖 General Prize Awarded · Third Prize

火 · 塘 Fire Place	022
风穿穿斗 Wind Blowing through Column and Tie	026
旭日三重 The Solar Trilogy	032
俯仰之间 · 隐山 Between Looking down and Looking up · Hidden Mountain	036
藏谷翎羽 Falling into the Valley	042
风穿折屋 Sun, Winds and Roofs	048

综合奖 · 优秀奖 General Prize Awarded · Honorable Mention Prize

绵 Continuous	052
林间关坝 · 沐日而生 Immersing in the Forest, Soaking up the Sunshine	056
东风知我欲山行，吹断檐间积雨声 Knowing that I Want to Go Hiking, the East Wind Blows off the Accumulated Rain between Eaves	062
山 · 望 Ridge · Telescopic	068
栖木 · 归尘 Live under Woods · Back to Dust	072
林遇 · 镜缘 Palpitating in the Forest · Forest in the Water	078

揽风·入垣　Touch the Wind	084
光栖坤灵　Growing in Soil	090
一川葳蕤　The Story between River and Trees	096
屋檐下　Under the Roof	102
细胞呼吸　Spring up Like Mushroom	108
聚·生　Gather and Live	114
溯·曦源　Look for the Place Where the Sun Rises	120
"在地""营造"——自然科学考察站设计　Design of Scientific Research Station	126
檐下·村落　Village under Eaves	130
辉隐驿事　Light in the Station	136
沐光山园　A Post Station in Guanba Village in the Sun	142
光遇　Light Encounters	148
山谷·漫步　Roam in the Valley	152
漂浮的森林　Floating Forest	158

综合奖·入围奖　General Prize Awarded · Finalist Award

光·弋林间　Light in the Forest	164
风隐山居　Hide on the Wind	170
林涧·驿　Springdale Station	176
土墙板屋　Earth Wall Board House	182
于山阆水 环蜂绕竹　Guanba Wonderland	188
光之滑梯　The Slide of Light	194
夏时冬语　Summer Blooming Winter Warming	200
流泉·聚旭　Flowing Spring · Warm Storage	206
共生·折驿　Symbiosis Folding Corridors	212
循院·浔源　Circulating Courtyards, Tracing the Source	218
木林·阳光山水驿　Tree Tree Tree	224
轻触·自然——装配式建筑理念下自然保护区科考站设计　Light Touch—Design of Scientific Research Stations in Nature Reserves under the Concept of Prefabricated Architecture	230
曲径通幽　Winding Path Leading to Seclusions	236
拾野　Return to Nature	242

隐山·观水·纳风	Lifting · Experiencing · Ventilating	246
落木·云台	Timber & Terrace	252
河谷林驿	Valley Villas at Guanba	258
凉厅新叙	Description of New Pavilion	264
蜀溪春	Spring Streams	270
生命·守望	Life · Care	276
山·风·舞	Mountain Breeze Dancing	282
处幽篁兮	The Bamboo Forest	288
万物生·归一	Living with Nature	292
画山水序	Preface of Painting Landscape	298
风游光塔	Roaming in the Sun Tower	304
檐下丘语	Hill Language under the Roof	310
山山而川·生生不息	The Circle of Life	316
山与火之歌	A Song of Mountain and Fire	322
山谷间的呼吸	Breathing in the Valley	328
归巢来兮	Back to Nature	332

技术专项奖　Technical Special Award

共生之径——光孕众生　The Path of Symbiosis—Light Gives Birth to Life	338
光泽·栖居　Sunshine · Habitat	344

有效作品参赛团队名单
Name List of All Participants Submitting Valid Works　　350

2022台达杯国际太阳能建筑设计竞赛办法
Competition Brief on International Solar Building Design Competition 2022　　358

2022台达杯国际太阳能建筑设计竞赛回顾
General Backgrond of International Solar Building Design Competition 2022

主题：阳光·山水驿

一、赛题设置

生态文明是人类文明发展的历史趋势，生态文明建设是关系中华民族永续发展的根本大计。坚持生态优先、科学确立绿色发展理念是重中之重。2022台达杯国际太阳能建筑设计竞赛以"阳光·山水驿"为题，以四川省平武县木皮藏族乡关坝村大熊猫栖息地自然保护区中建设科学考察站为赛题，探索生态保护新途径与新思路。

项目所在的四川省平武县关坝村自然保护区是重要生态保护区，有丰富的生物多样性资源，包括大熊猫、金丝猴等数十种珍稀动植物，2021年入选联合国《生物多样性公约》缔约方大会第十五次会议"生物多样性100+全球典型案例"。赛题将深入探寻人与自然和谐发展的理念，推动可再生能源在建筑上的应用，让科技惠及民生，协助推进环境保护与经济发展。

二、竞赛启动

2022年3月26日，2022台达杯国际太阳能建筑设计竞赛在北京启动。竞赛采用线上启动的形式，高效、广泛地传播竞赛理念，让更多感兴趣的网友们参

Theme: Sunshine & Station in the Scenery

I. Competition Preparation

Ecological conservation is a historical trend in the development of human civilization, and the development of ecological conservation is a fundamental plan for the sustainable development of the Chinese nation. It is of utmost importance to prioritize ecological conservation and the concept of green development. The International Solar Building Design Competition 2022, themed "Sunshine & Station in the Scenery", sets the subject as a scientific research station in the Giant Panda Habitat Nature Reserve in Guanba Village, Mupi Tibetan Township, Pingwu County, Sichuan Province, to explore new ways and ideas for ecological conservation.

The Guanbagou Nature Reserve in Pingwu County, Sichuan Province, where the project is located, is an important ecological reserve with rich biodiversity resources, including dozens of rare species of animals and plants such as giant pandas and golden monkeys. It was selected as one of the "100+ Biodiversity Positive Practices and Actions Around the World" at the 15th meeting of the Conference of the Parties to the Convention on Biological Diversity in 2021. The competition will delve into the concept of harmonious development of man and nature, promote the application of renewable energy in buildings, allow technology to benefit people's livelihoods and help promote environmental protection and economic development

II. Launch of the Competition

On March 26, 2022, International Solar Architecture Design Competition 2022 was launched in Beijing.The competition was still launched via video link, efficiently and widely spreading the concept of the competition and attracting more netizens to participate in the activities of the competition. Liu Zhihong, Vice President of China Construction Technology Consulting Co. Ltd. (CCTC) and Executive Director of Central Research Institute of

赛题场地
Project Site

启动会直播现场
Live from the Launch Event

竞赛海报（中、英文版）
Live from the Launch Event

与到竞赛的活动中。中国建设科技集团股份有限公司副总裁、中央研究院执行院长刘志鸿，中国建筑设计研究院有限公司董事、总经理、党委副书记马海，中国建筑设计研究院副总建筑师仲继寿，国家住宅与居住环境工程技术研究中心主任张磊，竞赛冠名单位台达集团创办人暨荣誉董事长郑崇华等嘉宾参加了本次竞赛的线上启动仪式，共同启动新一届竞赛。

三、竞赛宣传

自 2005 年第一届竞赛举办以来，竞赛组委会已先后前往清华大学、天津大学、东南大学、重庆大学、山东建筑大学等 60 多所建筑院校开展国际太阳能建筑设计竞赛巡讲活动，受到了高校师生的积极响应和好评。

2022 年 9 月 27 日，国际太阳能建筑设计竞赛组委会走进南京工业大学，与师生们围绕国际太阳能建筑设计竞赛进行了交流。

同时，竞赛组委会联合国家住宅与居住环境工程技术研究中心低碳建筑技术研究所，建立"迈向产能建筑系列云上讲堂"，组织国内知名专家向全社会传播建筑低碳技术发展与实践成果。内容涵盖了太阳能建筑技术应用趋势和现状、历届

CCTC; Ma Hai, Director, General Manager, and Deputy Secretary of the Party Committee of China Architecture Design & Research Group (CADG); Zhong Jishou, Deputy Chief Architect of CADG; Zhang Lei, Director of China National Engineering Research Center for Human Settlements and Zheng Chonghua, Founder and Honorary Chairman of Delta Electronics, attended the online launch of the competition and launched the new competition together.

III. Promotion Publicity

Since the first competition was held in 2005, the competition organizing committee has visited architecture departments of more than 60 universities, such as Tsinghua University, Tianjin University, Southeast University, Chongqing University, and Shandong Jianzhu University, to promote International Solar Building Design Competition, which has received positive response and praises from teachers and students.

线上宣讲会海报
Posters of Online Lectures

竞赛获奖作品分析和本届竞赛介绍。通过云上讲堂，师生们对竞赛和建筑技术有了更深入的了解，激发了参赛团队的设计灵感，对太阳能建筑应用技术进行了创新性的思考。在本届竞赛期间，"迈向产能建筑系列云上讲堂"邀请中国科学技术大学季杰教授、上海交通大学代彦军教授、湖南大学彭晋卿教授等领域专家，开展包括太阳能建筑一体化设计、建筑节能技术综合应用、建筑能效提升等内容的云上讲堂，提高了竞赛关注群体和观众的知识积累及相关技术素养的传播，使竞赛宣讲会成为太阳能建筑领域行业知识、技术、理念的重要科普和交流平台。

四、媒体宣传

组委会自竞赛启动以来通过多渠道开展媒体宣传工作，包括：竞赛官方网站

On September 27, 2022, the organizing committee of the International Solar Architecture Design Competition visited Nanjing Tech University and talked with teachers and students there about the competition.

In addition, the organizing committee and the Low Carbon Building Research Institute of China National Engineering Research Center for Human Settlements launched the "Online Lecture—Towards Productivity". They invited renowned domestic experts to promote the results of low carbon technology development and practice in construction to the whole society. They talked about the trends and status of solar building technology applications and the analysis of previous winning entries and then introduced this year's competition.

竞赛网站
Official Website

竞赛信息网络发布 1
Online Competition Information Release Ⅰ

After the lecture, students and teachers gained a deeper understanding of the solar architecture design competition and energy-saving technology. During this competition, participants were inspired to have more innovative ideas of technology applied to solar energy buildings. Besides, experts, including Prof. Ji Jie, Chair, University of Science and Technology of China; Prof. Dai Yanjun, Shanghai Jiao Tong University; and Prof. Peng Jinqing, Hunan University, were invited to conduct lectures via video, covering integrated solar building design, comprehensive application of building energy-saving technologies, and improvement of building energy efficiency. These activities, therefore, improved people's knowledge and promoted relevant technological literacy. As a result, the lecture has become an important platform for science popularization and communication of knowledge, technology, and ideas about solar energy buildings.

IV. Media Publicity

Since the launch of the competition, the organizing committee has carried out media publicity through various channels. For example, it reported the progress of the competition in real-time on the official website of the competition (bilingual) and popularized knowledge of solar architecture. Keywords can be searched on Baidu, which facilitated the public to log in to the official website. Over 50 domestic and international websites, including Xinhua Net, Tencent, Sina, AP, PRN, and CNET Japan, reported or put a link to the competition. The organizing committee contacted many foreign universities and media and released the information and process about the competition. The organizing committee also contacted many foreign universities and media to release information and news about the competition. Processes of the competition and online lectures were posted on the WeChat official account and Weibo. The introduction of cases and competition files were also provided for downloading. These measures increased the influence of the competition and the technical ability of the participating teams.

竞赛信息网络发布 2
Online Competition Information Release Ⅱ

（双语）实时报道竞赛进展情况并开展太阳能建筑的科普宣传；在百度设置关键字搜索，方便大众查询，从而更快捷地登录竞赛网站。在新华网、腾讯网、新浪网、美联社、德国财经网、CNET 日本等 50 余家国内外网站上报道或链接了竞赛的相关信息；组委会与多所国外院校和媒体取得联系并发布竞赛信息与动态。通过微信公众号、微博实施发布竞赛进展、云讲堂预告等动态，并提供竞赛相关资料下载与案例介绍等，有效地提高竞赛的影响力及参赛团队的技术能力。

五、竞赛注册及提交情况

本次竞赛的注册时间为 2022 年 3 月 26 日至 2022 年 8 月 15 日，共 654 组团队通过竞赛官网完成注册，其中，有包括来自德国、加拿大、西班牙等境外注册团队 10 组。截至 2022 年 9 月 30 日 24 时，竞赛组委会收到有效参赛作品 171 份。

六、作品初评

2022 年 10 月 1~21 日，组委会组织评审专家对通过形式筛查的全部有效作品开展初评工作。专家根据竞赛办法中规定的评比标准对每一件作品进行评审，经过合规性审查及初期评审成绩排名，选择分数前 100 名的作品进入中期评审。

V. Registration and Submission

Registration for this competition was open from March 26 to August 15, 2022. 654 teams registered through the competition website, including 10 teams from other countries such as Germany, Canada and Spain. By September 15, 2022, the competition organizing committee had received 171 valid entries.

VI. Preliminary Evaluation

From October 1 to 21, 2022, the Organizing Committee carried out the preliminary evaluation of all valid entries. Experts evaluated each entry according to the appraisal standard in the competition specification. After the compliance review, experts gave preliminary evaluations to those eligible entries and ranked them according to their scores. The top 100 entries were selected for the mid-term evaluation.

From October 24 to November 4, 2022, experts reviewed the 100 entries in the mid-term evaluation. After the strict evaluation, the top 60 scorers were awarded the general prize (First Prize, Second Prize, Third Prize, Honorable Mention Prize and Finalist Award), and the top 15 entered the final inquiry.

2022年10月24日至11月4日，组委会组织评审专家对进入中期评审的作品开展评审工作。经过专家组的严格评审，得分前60名获得综合奖（一、二、三等奖，优秀奖，入围奖），分数前30名将进入第二阶段评审，专家组经过二次评审和打分，前15名进入到现场答辩环节。

七、作品终评

竞赛终评会采用线上学生答辩与专家组集中研讨的复合形式，不仅让学生更好地展示作品设计内容，也能够使评审专家综合评价竞赛作品。2022年12月14日上午，进入决赛的15个团队在线上通过图像、汇报与视频等多元方式展示阐述作品，并回答评审专家组提问。专家们在关注作品设计与技术应用创意的同时，也注重设计与技术应用的适用性和可操作性。下午，崔愷院士、杨经文博士等九位国际评审专家通过视频会议的方式连线，历经3轮讨论和评选，最终选出竞赛一等奖1项、二等奖3项、三等奖6项、优秀奖20项、入围奖30项、技术专项奖2项，共计62项综合奖作品。

VII. Final Evaluation

The final evaluation of this year's competition was held online in the form of students´ defense and experts´ panel discussions. In this way, students better presented their designs, and experts could evaluate those entries comprehensively. On the morning of December 14, 2022, the 15 finalists presented their works online through images, texts, speeches, and videos, and answered questions from the judging panel. In the final evaluation, feasibility and operability were given the same importance as the design and technological ideas. On the afternoon of December 14, 2022, nine international experts, including Academician Cui Kai and Dr. Yang Jingwen, held a video conference. After three rounds of evaluation and discussion, the panel selected 1 First Prize, 3 Second Prizes, 6 Third Prizes, 20 Honorable Mention Prizes, 30 Finalist Awards and 2 Prizes for Technical Excellence Works, totaling 62 General Prizes.

终评专家组与答辩师生合影
Photo of the Final Evaluation Panel, Students and Teachers

线上终评会专家讨论
Expert Discussion at the Online Final Evaluation Meeting

2022台达杯国际太阳能建筑设计竞赛评审专家介绍
Introduction to Jury Members of International Solar Building Design Competition 2022

评审专家
Jury Members

杨经文，马来西亚汉沙杨建筑师事务所创始人、2016梁思成建筑奖获得者
Mr. King Mun YEANG: Founder of T. R. Hamzah & Yeang Sdn. Bhd (Malaysia), 2016 Liang Sicheng Architecture Prize Winner

Deo Prasad，澳大利亚科技与工程院院士、澳大利亚勋章获得者、澳大利亚新南威尔士大学教授
Mr. Deo Prasad: Academician of the Australia Academy of Technological Sciences and Engineering, Winner of the Order of Australia, and Professor of University of New South Wales, Sydney, Australia

Peter Luscuere，荷兰代尔伏特大学建筑系教授
Mr. Peter Luscuere: Professor of Department of Architecture, Delft University of Technology

陈绍彦，新加坡CPG集团首席创新官、新加坡建筑师注册局主席
Mr. Chen Shaoyan: Group Chief Innovation Officer of CPG Corporation Pte Ltd; President, Board of Architects Singapore

崔愷，中国工程院院士、全国工程勘察设计大师、中国建筑设计研究院有限公司总建筑师
Mr. Cui Kai: Academician of China Academy of Engineering, National Engineering Survey and Design Master and Chief Architect of China Architecture Design & Research Group (CADG)

仲继寿，中国建筑设计研究院有限公司副总建筑师、国家住宅科技产业技术创新战略联盟秘书长
Mr. Zhong Jishou: Chief Commissioner of Special Committee of Solar Buildings, CRES, and Deputy Chief Architect of CADG

宋晔皓，清华大学建筑学院建筑与技术研究所所长、教授、博士生导师，清华大学建筑设计研究院副总建筑师
Mr. Song Yehao: Director, Professor and Doctoral Supervisor of Institute of Architecture and Technology, School of Architecture, Tsinghua University, and Deputy Chief Architect of Architectural Design and Research Institute of Tsinghua University

钱锋，全国工程勘察设计大师，同济大学建筑与城市规划学院教授、博士生导师，高密度人居环境生态与节能教育部重点实验室主任
Mr. Qian Feng: National Engineering Survey and Design Master, Professor and Doctoral Supervisor of College of Architecture and Urban Planning Tongji University (CAUP), Director of Key Laboratory of Ecology and Energy-saving Study of Dense Habitat (Tongji University), Ministry of Education

黄秋平，华东建筑设计研究总院总建筑师
Mr. Huang Qiuping: Chief Architect of East China Architectural Design & Research Institute (ECADI)

冯雅，中国建筑西南设计研究院顾问总工程师、中国建筑学会建筑热工与节能专业委员会副主任
Mr. Feng Ya: Chief Engineer of China Southwest Architectural Design and Research Institute Corp. Ltd. and Deputy Director of Special Committee of Building Thermal and Energy Efficiency, Architectural Society of China（ASC）

获奖作品

Prize Awarded Works

光・影・巷
——四川关坝科学考察站

Design Concept

设计旨在回应场地的气候、环境、功能以及文脉。通过对当地气候和环境性能需求的分析，我们确定了建筑的总体布局，并通过适应性的体形和缓冲空间回应气候的问题，同时结合场地的高差变化，营造出丰富、错落有致的空间序列。

在节能策略上，以空间调节为理念，采取被动优先、主动优化的方式，从形体、表皮以及构造三个层面进行设计与应对。为了充分利用太阳辐射来解决冬季热舒适问题，通过建筑形态控制来获得更多有利的朝向感，并结合温室空间等设置蓄热体等被动式策略提升室内舒适性。材料上则主要采用当地易于获取且造价低廉的木材以及夯土以减少整个建筑生命周期的碳排放。在主动式应用方面，通过计算得出最佳的太阳能光伏板角度，并将其合理地与建筑形态相结合，确保了其功能和美学方面的统一。

The design aims to respond to the climate, environment, function and context of the site. Based on the analysis of the local climate and the requirements of environmental performance, we determined the overall layout of the building, and responded to the climate through the adaptive shape and buffer zone, while creating a rich and well-proportioned spatial sequence combined with the variation of the site's elevation.

In terms of energy strategy, with the concept of adaptive space and the principle that passive strategies are preferred and active strategies optimize result, the design are carried out from three levels: form, facade and details. In order to make full use of solar radiation to solve the problem of thermal comfort in winter, the building's form aims obtain more favorable orientation, and the passive strategies such as greenhouse, heat accumulator are set in combination with the exhibition space to improve indoor comfort. Locally available and inexpensive wood and rammed earth are used to reduce carbon emissions throughout the building's life cycle. For the active application, the optimal Angle of solar photovoltaic panels is calculated and rationally combined with the building form to ensure the unity of its functional and aesthetic aspects.

Form Analysis

综合奖・一等奖
General Prize Awarded・First Prize

注 册 号：100676
项目名称：光・影・巷
　　　　　Light・Shadow・Lane
作　　者：马志强、王凌豪、邹立君
参赛单位：浙江大学建筑设计研究院、新加坡国立大学、华东建筑设计研究院

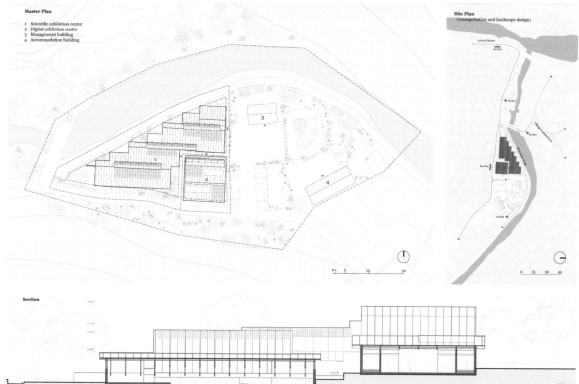

光·影·巷
——四川关坝科学考察站

专家点评：

该作品建筑总体布局体现了对当地气候和环境需求的呼应，结合场地高差变化，空间序列丰富有致，缓冲空间的设置与建筑功能结合良好。采用当地木材以及夯土等传统技术用现代的建筑语言进行建构，有效减少整个建筑生命周期的碳排放。光塔的设计比较巧妙，解决冬季室内日照不足的问题，以及在夏季利用热压通风加强室内空气流动。太阳能与建筑一体化程度高，建筑室内环境及建筑能耗分析较为深入。

The general layout of the building adapts local climate and environment. According to changes in site elevation, it has a diversified and appropriate space sequence and integrates transition spaces and building functions. Local timber and traditional techniques such as rammed earth are used to build this modern architecture, effectively reducing carbon emissions throughout the building's life cycle. The design of the light tower is ingenious, solving the problem of insufficient indoor sunlight in winter and enhancing indoor airflow in summer with thermal natural ventilation. Solar energy and the building are highly integrated. The building's indoor environment and energy consumption are analyzed in depth.

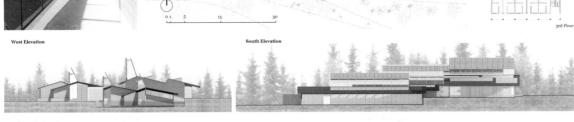

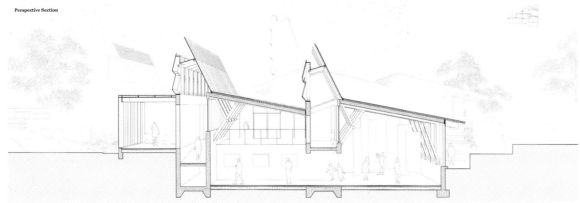

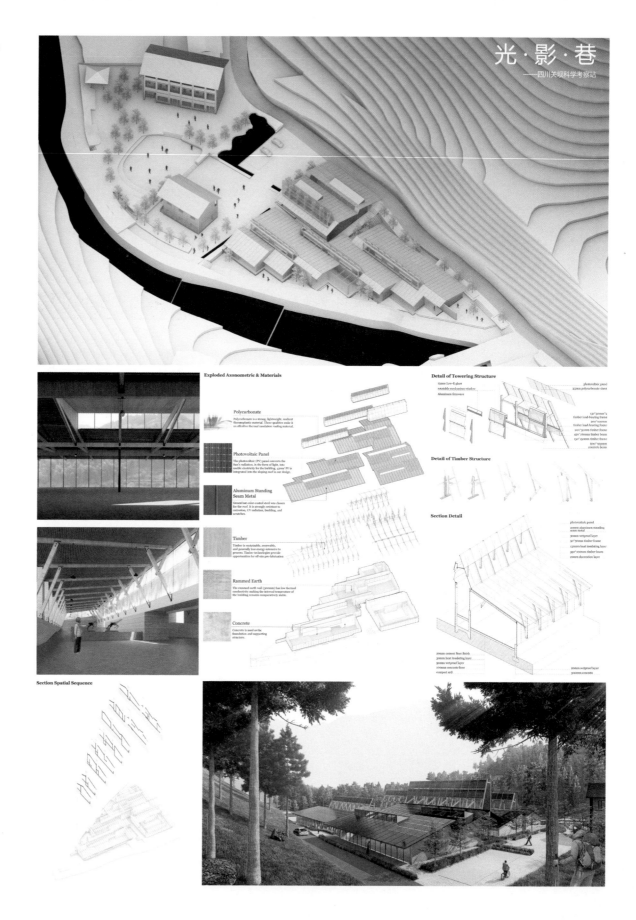

光·影·巷
—— 四川关坝科学考察站

Climate Analysis

The site is located in Pingwu County of Sichuan Mianyang City, which is located in the west edge of **hot summer and cold winter area** of our country and close to cold area. It is **cool and rainy in summer and cold and snowy in winter**, which requires special attention to insulation, heating and ventilation. At the same time, the design area is also located in the valley on both sides of the north and south, with a relatively unique microclimate environment.

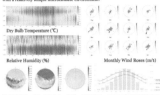

Dry Bulb Temperature (°C)
Relative Humidity (%)   Monthly Wind Roses (m/s)
Total Radiation   Direct Radiation   Diffuse Radiation   Average Precipitation (mm)

According to the visual chart analysis of meteorological data, the local climate environment needs to **enhance ventilation in summer** to take away moisture, while in winter, it is necessary to **make use of the heat gained by the sun** as much as possible, **improve the thermal performance**, and reduce indoor heat loss.

Passive Strategy Analysis

- Comfort time 1.73%
- Evaporative Cooling 0.28%
- Mass + Night Vent 0.28%
- Occupant Use of Fans 0.31%
- Capture Internal Heat 20.3%
- Passive Solar Heating 24.1%

The following passive strategies are suitable for the mediation of building environment in Guanba Village: The **Capture Internal Heat** and the **Passive Solar Heating**, which could enhance 20.3% and 24.% comfortable time in the whole year respectively.

Passive Design Strategy of Climate

In summer, the design can use a **mixed ventilation strategy combining wind pressure and heat pressure** to take away indoor heat and moisture, and the light reflected by the light pipe is also more suitable for the display of exhibits in the exhibition hall. In winter, the **greenhouse effect** is used to heat the air in the snoot, so as to relax and heat the core space.

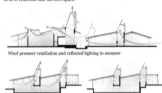

Wind pressure ventilation and reflected lighting in summer
Heat press ventilation in summer   greenhouse effect in winter with reflected light

Radiation Strategy

Direct Sun Hours   Incident Radiation

Due to the blocking of the mountain on the south side, the direct sun hours of the site shows a downward trend from north to south. In the design, we placed the solar panels on the north side of the sun as much as possible, and improved the insulation performance of the building on the north side.

Wind Simulation

Cold lane:Ventilation by wind pressure mechanism

Section 1-1: Ventilation by hot pressing mechanism

Section 2-2: Ventilation by hot pressing mechanism

Due to the existence of the mountains, there are much east-west winds in the site. By opening the doors and windows on the east and west sides, the wind pressure and the "cold lane" effect could be used to dissipate heat and dehumidify. In transition seasons,vertical ventilation based on thermal pressure will have the same effect.

Analysis of Active Energy Saving Strategies

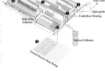

1. The toilets are flushed with rainwater.
2. The rooms are lit by daylight,which is reflected by light cube in the skylight.
3. Through 60 meters deep boreholes, the warmth from the underground is brought to the surface and used to heat the building.
4. The green roofs keep the building warm in winter and cool in summer.
5. Windows and doors are fitted with triple glazing.
6. Rammed erath is used to enhance the thermal-insulation performance of the wall.
7. Under the floor of the core space, there is an integrated pipeline that can be used for indoor heating in winter using a ground source heat pump.

Simulation Validation

From "**Air-conditioning**" to "**Space-conditioning**": The starting point of the design is based on the conditioning of "space" to promote the green building design from "**measure-oriented to effect-oriented**", and from "**technology-oriented to design-oriented**". Compared with the traditional "air conditioning", it will consume less energy, and at the same time realize the local creation and experience with the space itself.

The design realizes the regulation of the space on both sides by placing staggered "**buffer spaces**" in the horizontally arranged functional volumes. The towering "**light tube**" shape can not only bring soft reflected light, but also promote heat pressure and wind pressure ventilation through the opening and closing of skylights and east-west window sashes. In the cold season, it also acts as a "**greenhouse**" to form thermal insulation.

Contrast Model

In the design, we mediate the different external environment through the "**virtual & real**" and the "**open & close**" of the buffer space, and establish contrast experimental group for simulation and validation.

#1 Model without skylight   #2 Model with skylight

Energy Load (without skylight)

For the control group without skylight, the annual cooling load was 84kWh/m², heating load was 44kWh/m², and lighting load was 27Kwh/m². Compared with the skylight group, due to the lack of appropriate regulation, it will consume more energy.

Energy Load (with skylight)

The skylight group experienced a **23%** reduction in load throughout the year. In summer, you can open the shading curtains during the day and open the skylight at night to enhance ventilation. In winter, heat loss can be reduced through the greenhouse effect.

Active & Passive Energy Saving Strategy System

In this design, we selected appropriate technologies and strategies according to the local climate and environmental performance needs to create a green building with a sense of place. And the design concept has changed from the original "**energy consuming building**" to "**zero energy building**" and even "**positive energy building**". Adhering to the concept of "**passive first, active optimization**", we will be active technology combined. The passive strategy mainly includes three aspects: **shape**, **skin** and **construction**. Active technologies include solar energy, rainwater recycling, ground source heat pump and efficient lighting system.

Indoor Illuminance Analysis

Indoor Illuminance (With skylight)   Indoor Illuminance of plan (With skylight)
22 Jun 12:00   22 Jun 12:00

Indoor Illuminance (Without skylight)   Indoor Illuminance of plan (Without skylight)
22 Jun 12:00   22 Jun 12:00

The indoor illumination of the two control groups was simulated at 12 noon on the summer solstice. The skylight group could reflect the light reflected many times into the core space on both sides through the high side window, creating a softer and brighter environment. When special display is needed, the light can be shielded by shading curtain.

Spatial Climate Gradient

According to the functional and environmental performance requirements of different Spaces inside the building, a reasonable spatial climate gradient is adopted. Different Spaces are wrapped in layers like "**Temperature Onions**", and the spatial organization itself realizes the environmental regulation of the main used space. We selected multiple measuring points in a typical space to simulate the indoor operating temperature, and obtained the average value in chart.

■ Functional space with strict environmental mediation
■ Public space without strict environmental mediation
++ Climate buffer zone
● Test point 1, 2, 3
● Test point 4, 5, 6

Indoor OT at Summer Solstice

Through comparative study, the sun radiation can be effectively reduced by opening the shading curtain in the light cylinder and the north-south opening window, and the indoor temperature can be reduced by using the wind pressure ventilation. At night, the skylight can be opened to vent the indoor heat.

Indoor OT at Winter Solstice

In winter, the glass lighting tube acts as a greenhouse, wrapping outside the functional space like a cotton coat, reducing heat loss inside the main space and reducing heating energy consumption. Meanwhile, improve the indoor illumination.

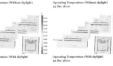

综合奖・二等奖
General Prize Awarded・
Second Prize

注　册　号：100929
项目名称：山间・山现
　　　　　Roof Hiding in the Mountain
作　　　者：何勇杰、徐欣然、魏钰丰、
　　　　　刘孙卓然
参赛单位：北京工业大学
指导教师：陈　喆

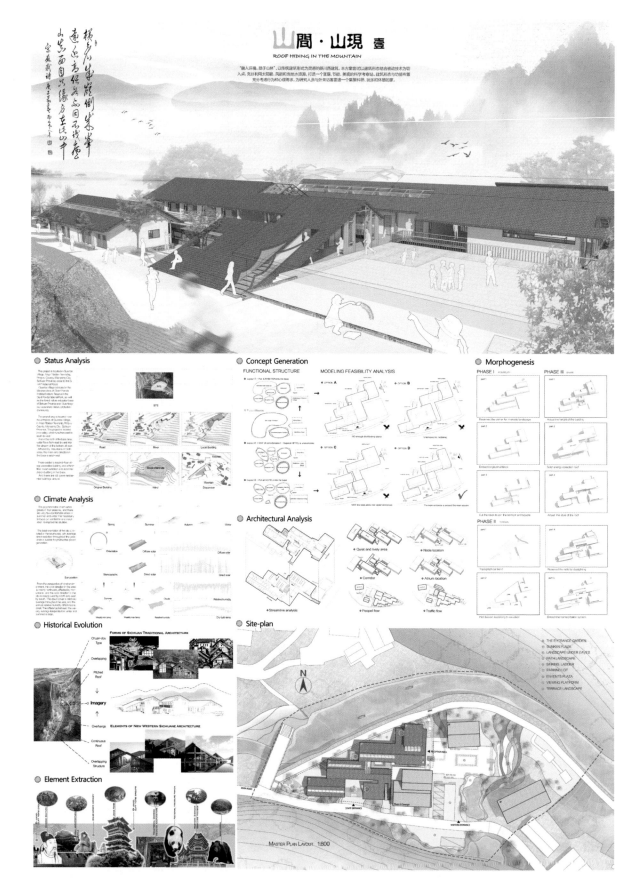

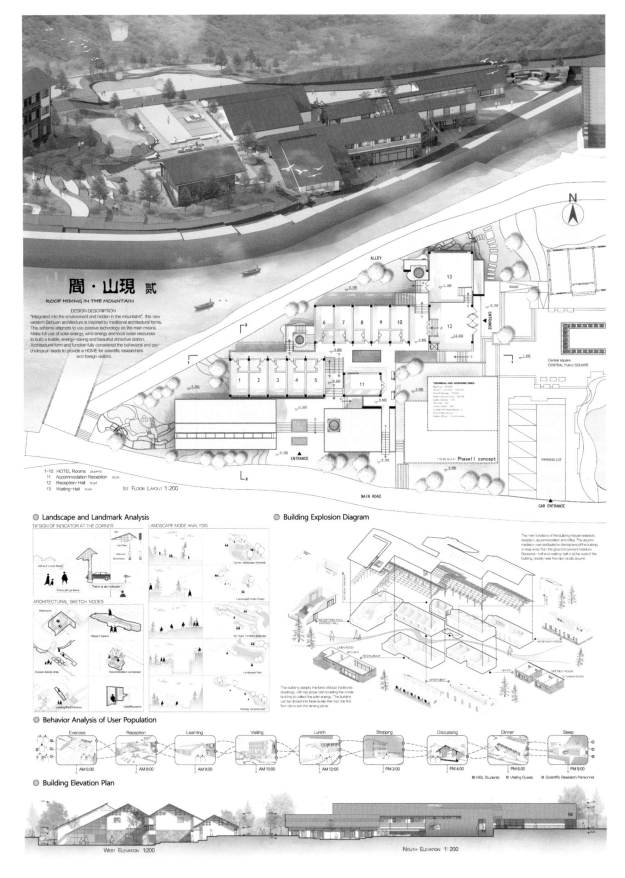

专家点评：

本方案以建筑形态结合被动技术为切入点，充分考虑了太阳能、风能和当地水资源的利用，同时充分考虑了所选基地的环境，利用地势尝试创造一个和谐丰富的展示性建筑空间，将规划、建筑形象与气候特点、文化特点及技术应用较好地结合在一起，建筑形象简洁且具有特色，丰富的屋面元素处理，使建筑融入了山的怀抱；总体规划布局合理、建筑空间组合既丰富又体现了各功能分区的特点；采用了太阳能利用技术，并将技术与建筑构件处理融为一体；技术应用具有可实施性；设计表达清晰、充分、深入。

This proposal takes the architectural form combined with passive design as the starting point and fully considers the use of solar energy, wind energy and local water resources. It also considers the chosen base's environment, using the terrain to create a harmonious and abundant architectural space for exhibition. Planning, architectural image, climate features, cultural characteristics and technology applications are properly integrated. The building's image is simple and distinctive, with a rich treatment of roofing elements, which integrates the building into the mountain. The general planning is well laid out, and the rich combination of building spaces reflects the characteristics of each functional partition. Solar energy technology is adopted. The technology is integrated with the treatment of building elements and is implementable. The design is clear, well-expressed, and in-depth.

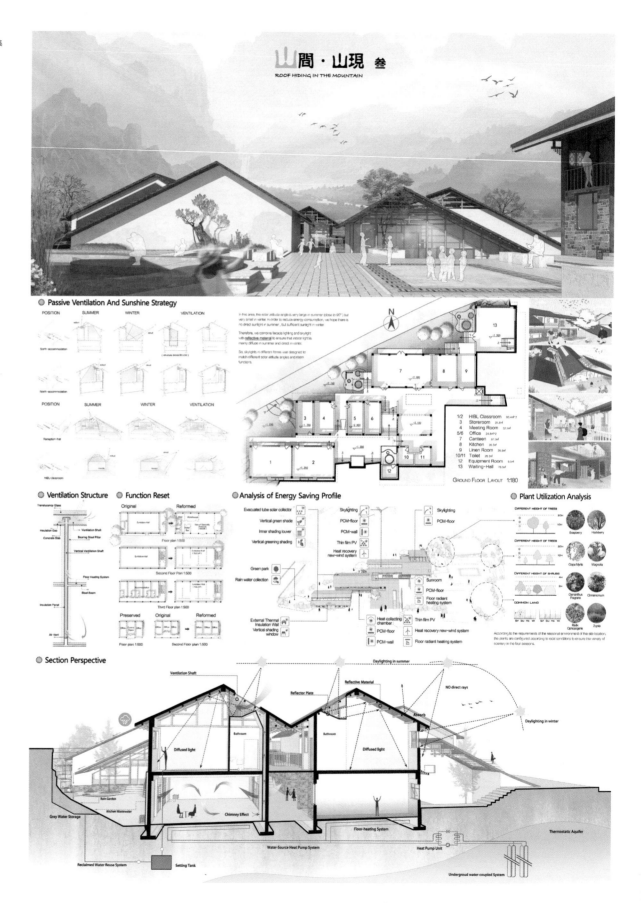

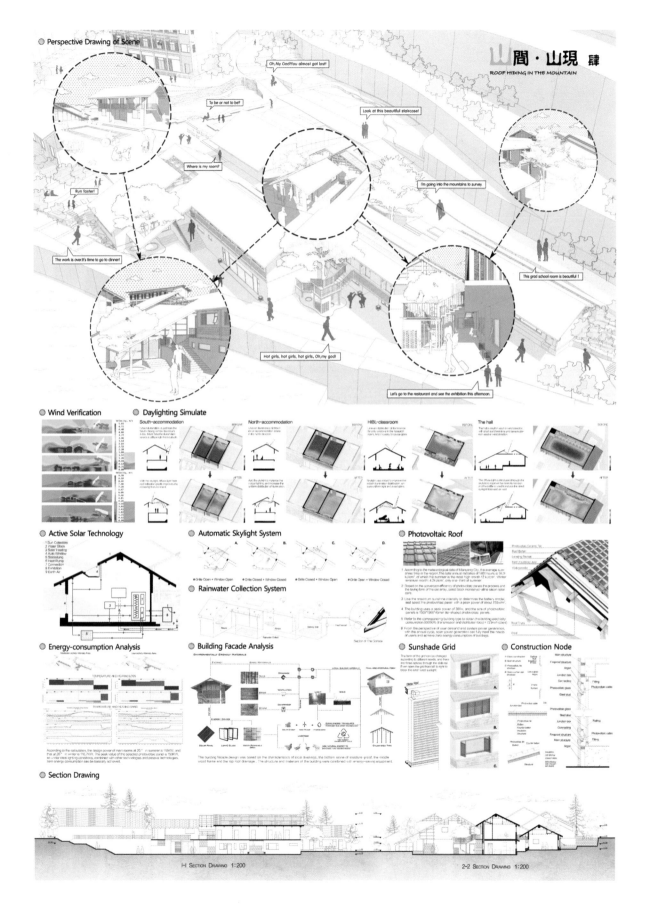

综合奖·二等奖
General Prize Awarded · Second Prize

注 册 号：100961
项目名称：光·构
　　　　　Light and Structure
作　　者：王占阳、张志达、李　阳、
　　　　　许人天
参赛单位：米兰理工大学
指导教师：Angelo Lorenzi

Design Description / 设计说明

1. 尊重场地，充分利用原有地形高差，建立规划系统，以达到最小土方量挖掘。
2. 尽量保留原场地树木及水景系统，将其纳入规划之中，达到微环境调节的作用。
3. 就地取材，建筑采用木材、生土、石头、茅草等当地材料作为建筑的主要材料。
4. 设计采用被动式及主动式结合的策略，采用太阳能光伏板、格构空腔天窗、可调节通风装置、特郎勃墙体、地道风系统、太阳能储能系统等可持续设计策略。

1. Respect the site, make full use of the original terrain height difference, and establish a planning system to achieve the minimum amount of earthwork for excavation.
2. Try to retain the original site trees and waterscape system and incorporate them into the planning to achieve the role of micro-environment regulation.
3. Make full use of local materials with contemporary technology, such as wood, raw soil, stone, thatch are used as the main materials for the construction.
4. The design adopts a combination of passive and active strategies, and adopts sustainable design strategies such as solar photovoltaic panels, lattice cavity skylights, adjustable ventilation devices, Trombo walls, tunnel wind systems, and solar energy storage system.

Design Strategy
1. Keep the original trees
2. Little intervention for original landscape and terrain elevation
3. Utilize local materials
4. Sustainable design

Functional Analysis

reception　　toilet　　sales and bar
exhibition　outdoor stage　sales for free
　　　　　　research room　office
　　　　　　　　　　　　　hotel and restaurant

Streamline Analysis

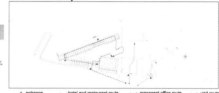

▲ entrance　　- hotel and restaurant route
　　　　　　-- managent office route　-·- visit route

Climate Analysis

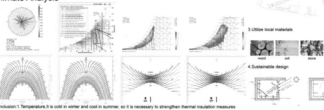

Conclusion: 1. Temperature, It is cold in winter and cool in summer, so it is necessary to strengthen thermal insulation measures.
2. Humidity, with high humidity and ventilation measures are required.
3. Radiation, the best solar angle for photovoltaic panel is between 30°-90°.

Master Plan 1:400

专家点评：

作品总平面顺应基地三角形周边布局，最大限度利用北侧河流及远处山势景观，中间利用地形自然高差围合成下沉式广场，形成与基地环境相适应的总体设计。

单体建筑设计创新性地采用现代木结构建筑语言，结合建筑功能有机融合建筑的自然采光通风以及太阳能等主被动技术，自然恰当；建筑材料采用木材、夯土墙、毛石以及茅草等自然材料，采用标准化构件技术，容易实施。方案回归建筑设计本源，具有科学考察站建筑性格特点和绿色节能性能。

The general plan of the work follows the triangular layout of the base, making maximum use of the river on the north side and the distant mountain landscape. It uses the natural height difference of the terrain in the middle to enclose a sunken square, forming a general design that is compatible with the base environment.

The work attempts to return to the origin of architectural design. The single building design innovatively adopts the modern wooden structure and organically integrates the natural lighting and ventilation of the building with active and passive technologies such as solar energy, which is naturally appropriate and reasonable. The building has the feature of a scientific research station. It uses natural materials such as timber, rammed earth walls, rough stones and thatch. The use of standardized component technology makes it easy to implement. The scheme returns to the original architectural design and has the character and green energy efficiency of a scientific research station.

Ground Floor 1 : 300

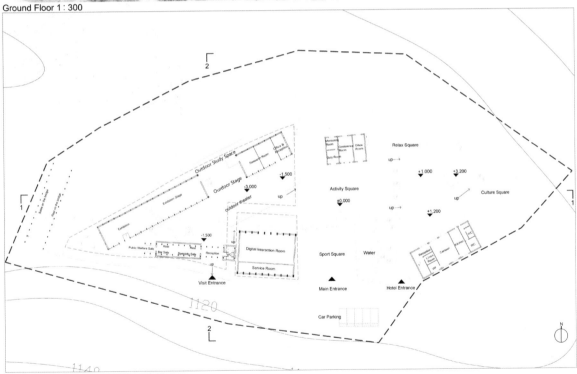

1-1 Section 1 : 300

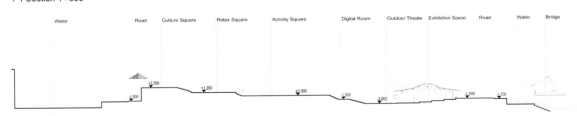

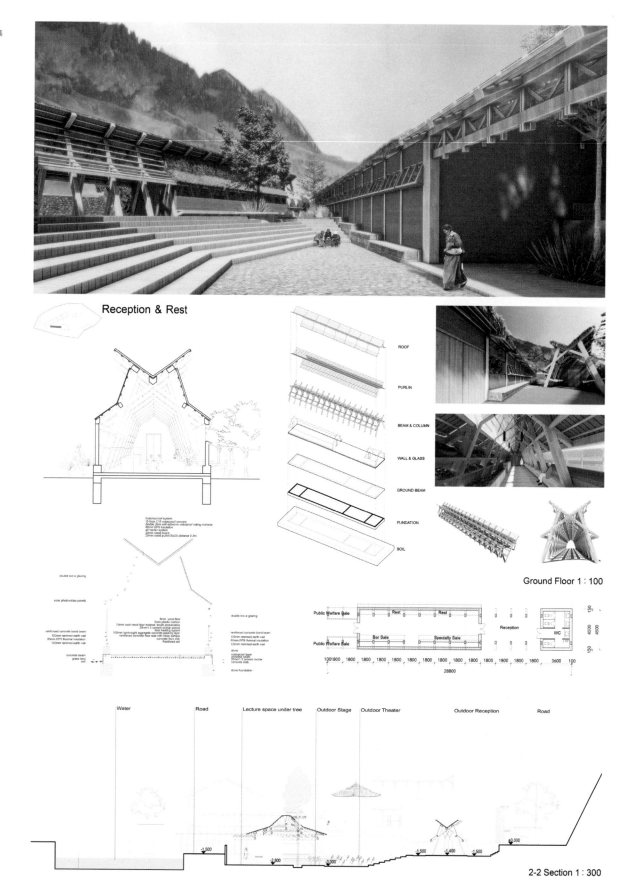

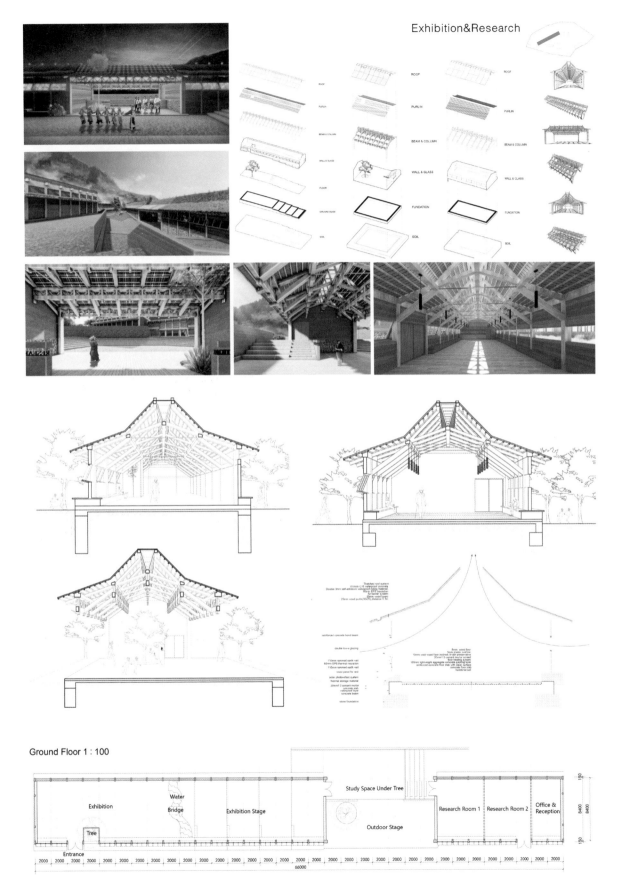

Exhibition&Research

Ground Floor 1 : 100

Digital Interaction

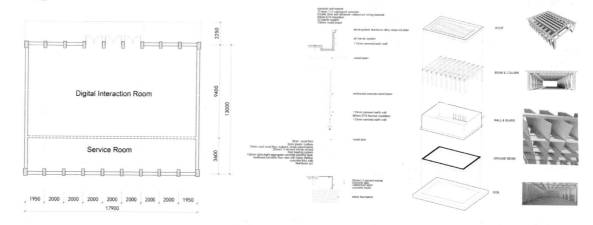

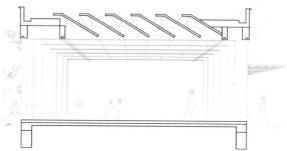

Hotel & Restaurant

Ground Floor 1 : 100

Second Floor 1 : 100

Third Floor 1 : 100

Management Office

Ground Floor 1 : 100

Second Floor 1 : 100

Bridge

Sustainable Technology

Trumbo Wall -(for the hotel south wall reform)

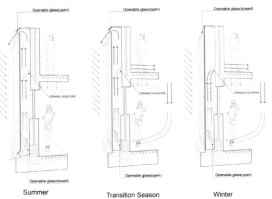

Summer | Transition Season | Winter

Adjustable Ventilation Devices

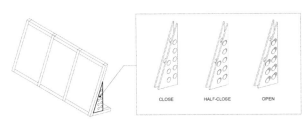

According to different season to adjust the open size of the ventilation hole

Solar Energy Storage System

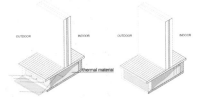

Winter Day-(Collect solar energy from outdoor) | Winter Night-(Release solar energy to indoor space)

Lattice Cavity Skylights

Utilize the triangular cavity space as temperature adjustment

Solar Photovoltaic Panels (for south roof)

According to climate analysis, the best solar angle for photovoltaic panel is between 30°-90°.

Ladybug Analysis for Radiation

Ladybug Analysis for Radiation

Wood Detail

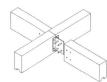

U-value Analysis

For better analysis the rammed earth wall thermal insulation properties, use the Openstudio software, selected different materials for comparison. The rammed earth wall has better insulation performance.

Material	D	λ	R
Rammed earth	11cm	0.76	0.381
EPS	8cm	0.33	1.51
Rammed earth	11cm	0.76	0.381

U=0.36

Material	D	R
Heavyweight concrete	11cm	0.08
EPS	8cm	1.75
Heavyweight concrete	11cm	0.08

U=0.44

Material	D	R
wood siding	4cm	0.44
EPS	8cm	1.25
Lightweight concrete	16cm	0.38
Tile	2cm	0.01

U=0.53

Material	D	R
Plaster	2cm	0.08
EPS	8cm	1.5
Brick	16cm	0.27
Terrazzo	4cm	0.02

U=0.54

Material	D	R
Plaster	2cm	0.12
Brick	16cm	0.08
EPS	8cm	1.5
Terrazzo	4cm	0.02

U=0.54

Marqutte

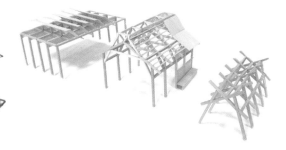

综合奖・二等奖
General Prize Awarded · Second Prize

注 册 号：101058
项目名称：风之谷
　　　　　Valley of the Winds
作　　者：施语林、杨　光、万沐霖、
　　　　　蒋旭亮、胡家皓
参赛单位：南京工业大学、东华大学
指导教师：薛　洁、刘　强、周紫昱

DESIGN SPECIFICATION

此设计以"山谷风"这一显著条件，围合风道空间，着重加强冬季保温，部分进行装配式建造。"风道"改良当地民居对山谷风的利用模式，保留风干储藏功能，促进通风除湿散热，兼作冬季阳光房，调配出适宜的空间感受和季节温度。建筑形体以不同坡向应对受限的山谷条件，配合平面功能集中布置太阳能光伏板系统，整合风能、水能、雨水收集和污水无害化处理措施实现资源再生和能源一体化供给。
风道组成风干储藏、售卖展示、连廊观星等不同空间，人与自然在以风道为界的不同层级下和谐共处。山间风过，且听风吟，这是我们的期盼，也是白熊沟的未来。

This design is based on the remarkable condition of "valley wind", enclosing the air duct space, focusing on strengthening winter insulation, and partially prefabricated construction. The "wind channel" improves the utilization mode of the valley wind in the local residence, retains the function of air drying and storage, promotes ventilation and dehumidification and heat dissipation, and has both a winter sun room and a suitable space feeling and seasonal temperature. The building shape responds to the limited valley conditions with different slopes, and the solar photovoltaic panel system is centrally arranged with the plane function, and the wind energy, hydropower, rainwater collection and sewage harmless treatment measures are integrated to achieve resource regeneration and energy integration.
The wind tunnel consists of different spaces such as air drying storage, sales display, and stargazing in the corridor, and man and nature coexist harmoniously under different levels bounded by the wind tunnel. The wind in the mountains passes and listens to the wind and sounds, which is our expectation and the future of White Bear Valley.

LOCAL LIVING CONDITION

Local villagers use the east-west valley winds throughout the year to design their crop storage space.

CLIMATE ANALYSIS

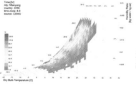

Valley winds prevail from east to west throughout the year.

Wind speeds are high in the valley, and it is often cloudy, which is bound to lead to a lack of sunshine.

Pay attention to heat preservation in winter and heat prevention in summer.

TOTAL DESIGN FRAME

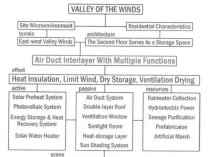

TECHNICAL-ECONOMIC INDICATOR

Land Area 2095 m²
Building Area 2860 m²
Plot Ratio 0.73
Building Height 9.1m

SITE PLAN 1:500

专家点评：

作品以"山谷风"构想为建筑形似创造的源泉，力图通过各种建筑空间，学习并改造当地民居对山谷风的利用模式，促进通风、除湿、散热，建筑形体有意识以不同坡向应对受限的山谷场地条件。此外，设计非常关注冬季保温隔热和供热问题，强调冬季舒适的重要性，而这些都是非常适应当地气候要求的有意义的设计对策。设计还力图表达对于装配式建造的关注，以保证更好的建造品质。除此之外，各种配合平面功能设计的太阳能光伏系统、整合风能、水能、雨水收集和污水无害化处理措施，实现资源再生和能源一体化供给等，均体现了设计的技术含量。

The work uses the idea of the 'valley wind' as a source of the architectural feature, seeking to learn from and adapt the use of the valley wind in local dwellings through various architectural spaces, to promote ventilation and dehumidification and heat dissipation. The building forms were consciously designed to respond to the restricted valley site conditions with different slope orientations. In addition, there is a strong focus on winter insulation and heating, emphasizing the importance of comfort in winter, which is clearly a meaningful design that is very much adapted to the requirements of the local climate. The design also seeks to express a concern for the use of assembled construction methods to ensure a better quality of construction. Besides, in terms of the use of technology, the technical aspects of the design are reflected in the application of various technologies such as solar photovoltaic systems designed to match the floor plan functions, the integration of wind and water energy, rainwater harvesting and harmless sewage treatment measures to achieve resource regeneration and integrated energy supply.

ENERGY STRATEGY

LOGICAL GENERATION OF THE SITE

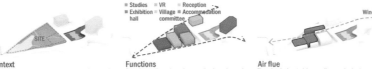

Context
The design includes new construction and renovation projects. The height difference in the first phase of the site is about 3.5 meters.

Functions
Based on the triangular terrain boundary, the building functions are arranged in a progressive manner.

Air flue
The site is located in a valley, and air ducts are inserted into the building according to local conditions.

Skylight
Skylights are set at the top of the building according to the specific needs of the interior, which also enriches the interior experience.

Light
The building is located at the foot of the hill, the sunshine duration is relatively insufficient, so it is appropriate to arrange roof photovoltaic panels.

View
The surrounding scenery is beautiful, and the window sash is arranged in a guided manner to bring the scenery indoors.

ENTRANCE MARKER

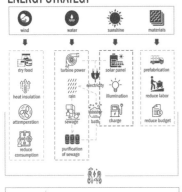

ARCHITECTURAL FORM STRSTEGY

Venue wind | North-south orientation | East-west orientation

Conclusion: By comparing the roof sunshine duration of the north-south and east-west directions, it is found that the east-west roof photovoltaic panel arrangement can convert solar energy more, and at the same time design the rotating body in the appropriate north-south direction to obtain the best south-facing sun.

SITE LAYOUT

Fitness Area

Rest Area

Bamboo Garden

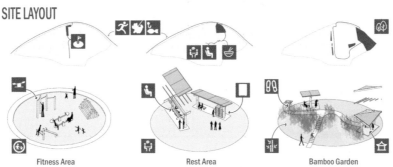

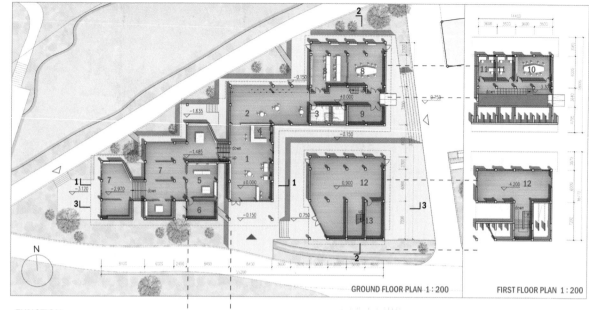

GROUND FLOOR PLAN 1:200 FIRST FLOOR PLAN 1:200

FUNCTION

1. Reception Lobby
2. Waiting Room
3. WC
4. Water Bar
5. Special Local Product Store
6. Storage
7. Exhibition Hall
8. Teaching & Research Office
9. Storage
10. Negotiating Room
11. Office
12. VR Room
13. Ancillary Room
14. Food Storage (experience space)

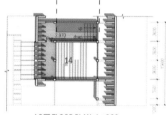

LOFT FLOOR PLAN 1:200

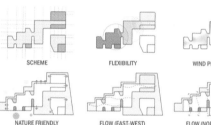

SCHEME · FLEXIBILITY · WIND PRESSURE
NATURE FRIENDLY · FLOW (EAST-WEST) · FLOW (NORTH-SOUTH)

WEST ELEVATION 1:200

REGENERATION

We keep the original functions of the village committee and move the reception center's exhibition to the observation station. So, the reception center only offers meals and accommodation. It means that we keep the relaxing functions together and so do the activity functions.

SECOND FLOOR OF THE RECEPTION CENTER 1:200

THE VILLAGE COMMITTEE

SOUTH ELEVATION 1:200 EAST ELEVATION 1:200 A-A SECTION 1:200

FIRST FLOOR OF THE VILLAGE COMMITTEE 1:200

FIRST FLOOR OF THE RECEPTION CENTER 1:200

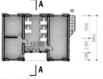

GROUND FLOOR OF THE VILLAGE COMMITTEE 1:200

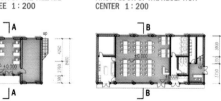

GROUND FLOOR OF THE RECEPTION CENTER 1:200

THE RECEPTION CENTER

NORTH-WEST ELEVATION 1:200 SOUTH-WEST ELEVATION 1:200 B-B SECTION 1:200

AIR DUCT STORAGE

EXHIBITION

STUDIES OFFICE

SPATIAL PROTOTYPE TRANSLATION

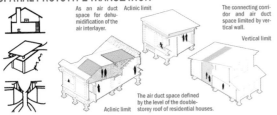

As an air duct space for dehumidification of the air interlayer.

Aclinic limit

The connecting corridor and air duct space limited by vertical wall.

Vertical limit

The air duct space defined by the level of the double-storey roof of residential houses.

Aclinic limit

ANALYSIS OF OPERATION FORMAT

Scientific research station model

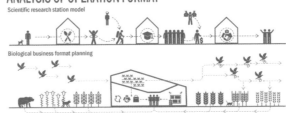

Biological business format planning

AIR DUCT SPACE LIMITED BY DOUBLE ROOF

Functioned as a collection of exhibition hall, sales, storage and stargazing platform.

In summer, it is ventilated by air vents, folding window sashes and sun shades. In winter, it is also used as a sun room through sunshine and active solar technology.

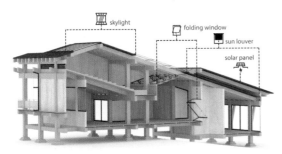

skylight · folding window · sun louver · solar panel

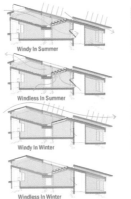

Windy In Summer

Windless In Summer

Windy In Winter

Windless In Winter

In summer, wind is channeled into the interior of the building through the double roofing, which is bounded by open and closed skylights.
In the wind ventilation dehumidification heat dissipation, in the absence of wind by hot pressure pull wind.

In winter, the building pays attention to the insulation effect, thereby closing the skylight and other vents.
Under the direct exposure of the western sunlight, the sun room with double roof and interlayer is formed to exchange heat radiation to other rooms.

AIR DUCT SPACE LIMITED BY THE WALL

For the most typical two-sided and one-corridor layout, the skylight in the corridor is opened to form an open space with ample light and shadow, connecting the research classrooms, auxiliary rooms and office space.

The roof of the building also uses solar panels, skylights, air intake grids and other structural ways to create a suitable use of space.

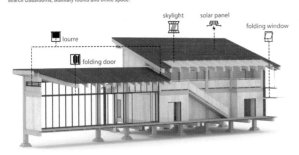

lourre · folding door · skylight · solar panel · folding window

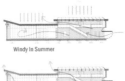

Windy In Summer

Windless In Summer

In summer, the building opens the air inlet and forms an air duct corridor limited by two pieces of wall in the inner corridor to take away the indoor heat.

Windy In Winter

In winter, the building closes the air vents and forms a sunroom on the inner porch through south-facing sunlight and skylights, as well as active solar technology.

Windless In Winter

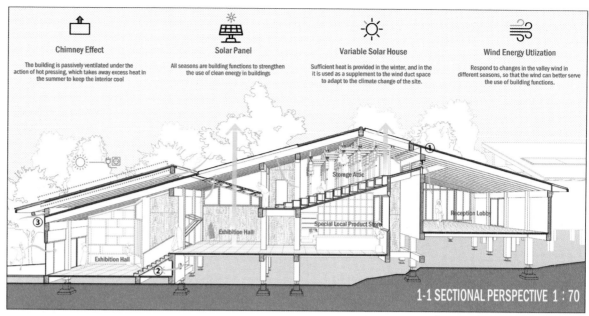

Chimney Effect
The building is passively ventilated under the action of hot pressing, which takes away excess heat in the summer to keep the interior cool.

Solar Panel
All seasons are building functions to strengthen the use of clean energy in buildings

Variable Solar House
Sufficient heat is provided in the winter, and in the it is used as a supplement to the wind duct space to adapt to the climate change of the site.

Wind Energy Utilization
Respond to changes in the valley wind in different seasons, so that the wind can better serve the use of building functions.

1-1 SECTIONAL PERSPECTIVE 1:70

MATERIAL SELECT

Cost of construction members of wood house

Stage	Material	Amount	Average	Total
Material income	Rebar	3120kg	4.17	13010
	Concrete	30m³	830	24900
	Specification wood	78m³	1000	78000
	Particleboard	24m³	2800	67200
	Antiseptic wood	3kg	4000	12000
	Steel connectors	1426kg		26300
	Small connectors	372m³		5700
Transport	Wood freight			
Field use	Power consumption	380kWh	0.8	304
	Scaffold			
Total				227414

TIMBER NODES

Sloped roof beam rafters junction nodes

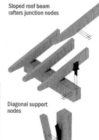

Diagonal support nodes

The column beam joins some nodes

Column base junction node

Concrete foundations

VENTILATED CONSTRUCTION

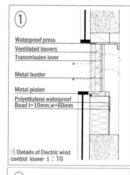

Waterproof press
Ventilated louvers
Transmission lever
Metal border
Metal platen
Polyethylene waterproof Bead t=10mm, w=60mm

① Details of Electric wind control louver 1:10

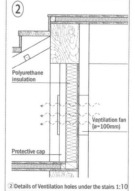

Polyurethane insulation
Ventilation fan (ø=100mm)
Protective cap

② Details of Ventilation holes under the stairs 1:10

③ WALL SECTION DETAILS 1:25

Photovoltaic panels
Magnesium aluminum alloy bracket
10mm Asphalt tile roofing
3mm SBS Waterproofing membrane
20mm Wooden lookoutboard
50mm Polyurethane insulation panel
20mm Wooden lookoutboard
220×30 Wooden upper edge board
Waterproof coating

Air spacer layer

20mm Wood finishes
120mm Polyurethane insulation panel
20mm Wood finish and protective layer

5+12A+5 Toughened glass

20mm Solid wood floors
Insulated cotton filling
20mm Solid wood floors
50×50 Wooden skeleton
220×30 Solid wood baffles

400×400 Weld metal plates
50×50×8 Angel
400×400 Weld metal plates
ø10mm Bolt the concrete

SOUTH ELEVATION 1:200

INTEGRATED ENERGY-SAVING TECHNOLOGY ANALYSIS

2-2 SECTIONAL PERSPECTIVE

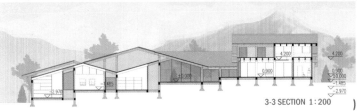

3-3 SECTION 1:200

BREAKDOWN VIEW OF SHOWROOM STRUCTURE

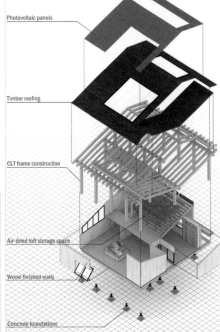

- Photovoltaic panels
- Timber roofing
- CLT frame construction
- Air-dried loft storage space
- Wood-finished walls
- Concrete foundations

THE GREEN BUILDING ASSESSING STANDARD OF CHINA

The result of assessing	pre-evaluation	self-evaluation		pre-evaluation	self-evaluation
Control basic score	400	400	Safty and durability	100	85
Healthy and comfortable	100	78	Convenience	100	65
Meterial saving	200	165	Environmentally livablity	100	83
Improvement and innovation	100	80			
Total	1100	956	Final	100	85
Grade			三星级		

ENERGY-SAVING DEVICE ANALYSIS

Rain water reuse system
Rainwater can be simply treated to realize the functions of flushing toilet, spraying road surface and greening watering, and reuse rainwater has significant water-saving efficiency.

Solar energy analysis
Photovoltaic power generation uses solar cells to convert solar energy directly into electricity.

Hydroelectric power
Hydroelectric power generation uses the water level drop, with the hydro-generator to generate electricity.

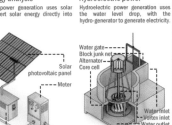
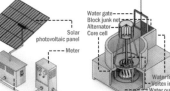

综合奖·三等奖
General Prize Awarded · Third Prize

注 册 号：100873
项目名称：火·塘
　　　　　Fire Place
作　 者：岳国威、刘 达、王 瑞、
　　　　　余浩然、高 溶
参赛单位：西安建筑科技大学
指导教师：孔黎明

火·塘
Fire · Place I

设计说明：
方案根据白马藏族传统民居建筑形式以及村落场地现状需求进行设计，结合火把节和火圈舞等风俗元素，结合场地地形，吸取火塘这一概念，将建筑塑造成一处既是科考中心同时又可以让村民举办节日的聚集性场所。以火塘为中心，形成环形看台。建筑功能环绕布置，形成建筑凝聚感和村民归属感；依地形而建，合理利用高差，错落有致，减少土方量。

Design Description:
The plan is designed according to the traditional Tibetan residential form of Baima and the current needs of the village site. Combining the elements of the torch festival and the fire circle dance, combined with the topography of the site, and absorbing the concept of the fire pond, the building is shaped into a gathering place that is both a scientific research center and a festival for villagers. With the fire pond as the center, a circular grandstand is formed, and the building function is arranged around the arrangement, forming a sense of architectural cohesion and a sense of belonging of the villagers; According to the terrain, rational use of height difference, staggered, reduce the amount of earth

■ Solar radiation analysis

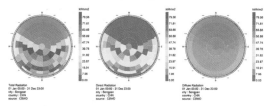

■ Climate analysis

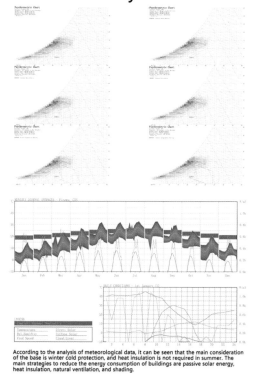

According to the analysis of meteorological data, it can be seen that the main consideration of the base is winter cold protection, and heat insulation is not required in summer. The main strategies to reduce the energy consumption of buildings are passive solar energy, heat insulation, natural ventilation, and shading.

■ Conceptual analysis

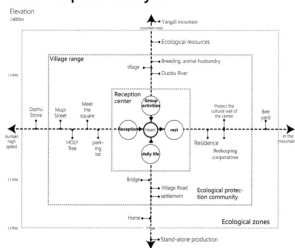

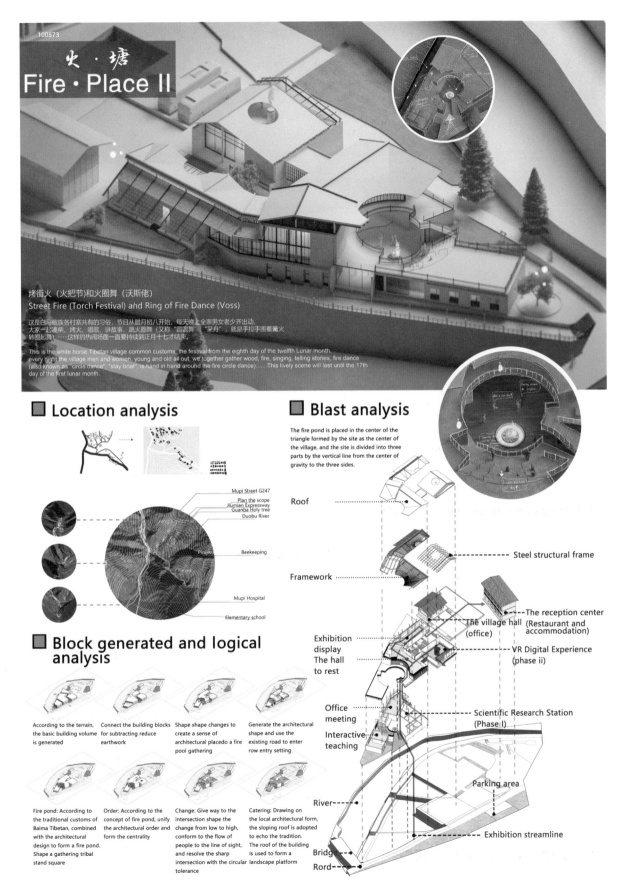

火・塘
Fire · Place II

烤街火（火把节）和火圈舞（沃斯佬）
Street Fire (Torch Festival) and Ring of Fire Dance (Voss)

这是白马藏族各村寨共有的习俗，节日从腊月初八开始，每天晚上全寨男女老少齐出动，大家一起凑乐、烤火、唱歌、讲故事、跳火圈舞（又称"圆圆舞"、"呆舟"，就是手拉手围着篝火转圈起舞）……这样的热闹场面一直要持续到正月十七才结束。

This is the white horse Tibetan village common customs, the festival from the eighth day of the twelfth Lunar month, every night the village men and women, young and old all out, we together gather wood, fire, singing, telling stories, fire dance (also known as "circle dance", "stay boat", is hand in hand around the fire circle dance)…… This lively scene will last until the 17th day of the first lunar month.

Location analysis

Blast analysis

The fire pond is placed in the center of the triangle formed by the site as the center of the village, and the site is divided into three parts by the vertical line from the center of gravity to the three sides.

Block generated and logical analysis

Fire pond: According to the traditional customs of Baima Tibetan, combined with the architectural design to form a fire pond. Shape a gathering tribal stand square

Order: According to the concept of fire pond, unify the architectural order and form the centrality

Change: Give way to the intersection shape change from low to high, conform to the flow of people to the line of sight, and resolve the sharp intersection with the circular tolerance

Catering: Drawing on the local architectural form, the sloping roof is adopted to echo the tradition. The roof of the building is used to form a landscape platform

---

专家点评：

该作品在分析场域主导风向、太阳辐照、场地高差等资源条件下，结合地方生活习俗、"火塘"特色、建筑文化，将驿站功能与太阳能资源利用巧妙地结合起来，营造出一处兼顾当地居民交流集会、科考旅游歇息的公共场所。

因此，利用场地地形东西向山谷风廊和沿河景观引入与"火塘"为中心的室内外集会广场的结合成为项目最大的亮点。当然，在处理圆形广场与屋面形式时，还可以采用更加经济并符合地方传统的屋顶形式。

This work combines local living customs, the characteristics of the 'fire place' and architectural culture in an analysis of the prevailing wind direction, solar irradiation, site height difference and other resources. It also integrates the function of stations and solar energy to create a public place for local residents to communicate and meet, and for scientific research and tourism to rest.

Given this, the combination of the east-west valley breezeway and the river landscape with the 'fire pit' as the centerpiece of the indoor and outdoor gathering plaza is the highlight of the project. Of course, a more economical and traditional roof form can be used when dealing with the circular square and the roof form.

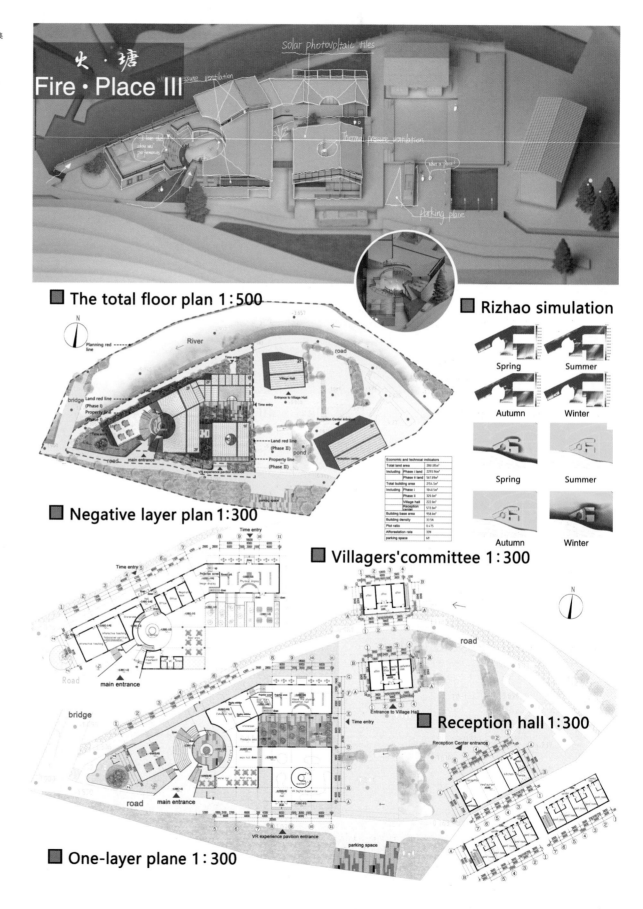

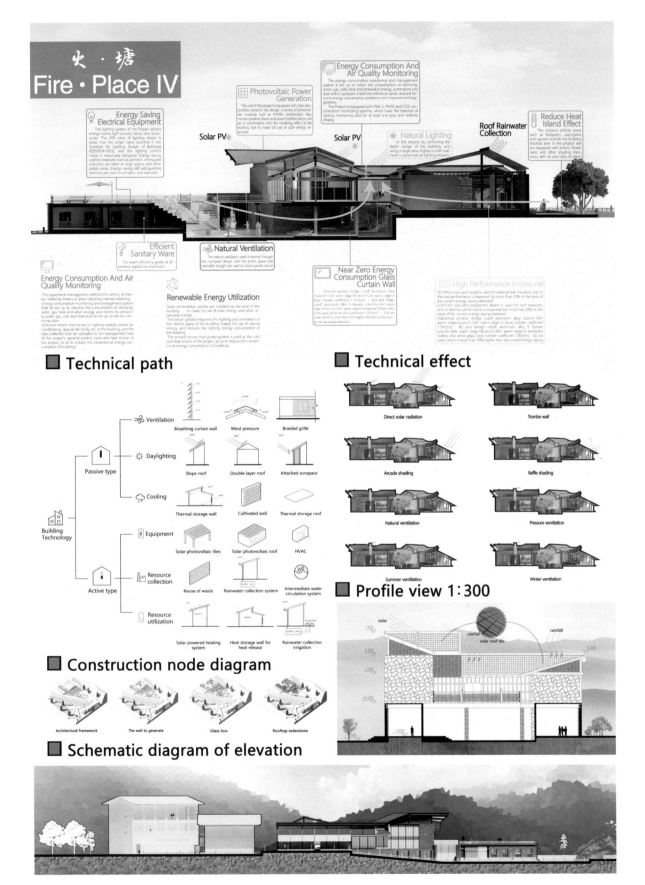

综合奖·三等奖
General Prize Awarded · Third Prize

注 册 号：100936
项目名称：风穿穿斗
　　　　　Wind Blowing through
　　　　　Column and Tie
作　　者：杜天慧、武云杰、刘芳鸣、
　　　　　袁梓飞
参赛单位：华北理工大学、厦门大学、浙大
　　　　　宁波理工学院、华北理工大学
指导教师：檀文迪

风穿穿斗·01
WIND BLOWING THROUGH COLUMN AND TIE

Design Desription

设计源于四川穿斗式特色民居，依据"青瓦出檐长、特色格子墙"的特点，运用冷巷、敞厅、抱厅、天井、等传统建筑布局手法达到自然通风采光。将传统穿斗式建筑与现代钢木结构融合，传承的基础上注重创新。场地中的树与建筑有机结合，可变化的悬挂花架，人们可从不同高度与植物互动。底层架空既符合防潮要求，又为植物留出空间。景观部分结合原有场地，设计叠水景观，与建筑部分相呼应。

This design originated from Sichuan's Column and Tie Construction. According to the characteristics of "green tiles with long eaves and characteristic lattice walls", it uses traditional architectural layout techniques such as cold alleys, open halls, holding halls, patios and so on to achieve natural ventilation and lighting. Combining traditional bucket-through architecture with modern steel-wood structure, and paying attention to innovation on the basis of inheritance. The trees and buildings in the site are organically combined, and the changeable hanging flower stands allow people to interact with plants from different heights. The bottom layer not only meets the moisture-proof requirements, but also leaves space for plants. Part of the landscape is combined with the original site, and the overlapping water landscape is designed, which echoes the architectural part.

风穿穿斗 · 02
WIND BLOWING THROUGH COLUMN AND TIE

2022 台达杯国际太阳能建筑设计竞赛获奖作品集

专家点评：

作品建筑形态采用化整为零的手法，具有山区民居村落的特点，与山体环境有较好的融合，变化丰富的剖面处理，结合被动技术，有利于通风采光。总体规划布局合理、建筑空间丰富，体现了各功能分区的特点；采用了太阳能利用技术，技术应用具有可实施性；设计表达清晰、深入。

The architectural form of the work adopts the technique of turning the whole into pieces. It has the features of a mountainous residential village and a good integration with the mountain environment. The overall planning layout is reasonable and the architectural space is rich. The varied section combined with passive technologies is conducive to natural ventilation and lighting. All of these reflect the features of each functional partition. Solar energy technology is used and the technical application is implementable. The design is clear and in-depth.

Building form Generation

Optimization of Main Ventilation

Solar Radiation Intensity

Terrain Analysis

Locial Climate Analysis

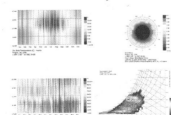

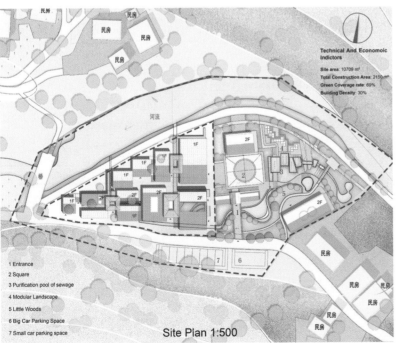

Site Plan 1:500

1 Entrance
2 Square
3 Purification pool of sewage
4 Modular Landscape
5 Little Woods
6 Big Car Parking Space
7 Small car parking space

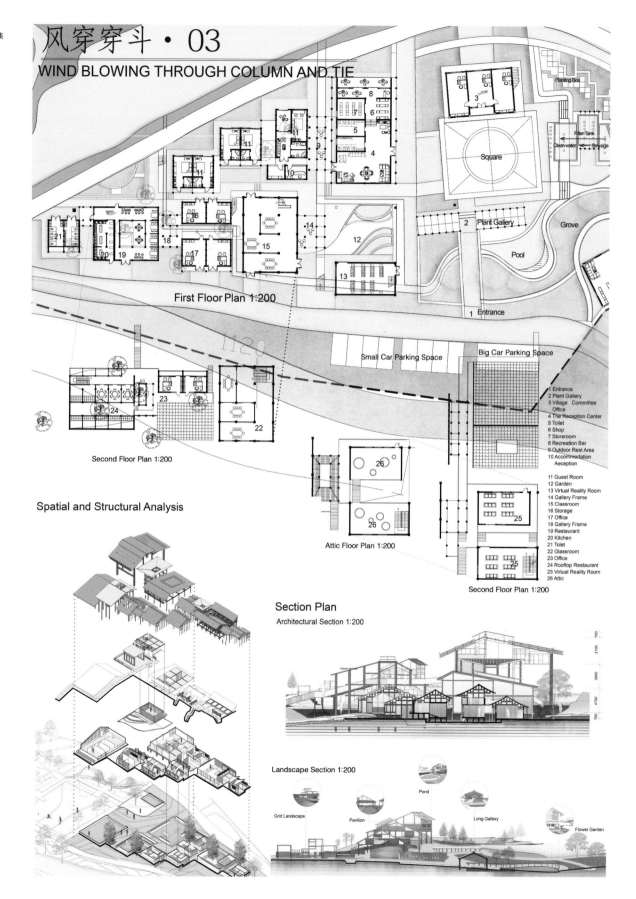

风穿穿斗 · 04
WIND BLOWING THROUGH COLUMN AND TIE

Manual of Key Construction Schemes

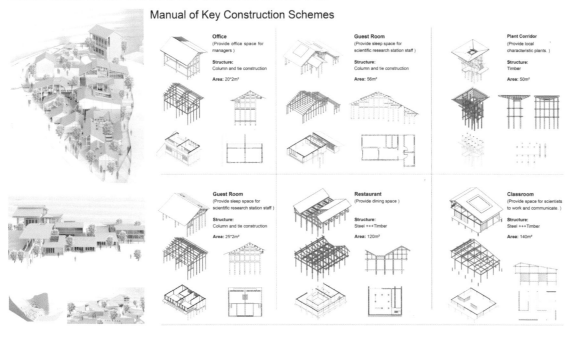

Multifunctional Classroom and Office

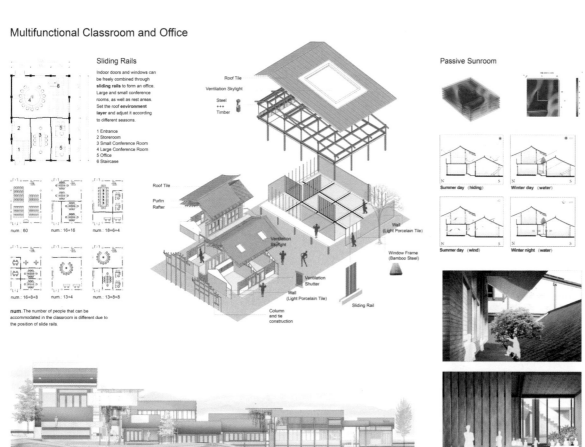

风穿穿斗 · 05
WIND BLOWING THROUGH COLUMN AND TIE

Plant Corridor

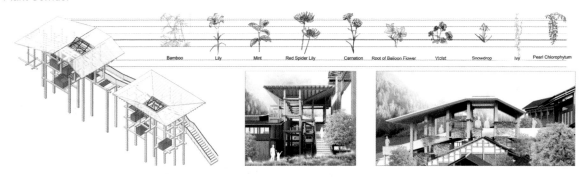

Tree House Restaurant

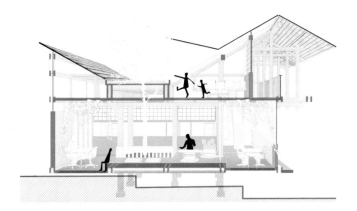

Plant Corridor

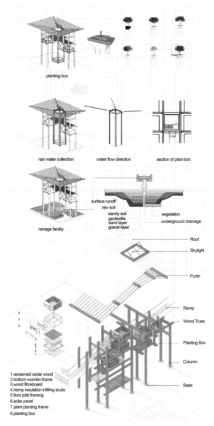

Atrium: Inherit the layout of the traditional fire pond, and form an open space for dining around trees.

Roof Greening: Suitable for private dining. Appreciate the leaves on the roof.

Passive Sunroom

Summary: Through the adjustment of various building envelope structures, different opening methods can be adopted in different seasons to achieve the passive solar energy strategy.

Summer day (hiding)
1. **Sunshading Board**: Tilt to block the sun.
3. **Shutter**: The angle is perpendicular to the sunlight and blocks the sunlight.

Winter day (take in sunlight)
1. **Sunshading Board**: Close
2. **Skylight**: Close
3. **Shutter**: The angle is parallel to sunlight to absorb sunlight.

Winter day (water)
4. **Atrium**: Set plants and water, which absorbs solar radiation and stores heat during the day.

Summer day (wind)
1. **Sunshading Board**: Open all
2. **Skylight**: Open all
3. **Shutter**: Open all

Winter night (Stop the cold wind)
1. **Sunshading Board**: Close
2. **Skylight**: The movable sliding door closes to block the cold wind.

Winter night (water)
4. **Atrium**: plants and water are set, and the water releases solar radiation, which causes the indoor temperature to rise at night.

South Elevation

风穿穿斗 · 06
WIND BLOWING THROUGH COLUMN AND TIE

Reception Center and VR

Multifunctional Classroom and Guest Room

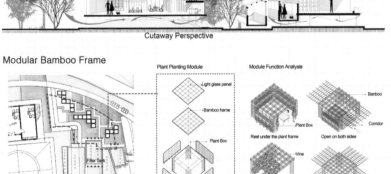

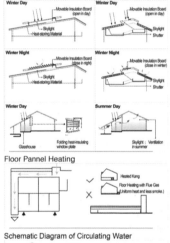

综合奖·三等奖
General Prize Awarded · Third Prize

注　册　号：100995
项目名称：旭日三重
　　　　　The Solar Trilogy
作　　　者：李春颖、贺　川、颜廷旭
参赛单位：多伦多大学

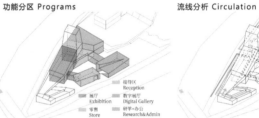

旭日三重
The Solar Trilogy

本项目位于山谷之中，北侧临水，南面傍山。主要广场设在乡镇公路和临水人行道路的交界处，提供了明显的视觉中心。景观坡地依托原有地势升起，增加了景观的层次性。除了对外的广场，建筑体块合成对内开放的庭院，形成了多种空间秩序。主入口设在临水的人行道路一侧，远离交通干扰，拉近了人与建筑的尺度。

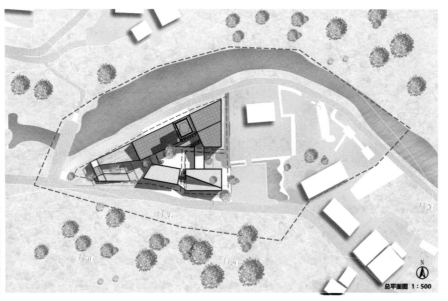

总平面图 1:500

专家点评：

方案总体围绕场地周边布局，较好地利用了场地环境景观资源。方案利用山谷风这一场地资源，趣味性地引入了捕风筒发电装置，装置结合夏季拔风、冬季太阳能储热等主被动技术，形成了较有特征性或标志性的建筑形象。

The scheme is generally laid out around the perimeter of the site, making good use of the site's environmental landscape resources. The scheme makes use of the valley wind as a site resource and introduces a wind catcher power generation device, which combines active and passive technologies such as draft in summer and solar thermal storage in winter to form a more characteristic or iconic architectural image.

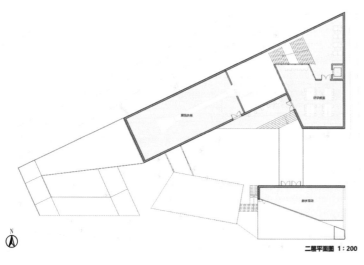

二层平面图 1:200

一层平面图 1:200

旭日三重
The Solar Trilogy

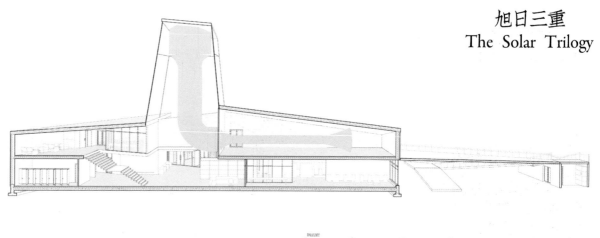

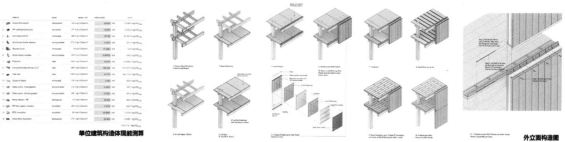

单位建筑构造体现能测算　　　　　　　　　　　　　　　　　　　　　　外立面构造图

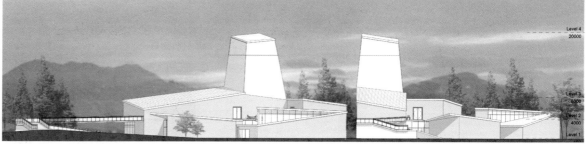

南立面 1:200　　　　　　　　　　　　西立面 1:200

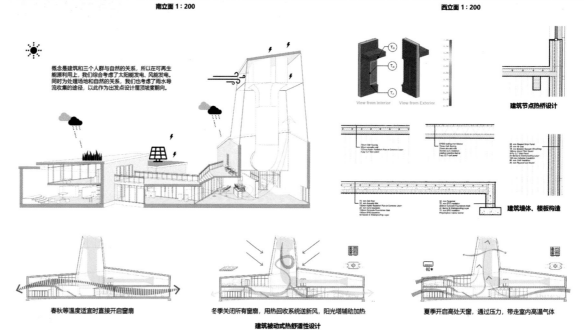

概念是建筑和三个人群与自然的关系，所以在可再生能源利用上，我们综合考虑了太阳能发电、风能发电。同时为了处理场地和自然的关系，我们也考虑了雨水导流收集的途径，以此作为出发点设计屋顶坡度朝向。

建筑节点热桥设计

建筑墙体、楼板构造

春秋等温度适宜时直接开启窗扇　　　冬季关闭所有窗房，用热回收系统送新风，阳光塔辅助加热　　　夏季开启高处天窗，通过压力，带走室内高温气体

建筑被动式热舒适性设计

旭日三重
The Solar Trilogy

微风发电系统原理及发电量测算

场地位于山谷，日照使临近山坡上下产生温差，形成山谷风。在获得的数据中得知，场地在下午会形成三到五级山谷风，并日落消失。

取上午11点到下午4点为山谷风形成时间，每天平均5小时，风廊中设置4个微风发电装置，在额定风速下，额定功率每个可达1000w。

最终计算一年风力发电量为 14600kW·h

光伏发电系统原理及发电量测算

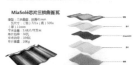

屋面采用光伏瓦技术，形体设计让屋面向南倾斜，以直接铺设更大面积光伏瓦片。

光伏瓦平均每平米发电功率为100w。
屋顶可铺设光伏瓦面积为965㎡，考虑到铺设技术，按修率问题，以0.8作为面积修正系数。

当地高峰太阳小时数经软件模拟，约为2.22h每天。

最终计算一年太阳能发电量为 50025.6kW·h

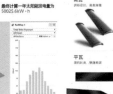

建筑总能耗模拟及热舒适性模拟

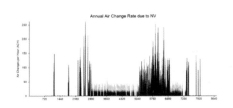
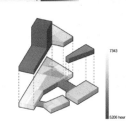

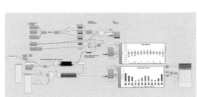
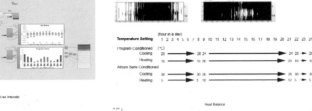
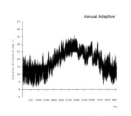

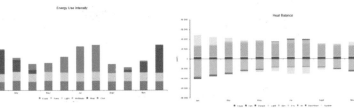

建筑能耗模拟结果：
建筑平均能耗EUI：54.5 k·W/h/yr，分解结果如下：

制热：14.3　　设备：12.81
制冷：12.69　　风扇：1.21
热水：0.72　　照明：12.71

一年总能耗为 85641.3kW·h
其中可再生能源利用量为 64625.6kW·h
约占总能耗 75.46%

综合奖·三等奖
General Prize Awarded · Third Prize

注 册 号：101031
项目名称：俯仰之间·隐山
Between Looking down and Looking up · Hidden Mountain
作　　者：郑龙纪、肖　瑞、李霁玮、季乐宇
参赛单位：南京工业大学
指导教师：舒　欣、薛春霖

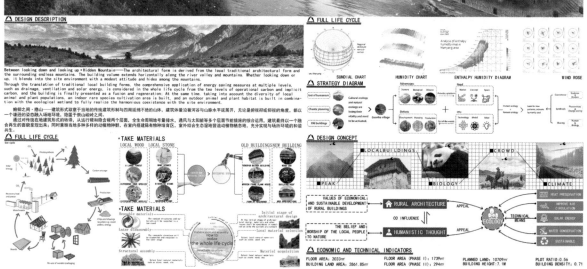

ARCHITECTURAL DESIGN
SCIENTIFIC INVESTIGATION STATION

N 39° 33' 18"
E 104° 33' 58"

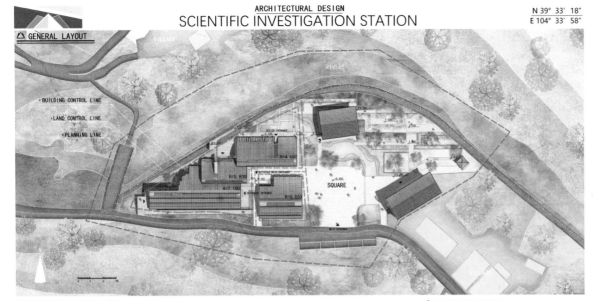

△ GENERAL LAYOUT

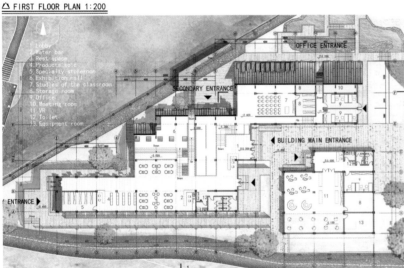

△ FIRST FLOOR PLAN 1:200

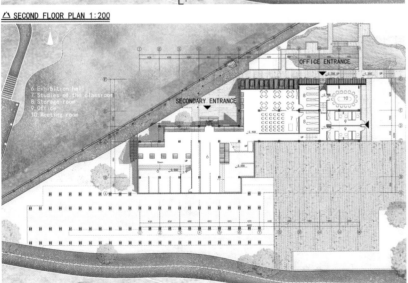

△ SECOND FLOOR PLAN 1:200

△ BLIND AREA SIGN

△ ANALYSIS OF BASIC NATURAL CLIMATE DATA

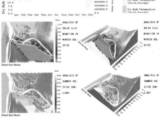

△ BLOCK GENERATED

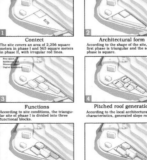

专家点评：

通过三角形断面的变化，形成整体感较强的、近乎连续的地景建筑，获得了足够多的可以利用太阳能的屋面面积。同时三角形的断面形式，使得建筑可识别性较强，木结构特征鲜明，具有较鲜明的建筑形式效果。但由于形式个性过于鲜明带来一些类似如何体现当地适应气候的传统建筑特色以及如何适应场地空间尺度等问题，这种疑问尤其会产生在对于四川气候和建筑更为熟悉的建筑师心中。同时略显夸张的形式会导致很多特殊的空间，有可能会产生建筑运维过程中保持足够舒适度时过高耗能的问题。

The triangular section forms an almost continuous landscape architecture with a large enough roof area for solar energy. At the same time, the triangular section makes the building more recognizable and identifiable. The timber structure makes it more distinctive and has a more distinctive architectural form effect. However, because the form is so distinctive, it brings some questions like how to reflect the traditional architectural features adapting to local climate and how to adapt to the spatial scale of the site. Such questions may arise particularly in the minds of architects who are more familiar with the climate and architecture of Sichuan. At the same time, the slightly exaggerated form will result in a number of special spaces, which may become a challenge to control energy consumption in the context of creating high energy consumption to maintain comfort during the operation and maintenance of the building.

ARCHITECTURAL DESIGN
SCIENTIFIC INVESTIGATION STATION

N 39° 33' 18"
E 104° 33' 58"

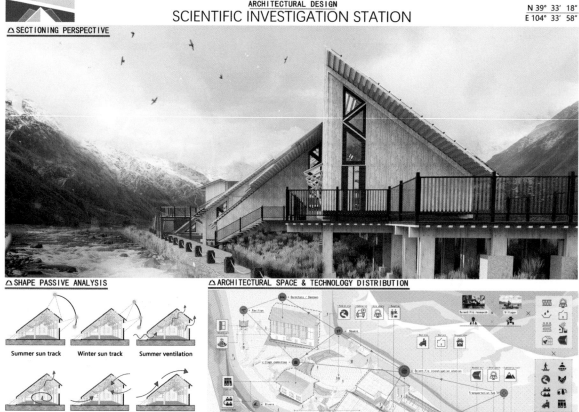

△ SECTIONING PERSPECTIVE

△ SHAPE PASSIVE ANALYSIS

Summer sun track | Winter sun track | Summer ventilation
Winter ventilation | Summer indoor airflow | Winter indoor airflow
Summer heat transfer | Winter heat transfer | Rainwater irrigation

△ ARCHITECTURAL SPACE & TECHNOLOGY DISTRIBUTION

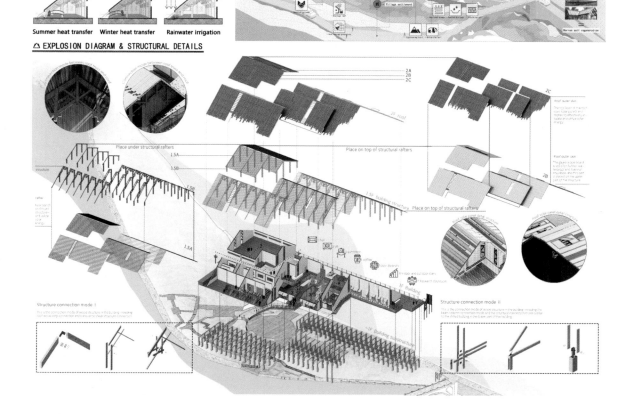

△ EXPLOSION DIAGRAM & STRUCTURAL DETAILS

Structure connection mode I | Structure connection mode II

ARCHITECTURAL DESIGN
SCIENTIFIC INVESTIGATION STATION

N 39° 33′ 18″
E 104° 33′ 58″

△ SECTIONING PERSPECTIVE

MAIN CONTENTS:
The passive and active techniques used in the design are displayed by using the expression technique of cross-sectional perspective.
At the same time, it represents the space, structure and use scene of the building.

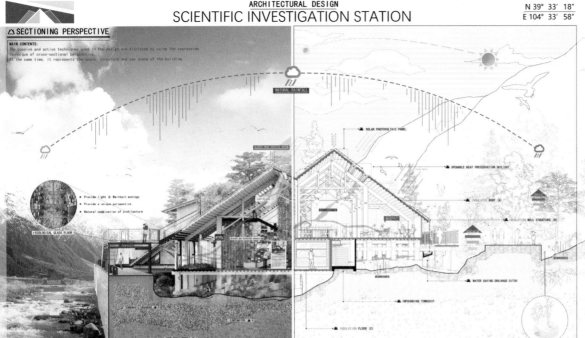

△ THERMAL INSULATION AND HEAT STORAGE DESIGN

(A) LIGHT TIMBER INSULATION ROOF STRUCTURE
- Roof shop
- Protective layer
- Waterproof layer
- Spacer layer
- Insulating layer
- OSB board
- Heavy wood
- Purline

(B) LIGHT WOOD INSULATION WALL STRUCTURE
- Interior trim
- Gypsum board
- Heat preservation cotton
- Main keel
- OSB board
- Keel
- Unidirectional waterproof breathing paper
- Anti-corrosion board for exterior wall

(C) LIGHT TIMBER INSULATION FLOOR CONSTRUCTION
- Floor decoration
- Cement mortar
- Spacer layer
- Noise reduction insulation layer
- OSB board
- Floor beam
- Dust separator layer
- Noise reduction spring
- Purline
- Floor beam

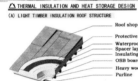 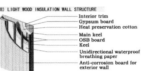

The site is located in guanba village, Pingwu County, Sichuan Province, China. It has a humid monsoon climate in the northern subtropical mountains. It belongs to the cold area in the building climate zoning, and the building must fully meet the requirements of thermal insulation in winter.

Therefore, the thermal insulation and heat storage performance of the building is fully considered in the design process. Lay insulation materials on roof, wall and floor.

△ SOLAR ENERGY TECHNOLOGY

• In order to meet the special needs of the site environment, two types of photovoltaic materials are used in this design. Polycrystalline silicon materials meet the conventional requirements. CdTe has good weak light power generation performance and low temperature coefficient (-0.25°C) which is suitable for the natural environment of the site. At the same time, traditional colors are extracted to echo the cultural elements of the site.

(1) Transparent CdTe thin film photovoltaic
(2) Polycrystalline silicon photovoltaic

△ ROOF PHOTOVOLTAIC INTEGRATION

(1) TRANSPARENT CdTe THIN FILM PHOTOVOLTAIC
In order to improve the efficiency of solar energy, semi transparent photovoltaic glass technology is introduced in the design. The technology has many advantages, such as good weak light power generation performance. The temperature coefficient is low, which is more effective in cold areas and can adapt to the site environment.

ADVANTAGE:
1. ENERGY GENERATION
2. UV & IR PROOF
3. THERMAL INSULATION
4. NATURAL ILLUMINATION
5. DESIGN
6. REDUCE

CONSTRUCTION DETAILS:
1. GLASS SUBSTRATE
2. PVB FILM
3. CRYSTALLINE SILICON WAFER
4. PVB FILM
5. GLASS SUBSTRATE

(2) POLYCRYSTALLINE SILICON PHOTOVOLTAIC
Photovoltaic curtain wall uses special resin to paste the solar cell on the glass and inlay it between two pieces of glass. Through the battery, light energy can be converted into electric energy. The back of the photoelectric template can also set off the color that the designer likes to adapt to different architectural styles.

△ GREENHOUSE

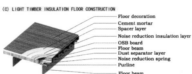

Heat massive house (summer)
Sunshine passive house (winter)

△ DESIGN SKETCH

EXHIBITION HALL

MAIN ENTRANCE

RECEPTION / REST

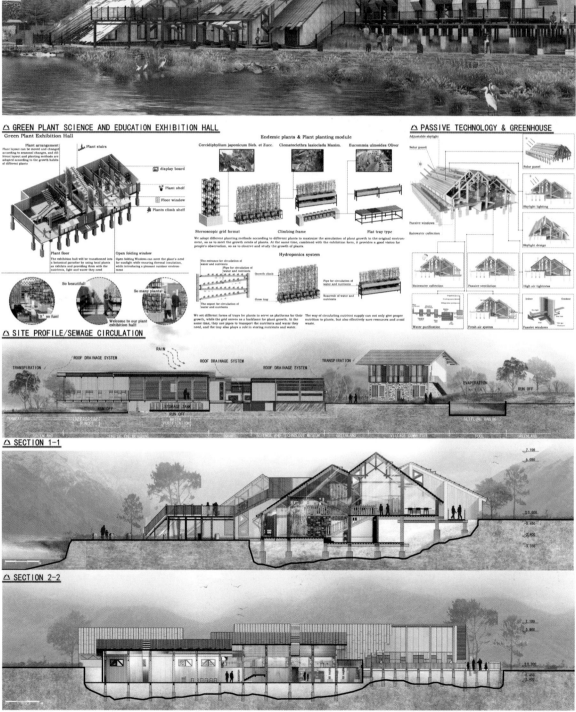

ARCHITECTURAL DESIGN
SCIENTIFIC INVESTIGATION STATION

N 39° 33' 18"
E 104° 33' 58"

△ NORTH ELEVATION

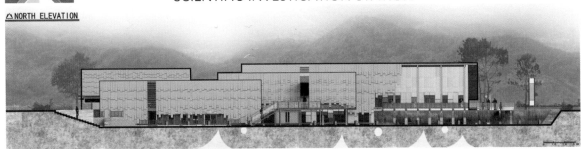

△ SOUTH ELEVATION

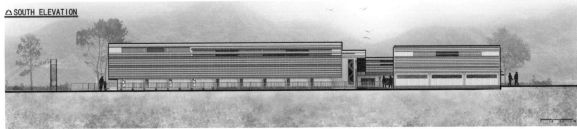

△ THERMAL INSULATION AND HEAT STORAGE DESIGN

Drainage on the same floor of the existing building toilet

Layout diagram   1-1 Section

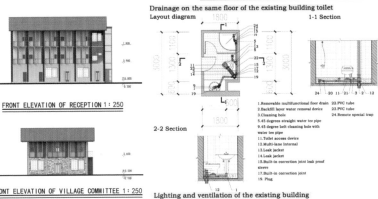

2-2 Section

1. Removable multifunctional floor drain
2. Backfill layer water removal device
3. Cleaning hole
5. 45 degrees straight water tee pipe
9. 45 degree belt cleaning hole with water tee pipe
11. Toilet access device
12. Multi-lane internal
13. Leak jacket
14. Leak jacket
15. Built-in correction joint leak proof sleeve
17. Built-in correction joint
19. Plug
22. PVC tube
23. PVC tube
24. Remote special trap

Lighting and ventilation of the existing building

When the Windows are closed, they can effectively block sunlight by using the external louvers.

Window closed condition

Windows in the open state, depending on the Angle of the window to get different light and ventilation, as well as light and shadow effects.

Window open condition

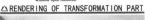

1. Canteen
2. Lobby
3. Laundry room
4. Storeroom
5. Disinfection washing
6. Food library
7. Food processing
8. Catering room
9. Accessible toilet
10. Male toilet
11. Ladies' room
12. Barrier-free elevator
13. Guest room
14. Toilet

GROUND FLOOR PLAN 1:250

SECOND FLOOR PLAN 1:250

THIRD FLOOR PLAN 1:250

FRONT ELEVATION OF RECEPTION 1:250

FRONT ELEVATION OF VILLAGE COMMITTEE 1:250

1-1 PROFILE 1:250   2-2 PROFILE 1:250

1. Lobby
2. Duty room
3. Monitor room
4. Office
5. Meeting room

GROUND FLOOR PLAN 1:250

SECOND FLOOR PLAN 1:250

△ RENDERING OF TRANSFORMATION PART

MODEL PHOTOS

MODEL PHOTOS

RECEPTION AND ACCOMMODATION

MODEL PHOTOS

MODEL PHOTOS

THE VILLAGE COMMITTEE

综合奖·三等奖
General Prize Awarded · Third Prize

注 册 号：101105
项 目 名 称：藏谷翎羽
　　　　　　Falling into the Valley
作　　　者：杨汶瑾、陈俊豪、古津铭、
　　　　　　林涵晔、叶家瑞、赵晓婷
参 赛 单 位：重庆大学
指 导 教 师：周铁军、张海滨、李　骏

藏谷翎羽
FALLING INTO THE VALLEY

四川省平武县关坝沟流域自然保护小区科学考察站设计 02
Design of Scientific Investigation Station for Guanbagou Nature Conservation Community in Pingwu County, Sichuan Province

2022 台达杯国际太阳能建筑设计竞赛获奖作品集

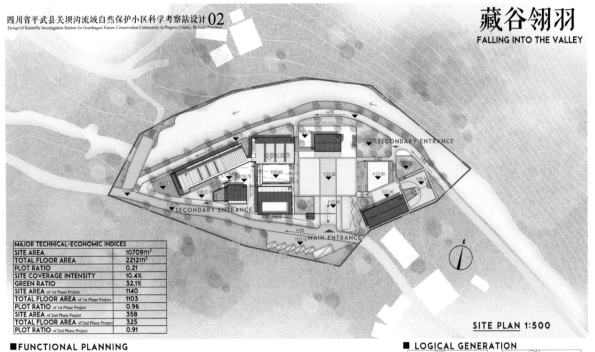

MAJOR TECHNICAL-ECONOMIC INDICES	
SITE AREA	10709m²
TOTAL FLOOR AREA	2212m²
PLOT RATIO	0.21
SITE COVERAGE INTENSITY	10.4%
GREEN RATIO	32.1%
SITE AREA of 1st Phase Project	1140
TOTAL FLOOR AREA of 1st Phase Project	1103
PLOT RATIO of 1st Phase Project	0.96
SITE AREA of 2nd Phase Project	358
TOTAL FLOOR AREA of 2nd Phase Project	325
PLOT RATIO of 2nd Phase Project	0.91

SITE PLAN 1:500

■ FUNCTIONAL PLANNING

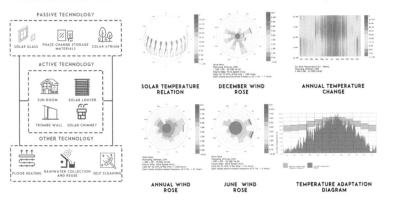

The site is divided by the east-west main axis, and buildings are arranged along both sides of the main axis. Crowd collecting and distributing nodes are set on the axis in series to create a sense of transparency of the site. The original functional positions of the village committee and reception are replaced, and the villagers' activity areas are concentrated. A secondary axis is set around the optimized reception area to divide the activity areas of villagers and students.

■ LOGICAL GENERATION

■ TECHNOLOGY SYSTEM

PASSIVE TECHNOLOGY
- SOLAR GLASS
- PHASE CHANGE STORAGE MATERIALS
- SOLAR ATRIUM

ACTIVE TECHNOLOGY
- SUN ROOM
- SOLAR LOUVER
- TROMBE WALL
- SOLAR CHIMNEY

OTHER TECHNOLOGY
- FLOOR HEATING
- RAINWATER COLLECTION AND REUSE
- SELF CLEANING

■ CLIMATIC SIMULATION

- SOLAR TEMPERATURE RELATION
- DECEMBER WIND ROSE
- ANNUAL TEMPERATURE CHANGE
- ANNUAL WIND ROSE
- JUNE WIND ROSE
- TEMPERATURE ADAPTATION DIAGRAM
- DIRECT EVAPORATIVE
- INDIRECT EVAPORATIVE
- NATURAL VENTILATION
- NIGHT PURGE VENTILATION
- PASSIVE SOLAR HEATING
- THERMAL MASS EFFECTS

专家点评：

建筑充分体现了当地夏凉冬冷气候的特点，建筑形态与立面具有川西特色，建筑布局和功能合理，交通流线流畅，将阳光间、太阳能光电与光热集热板、建筑双层空腔墙体等与建筑屋顶、立面造型紧密结合，有利于提高太阳能利用率。采用高性能的围护结构、电动百叶窗调节室内微气候、相变储能、多级缓冲空间的被动式建筑设计策略，减少建筑能量损耗；结合场地光照条件，合理地采用了光伏一体化设计、直流供电空调、照明及石墨烯电热膜供暖、太阳能热水系统，加热自来水、地源热泵系统等主动系统，实现了建筑的零碳运行。

The building fully reflects the features of the local cool summer and cold winter climate. The building form and facade have western Sichuan features, and the building layout and functions are reasonable. Its traffic flow is smooth. The winter garden, photovoltaic panels solar photovoltaic and solar thermal collector panels, and the cavity wall are closely integrated with the building's roof and facades, which can improve solar energy utilization. The passive building design strategies of high-performance envelope structure, electric shutters to regulate indoor microclimate, phase change energy storage and multi-stage transition space are adopted to reduce building energy loss. Combined with the light conditions of the site, the design reasonably adopts an integrated photovoltaic design, DC-powered air conditioning, lighting and graphene electric heating film, solar hot water system, heated tap water, ground source heat pump system and other active systems to achieve low zero carbon operation of the building.

四川省平武县关坝沟流域自然保护小区科学考察站设计 03

Design of Scientific Investigation Station for Guanbagou Nature Conservation Community in Pingwu County, Sichuan Province

藏谷翎羽
FALLING INTO THE VALLEY

■ CONCEPT

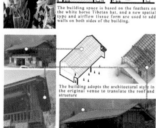

The building space is based on the feathers on the white horse Tibetan hat, and a new spatial type and airflow tissue form are used to add walls on both sides of the building.

The building adopts the architectural style in the original venue to translate the roof and structure.

The steel bars on the roof protrude upwards, imitate the green tile style of the roof of the local building, and the eaves of the corridor simulation of the eaves of the building.

The fire pond in the residential house — the middle column space of the pillar obtains another temperament in the air translation.

■ GROUND PLANE

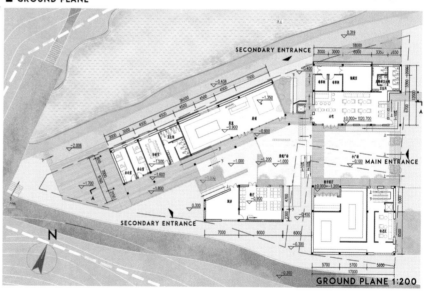

GROUND PLANE 1:200

■ VOLUME GENERATION

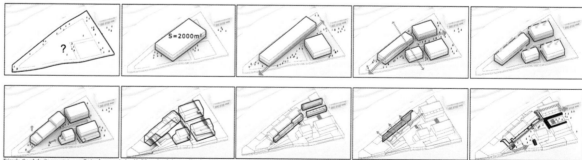

Triangle, the whole site presents two small triangle area can't placed buildings, the whole area of less than 2000 square meters, building to a layer of occupying space words appear abrupt area is not enough at the same time, the design of the building decorates two layer meet area is required, along the north decorate building long side, through things and two north-south axis will be divided into four whole individual piece piece of house. Three of them are arranged orthogonal to the village committee, and one is arranged to maximize the area along the long side. Ventilation and lighting can be satisfied through the built-in warm corridor and cold alley, which is specifically manifested as the double-layer wall in the north and the sunshine corridor in the south. The site is mainly divided into three terraces, which form an inward courtyard in the middle, and the building corridor forms a whole, making the streamline and the sight connected.

四川省平武县关坝沟流域自然保护小区科学考察站设计 04
藏谷翎羽
FALLING INTO THE VALLEY

■ SECOND FLOOR PLAN

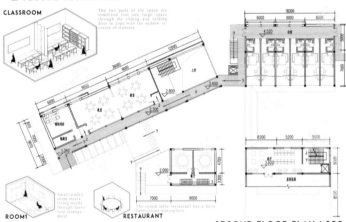

SECOND FLOOR PLAN 1:200

■ RETROFIT FLOOR PLAN

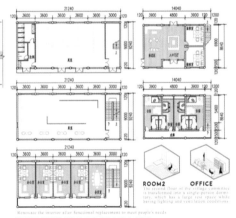

RETROFIT FLOOR PLAN 1:200

■ TECHNIQUE

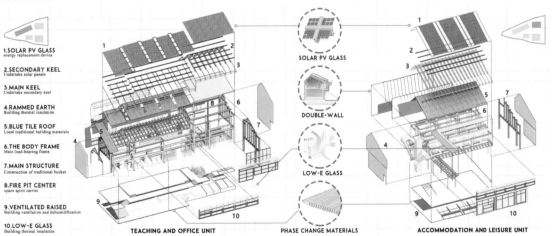

1. SOLAR PV GLASS — energy replacement device
2. SECONDARY KEEL — Undertake solar panels
3. MAIN KEEL — Undertake secondary keel
4. RAMMED EARTH — Building thermal insulation
5. BLUE TILE ROOF — Local traditional building materials
6. THE BODY FRAME — Main load-bearing frame
7. MAIN STRUCTURE — Construction of traditional bucket
8. FIRE PIT CENTER — space spirit carrier
9. VENTILATED RAISED — Building ventilation and dehumidification
10. LOW-E GLASS — Building thermal insulation

TEACHING AND OFFICE UNIT | PHASE CHANGE MATERIALS | ACCOMMODATION AND LEISURE UNIT

四川省平武县关坝沟流域自然保护小区科学考察站设计 05

藏谷翎羽
FALLING INTO THE VALLEY

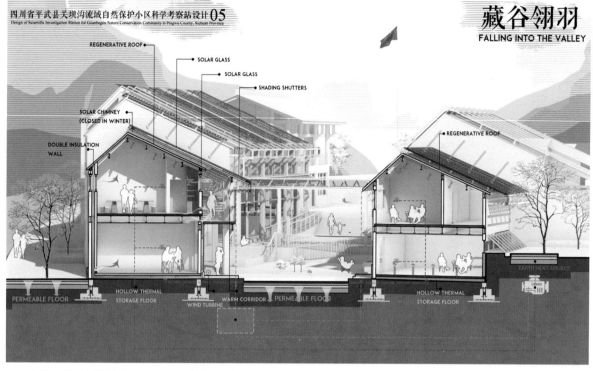

■ STRUCTURE DETAILS

■ SIMULATION OF LIGHTING AND VENTILATION

WINTER SUNSHINE / GROUND WIND SIMULATION / FIRST FLOOR WIND SIMULATION

SUMMER SUNSHINE / SECOND FLOOR WIND SIMULATION / ROOF WIND SIMULATION

■ ENERGY CALCULATION

According to the provision of the national regulation, the electricity consumption index of the accommodation space and the office and teaching space is about 30W/㎡, while the exhibition room rises up to 50W/㎡, with the 40W/㎡ of the kitchen room and 45W/㎡ of the leisure space.

As for the area, both the accommodation space and the leisure space have the area of 330㎡, and the office and teaching space is 644㎡, while the exhibition room covers an area of 736㎡, along with the smallest dining space which has the area of 172㎡. In our estimation, the power consumption of the accommodation space and the leisure space is about 16 hours a day, while the office and teaching space and the exhibition room drop to 10 hours a day , the last dining space has the shortest time of energy consumption, which is only used for 6 hours a day. Finally, the calculation shows that the total electricity consumption of the building is about 988.16kW·h a day.

(330㎡×30W/㎡＋330㎡×45W/㎡)×16h＋(644㎡×30W/㎡＋736㎡×50W/㎡)×10h＋172㎡×30W/㎡×6h=988.16kW·h

The energy production of each square of solar glass panels varies from 0.3kW·h to 10kW·h a day, for costs saving, we choose the type of panels that produces 2kW·h a day for the building, with the area of the 608㎡ of the panels, we can produce 1216kW·h of electricity a day, which is more than the waist of the building itself.

608㎡×2kW·h=1216kW·h＞988.16kW·h

■ TECHNOLOGY STRATEGY

DIRECT SOLAR RADIATION / SUMMER DAY CLASSROOM SIDE VENTILATION / DIRECT SOLAR RADIATION / SUMMER DAY HOTEL SIDE VENTILATION

WINTER DAY CLASSROOM SIDE MAINLY HOT PRESS CYCLE / WINTER NIGHT CLASSROOM SIDE PHASE CHANGE EXOTHERM / WINTER DAY HOTEL SIDE MAINLY HOT PRESS CYCLE / WINTER NIGHT HOTEL SIDE PHASE CHANGE EXOTHERM

■ SOUTH ELEVATION

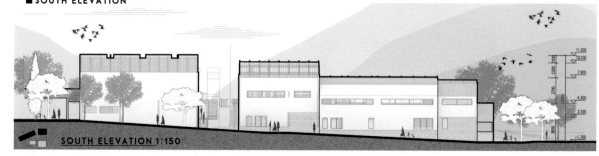

SOUTH ELEVATION 1:150

四川省平武县关坝沟流域自然保护小区科学考察站设计 06

藏谷翎羽
FALLING INTO THE VALLEY

■ STRUCTURE ANALYSIS

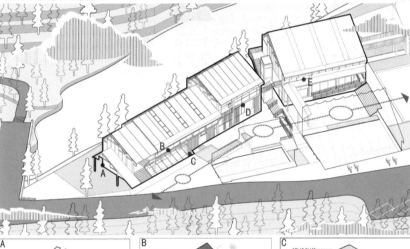

■ CARBON EMISSION

As is calculated in the last page, the total electricity consumption that the building has waist a day is about 988.16Kw·h, and according to the provision of the national regulation, the emission factor of carbon dioxide(CO_2) for the electricity is 0.5810tCO$_2$/MWh, we can come to a result of about 12.97 kilogram of CO_2 per square.

$(988.16Kw·h \times 0.5810tCO_2/MWh·0) \times 50/(330 m^2 + 330 m^2 + 644 m^2 + 736 m^2 + 172 m^2) = 12.98 kgce/(m^2·a)$

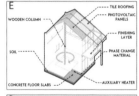
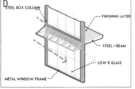

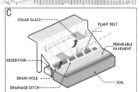

SECTION A-A 1:150

2022 台达杯国际太阳能建筑设计竞赛获奖作品集

风穿折屋——折屋伏地清风过 01

该设计从尊重地域文化、顺应山地形态、适应当地气候出发，下部采用紧贴地形、敦实厚重的错层形式。从传统建筑中转译而来的木屋架屋顶作为特色，其下方的灰空间与上部的天窗起到调节室内舒适度的作用。形体顺应由东至西的山谷风，东边分散西边聚集，内部形成利于山风穿过的通廊，充分利用了自然通风。在太阳能利用方面，朝向西南的屋面满铺单晶硅光伏，天窗则采用彩色薄膜光伏，兼顾采光、遮阳及室内空间氛围感的营造。

No.	NAME	QUANTITY	
1	total planning area	10709.02㎡	
2	total site area	2861.85㎡	the first phase 2293.96㎡
			the second phase 567.89㎡
3	building area	2195.28㎡	scientific research and exhibition 922.43㎡
			hotel 558.76㎡
			office 225.32㎡
			VR mall 208.77㎡
4	area of roads and places	5231.55㎡	
5	floor area ratio	0.2	
6	building density	0.14	
7	greening rate	44.65%	
8	parking lot	1 bus and 5 cars	

综合奖 · 三等奖
General Prize Awarded · Third Prize

注 册 号：101224
项目名称：风穿折屋
　　　　　Sun, Winds and Roofs
作　　者：路　易、梁伟豪、白盛鸿、
　　　　　李景秀、孙天伊、侯亚玲
参赛单位：天津大学、北京林业大学
指导教师：郭娟利、李　伟

The design starts from respecting the culture, adapting to the mountain form and adapting to the local climate. The lower part adopts the split-layer form which is close to the terrain and thick. The wooden frame roof, translated from the traditional building, is featured by the gray space below and the skylight above to adjust the indoor comfort.

Diagram of Design Process

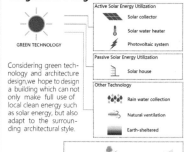

Site Analysis

Guanba Village is a forest nature education base in Sichuan Province and a nature conservation community in the Guanbagou River Basin. The total area of woodland in the village is 6499.3 mu, rich in biological resources and high quality of ecological environment, which is an important ecological hinterland and water conservation area of Mianyang City.
The main population of Guanba Village is the Baima Tibetan people. Their houses are built on the mountain, and the costumes of the white horse people who live on farming, animal husbandry, hunting and gathering are mainly white, black and flowers, and the colors are gorgeous. The primitive production and way of life have formed the worship of nature and the worship of mountains and rivers by the Baima people, which has also created their industrious and brave character.

Site
Surroundings, Buildings, Road, Transportation

Culture
Dance, Mask, Hat, Praying Dance

Site Plan 1:1000

Logical Generation

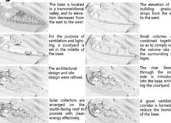

风穿折屋——折屋伏地清风过 02

FIRST FLOOR PLAN 1:300

1. Lobby
2. Water Bar
3. Specialty Shops
4. Storage Room
5. Toilet
6. Lounge
7. Exhibition Hall
8. Office
9. Conference Room
10. Classroom
11. Digital Exhibtion Hall
12. VR Experience Area
13. Security Room
14. Monitor Room
15. Office area
16. Restaurant
17. Kitchen

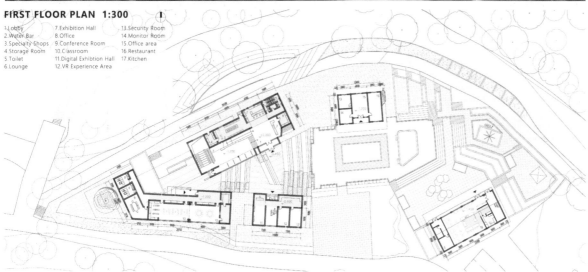

RESEARCH STATION OF NATURE PROTECTION AREA

GROUND FLOOR PLAN 1:300
1. Exhibition Hall
2. Media Exhibition

SECOND FLOOR PLAN 1:300
1. Activity Room
2. VR Experience Area

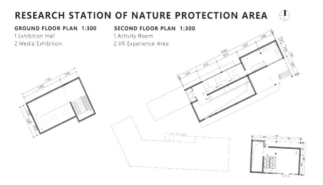

ACCOMMODATION AND RECEPTION

SECOND FLOOR PLAN 1:300
1. Guest Room
2. Lavatory

THIRD FLOOR PLAN 1:300
1. Guest Room
2. Lavatory

VILLAGE COMMITTEE

SECOND FLOOR PLAN 1:300
1. Conference Room
2. Office

ENTRANCE MARKING

专家点评：

作品从尊重地域文化、顺应山地形态、适应当地气候的角度出发，建筑采用生态特色的木结构形式，将川西传统特色与现代木结构建筑结合，建筑形态丰富，立面美观，平面布局和功能合理，交通流线流畅。采用天窗与自然通风走廊形成舒适空调。屋面和天窗采用单晶硅光伏和彩色薄膜光伏一体化，兼顾采光、遮阳，合理采用了太阳能热水系统、光伏直流系统、地源热泵系统等主动系统；生土墙体等与建筑立面造型紧密结合，有利于提高太阳能储能的被动式建筑设计策略，减少建筑能量损耗，实现了建筑低碳运行的目标。

The work is based on respect for local culture, the adaptation of local mountainous form and the climate. The building adopts a wooden structure with ecological characteristics, combining traditional western Sichuan characteristics with modern wooden architecture. The building has rich architectural forms, beautiful facades, reasonable layout and functions, and smooth traffic flow. Skylights and natural ventilation corridors are used to make people feel comfortable. The roof and skylights are integrated with full-length monocrystalline silicon photovoltaic cells and colorful thin-film photovoltaic cells, taking into account light and shade. Active systems such as solar hot water system, photovoltaic DC system and ground source heat pump system are reasonably adopted; the rammed earth walls and so on are closely integrated with the building facades, which is conducive to passive building design strategies that improve solar energy storage, reduce building energy losses and achieve the goal of low carbon operation of buildings.

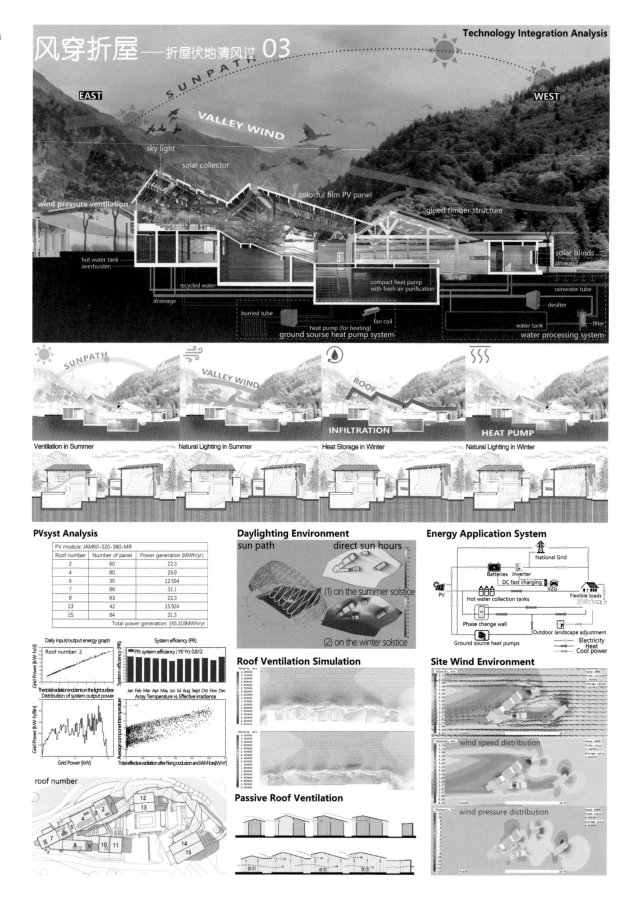

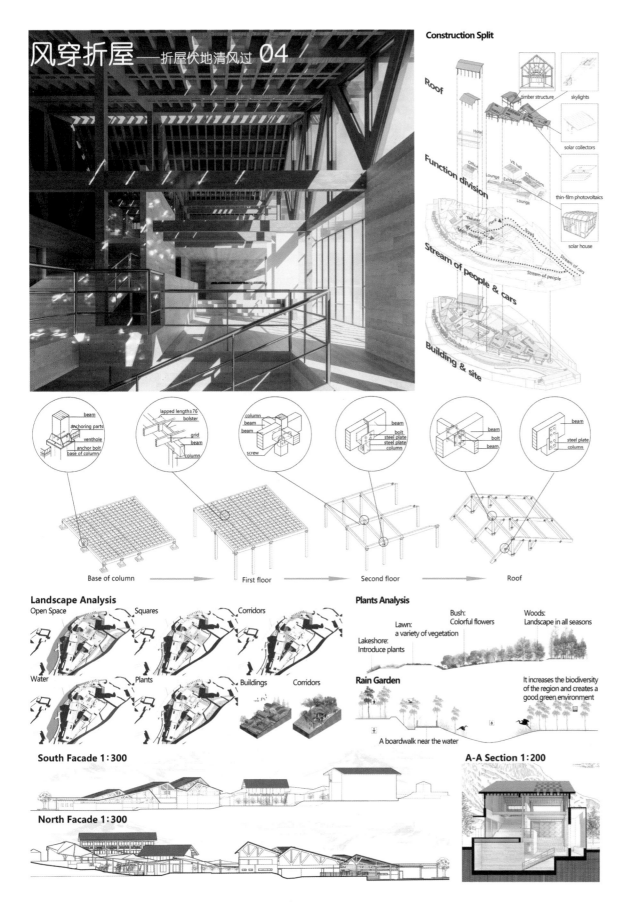

综合奖・优秀奖
General Prize Awarded · Honorable Mention Prize

注 册 号：100682
项目名称：绵
　　　　　Continuous
作　　者：郑新杰、李润霖、陈怡凤、
　　　　　郑颖茵、张明斌
参赛单位：广州大学
指导教师：庞　玥、李　丽

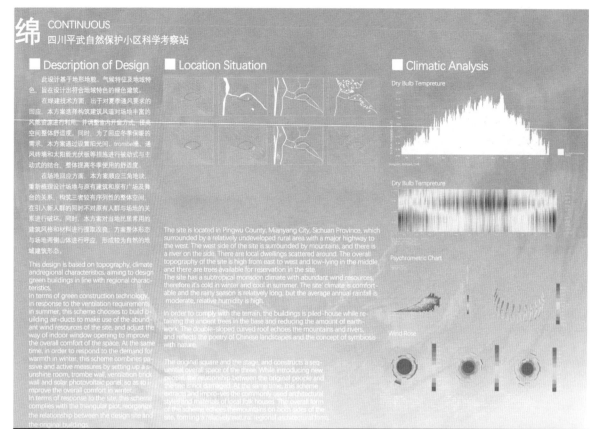

绵 CONTINUOUS
四川平武自然保护小区科学考察站

■ Description of Design

此设计基于地形地貌、气候特征及地域特色，旨在设计出符合地域特色的绿色建筑。
在绿建技术方面，出于对夏季通风要求的回应，本方案选择构筑建筑风道对场地丰富的风能资源进行利用，并调整室内的开窗方式，提高空间整体舒适度。同时，为了回应冬季保暖的需求，本方案通过设置阳光间、trombe墙、通风砖墙和太阳能光伏板等措施进行被动式与主动式的结合，整体提高冬季使用的舒适度。
在场地回应方面，本方案顺应三角地块，重新梳理设计场地与原有建筑和原有广场及舞台的关系，构筑三者较有序列性的整体空间，在引入新人群的同时不对原有人群与场地的关系进行破坏。同时，本方案对当地民居常用的建筑风格和材料进行提取改良，方案整体形态与场地两侧山体进行呼应，形成较为自然的地域建筑形态。

This design is based on topography, climate and regional characteristics, aiming to design green buildings in line with regional characteristics.
In terms of green construction technology, in response to the ventilation requirements in summer, this scheme chooses to build building air-ducts to make use of the abundant wind resources of the site, and adjust the way of indoor window opening to improve the overall comfort of the space. At the same time, in order to respond to the demand for warmth in winter, this scheme combines passive and active measures by setting up a sunshine room, trombe wall, ventilation brick wall and solar photovoltaic panel, so as to improve the overall comfort in winter.
In terms of response to the site, this scheme complies with the triangular plot, reorganize the relationship between the design site and the original buildings.

■ Location Situation

The site is located in Pingwu County, Mianyang City, Sichuan Province, which surrounded by a relatively undeveloped rural area with a major highway to the west. The west side of the site is surrounded by mountains, and there is a river on the side. There are local dwellings scattered around. The overall topography of the site is high from east to west and low-lying in the middle, and there are trees available for reservation in the site.
The site has a subtropical monsoon climate with abundant wind resources, therefore it's cold in winter and cool in summer. The site' climate is comfortable and the rainy season is relatively long, but the average annual rainfall is moderate, relative humidity is high.

In order to comply with the terrain, the buildings is piled-house while retaining the ancient trees in the base and reducing the amount of earthwork. The double-sloped curved roof echoes the mountains and rivers, and reflects the poetry of Chinese landscapes and the concept of symbiosis with nature.

The original square and the stage, and constructs a sequential overall space of the three. While introducing new people, the relationship between the original people and the site is not damaged. At the same time, this scheme extracts and improves the commonly used architectural styles and materials of local folk houses. The overall form of the scheme echoes the mountains on both sides of the site, forming a relatively natural regional architectural form.

■ Climatic Analysis

Dry Bulb Tempreture

Dry Bulb Tempreture

Psychrometric Chart

Wind Rose

绵 CONTINUOUS
四川平武自然保护小区科学考察站

■ Sunlight Hours Analysis   ■ Analysis Solar Radiation on Buildings Surface   ■ Design Strategy

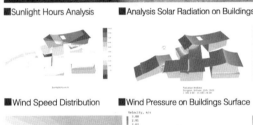

■ Wind Speed Distribution   ■ Wind Pressure on Buildings Surface

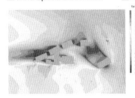
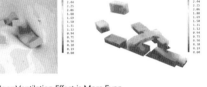

■ The Overall Indoor Ventilation Effect is More Even   ■ Reception Room Energy-Saving Technology

Meeting Room   Before   After

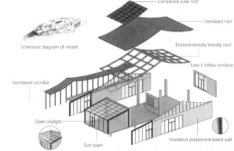

■ Indoor Ventilation Effect

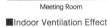

Elevation 1:200

绵 CONTINUOUS
四川平武自然保护小区科学考察站

Land area	2860.9 m²
Building area	1991.6 m²
Building density	41.20%
Number of floors	2

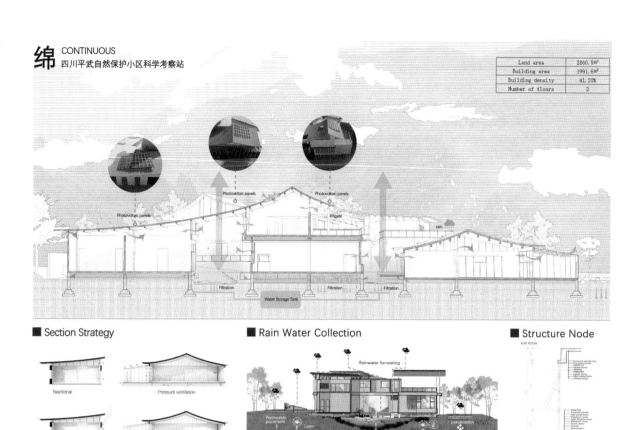

■ Section Strategy

■ Rain Water Collection

■ Structure Node

■ Sun Room Function

In winter days, trube wall, heavy heat storage floor, roof heat storage and meaned earth wall are mainly used for heat storage.

At night in winter, the energy stored in the sun room is transmitted to the interior of the main building through the glass curtain wall, raising the temperature at night in winter.

In summer, open the Windows of the sunroom to bring in natural air and provide good ventilation for the interior.

■ Local Perspective View

Elevation 1:200

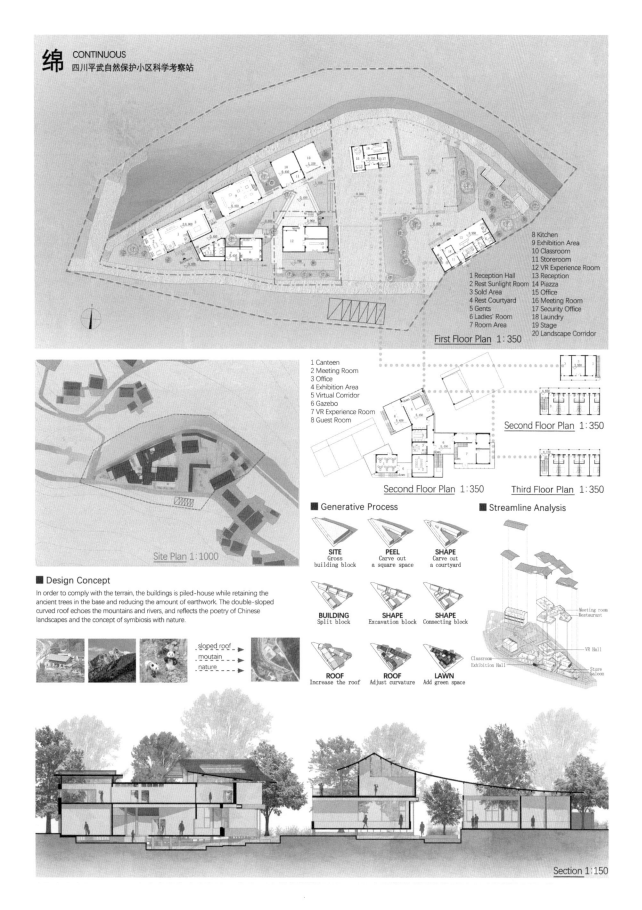

林间关坝·沐日而生
Immersing in the Forest, Soaking up the Sunshine

四川平武关坝自然保护区科学考察站设计

01

综合奖·优秀奖
General Prize Awarded·Honorable Mention Prize

注 册 号：100684
项目名称：林间关坝·沐日而生
Immersing in the Forest, Soaking up the Sunshine
作　　者：周可伊、陈　言、郭奕岑、赵庆卓、过翔天、毛思异、陈思妍
参赛单位：重庆大学
指导教师：黄海静

■ Design Specification

在双碳政策和可持续发展理念的背景下，本方案基于自然低碳、研学展示、民族文化三大场地特征，对场地进行湿地系统、雨水花园滤水系统、溪水利用系统、乡村民族文化展示、休闲小憩等多功能规划设计。

在空间功能上，提炼当地传统民居语言，形成以坡屋顶为单元的聚落形式模拟起伏山势，并以中央庭院、阳光走廊、天井、阳光间等串联建筑体块，引入室外绿化，让人有穿梭于林间之感。

在对气候的回应及被动式太阳能技术方面，本方案充分利用当地太阳辐射资源，通过设置光伏玻璃板、彩色太阳能光伏板、阳光间、夯土材料为基础的围护结构解决冬季保温问题。通过设置阳光走廊、通风式特朗勃墙、可开启的太阳外表皮促进建筑在过渡季节的自然通风。

此外，作为对地域文化的回应，本方案利用场地填挖剩余土方进行夯土墙制造，充分利用当地丰富的土、石、竹资源进行建设，在建设过程中控制碳排放量，达到低碳减碳目的。

In the context of carbon peaking and carbon neutrality goals and sustainable development, this project, based on the distinct site characteristics (natural ecology, research showcase and the minority culture), is targeted at the multifunctional design of the site, including constructed wetland system, rainwater garden filter system, stream water utilization system, rural minority culture display and repose space.

As for the spatial design, it is a refinement of the local traditional dwellings, forming unitized sloping roofs as symbol of traditional settlement and simulation of rolling ranges (or wavy terrain). It also features in threading different volumes with the central courtyard, solar corridor, patios and solar room as well as bringing in the outdoor greenery, leading to a sense of wandering in the woods.

Additionally, in regard to the climate response(or climatic concern) and passive solar energy application, this project takes full advantage of the local solar radiation resources to solve the problems in winter thermal insulation with the utilization of photovoltaic panel, coloured photovoltaic panel, solar room and rammed earth materials in the envelope enclosure. The natural ventilation of the structure is greatly improved by the setting of solar corridors, ventilated Trombe Walls and operable solar surfaces.

Besides, as another response to local cultures, with the recycle of the dug earthwork for rammed earth wall fabrication and the abundant local natural resources (soil, stone and bamboo) utilization for construction, controlling the carbon emissions in the construction process, the goal of reducing carbon emissions can be achieved.

■ Block Generation

01 Surroundings — the connection between the site and surroundings
02 Original Block — the connection between the block and site
03 Block Yield — the changes to adjust to the site edge line
04 Block Lift — the changes to adapt to the height difference
05 Block Piece of Cutting — block cut for the requirement of function
06 Roof Generation — exact-angle roof designed to be covered by solar panels
07 Courtyard — to take in the natural light and advance the conditions of ventilization
08 Trombe Wall — to make the complex energy-efficient

林间关坝·沐日而生
四川平武关坝自然保护区科学考察站设计

Immersing in the Forest, Soaking up the Sunshine

Economic and Technological Indexes
- Total area:
  - first phase: 2293.96m²
  - second phase: 567.89m²
  - total: 2861.85m²
- Overall floorage: 2130.39m²
- Building area: 1310.00m²
- Plot ratio: 0.74
- Square road paving area: 462.09m²
- Green area: 1089.77m²
- Greening rate: 38%
- Parking space: 6

Site Plan 1:300

■ Site Analysis

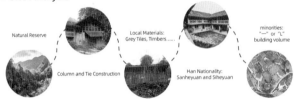

The site is situated in Pingwu County, Mianyang City, Sichuan Province. With the foundation of natural reserve for a long time, the village of Guanba has put forward a series of efficient ways to achieve the balanced relationship between nature and humans. Lying in such natural reserve, the site is low in the northwest and high in the southeast with several tall trees to protect.

Addtionally, in the southern side of the east - west stream right lies the site. The geographical location of Guanba village provides the site with the pleasant climate that makes it cool and rainy in summer, cold and snowy in winter. The annual average temperature is 13.9 ℃ and that of January is 4 ℃ which still remains above 0 ℃.

 site   surroundings   road
 transportation   sunlight   wind
  noise   buildings vegetation

■ Technology System

■ Climate Simulation

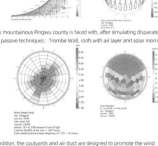

Concerning the major problem, the thermal insulation in winter, which mountainous Pingwu county is faced with, after simulating disparate architectural design strategies, this project has drawn the appropriate passive techniques: Trombe Wall, roofs with air layer and solar room.

According to the local information of wind direction and the land condition, the coutyards and air duct are designed to promote the wind environment of the buildings based on the soft ware simulation. According to the average daily temperature and solar radiation conditions, the distribution and the exact angle of solar panels are determined.

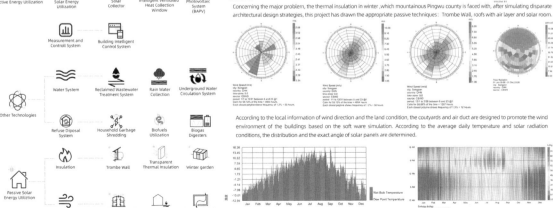
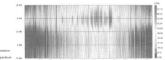

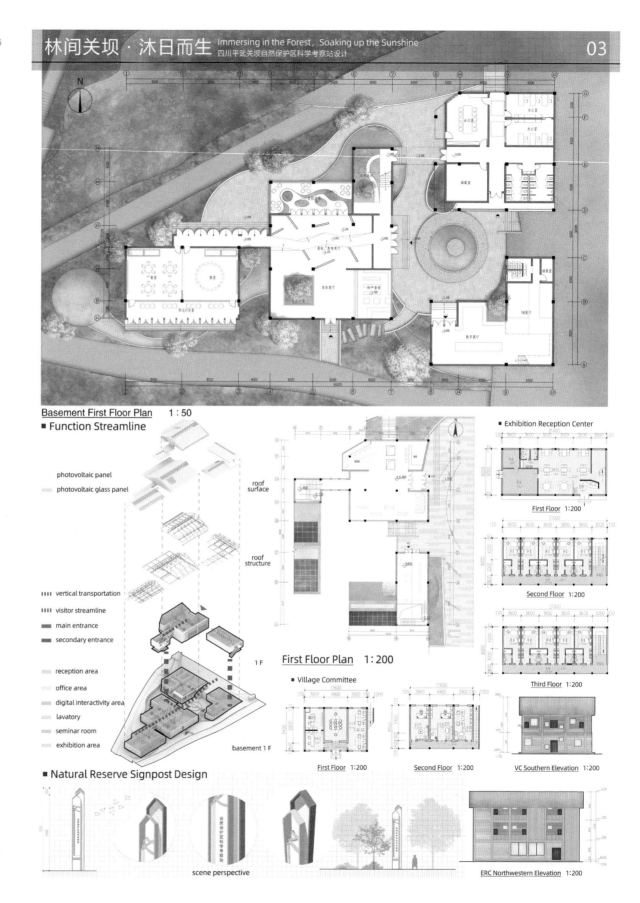

林间关坝·沐日而生
Immersing in the Forest, Soaking up the Sunshine
四川平武关坝自然保护区科学考察站设计

04

■ Active Solar Technology

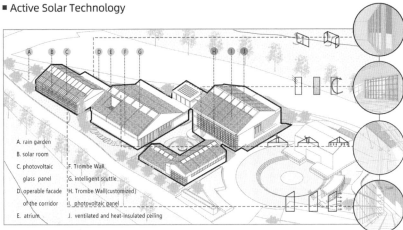

A. rain garden
B. solar room
C. photovoltaic glass panel
D. operable facade of the corridor
E. atrium
F. Trombe Wall
G. intelligent scuttle
H. Trombe Wall (customized)
I. photovoltaic panel
J. ventilated and heat-insulated ceiling

■ Energy Calculation

Electricity Consumption

According to Electricity consumption index table for all types of buildings in 19DX101-1 Common data for building electrical, electricity consumption of exhibition hall is 50W/m². The GFA of Phase I and Phase II is 1117.10m². If the working hours are 8h per day, the building's electricity daily consumption is:

$50W/m² \times 1117.10m² = 55855W = 55.855 kW \cdot h$

Electricity Production

The area of photovoltaic panels and photovoltaic glass on roof is 712.69m². Current module power of conversion efficiency of crystalline silicon cells is 100~120W/m² and the system efficiency is 70%. Thus, the amount of electricity generated on the winter solstice with a daylight time of 8 hours is:

$100W/m² \times 70\% \times 8h \times 712.69m² = 57012W \cdot h = 57.012 kW \cdot h$

Conclusion: $57.012 kW \cdot h > 55.855 kW \cdot h$

At the same time, the electricity production from PV façade is not counted in. It can be concluded that the complex's own electricity consumption can basically be covered by solar power generation.

■ Rain Garden Circulation System

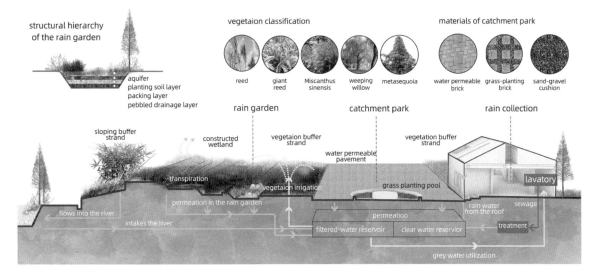

林间关坝·沐日而生 — Immersing in the Forest, Soaking up the Sunshine

四川平武关坝自然保护区科学考察站设计

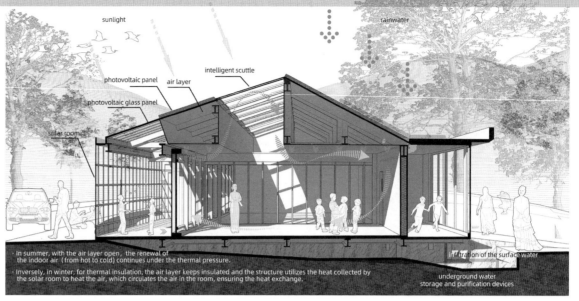

- In summer, with the air layer open, the renewal of the indoor air (from hot to cold) continues under the thermal pressure.
- Inversely, in winter, for thermal insulation, the air layer keeps insulated and the structure utilizes the heat collected by the solar room to heat the air, which circulates the air in the room, ensuring the heat exchange.

■ Details of Ventilators   ■ Analysis of Interior Wind Environment   ■ Section Strategy

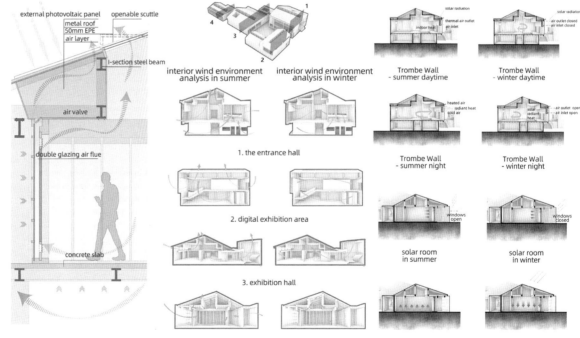

■ Scene Perspective

Solar Room

Hall Space

Wetland Exhibition Area

林间关坝·沐日而生
Immersing in the Forest, Soaking up the Sunshine
四川平武关坝自然保护区科学考察站设计

■ Indoor Ventilation Simulation

First Floor in Summer   First Floor in Summer   Second Floor in Summer

First Floor in Winter   First Floor in Winter   Second Floor in Winter

■ Sunlight Simulation

Winter Solstice Sunlight Period (N)   Summer Solstice Sunlight Period (N)   Summer Solstice Midday Shadow

Winter Solstice Sunlight Period (S)   Summer Solstice Sunlight Period (S)   Winter Solstice Midday Shadow

■ Energy Consumption Calculation

能耗分项	需求量(kW·h/㎡)	可再生能源利用	利用量(热量)(kW·h/㎡)
耗冷量 Qc	0.00		
耗热量 Qh	17.02	地源\空气源热泵 EPh	12.76
生活热水耗热量 Qw	1.62	太阳能\空气源热泵	1.62
照明能耗 Ql	29.00	光伏发电 Er	368.87
电梯能耗 Qe	0.00	风力发电 Ew	0.00
合计	47.64		383.25

■ Carbon Emissions Calculation

类别	年碳排放量(kgCO$_2$/㎡·a)	碳排放量(kgCO$_2$/㎡·a)
建筑材料生产	77.62	3881.39
建筑材料运输	0.00	0.00
建筑建造	0.01	0.40
建筑拆除	7.76	388.18
建筑运行	-53.60	-2680.06
碳汇	-5.69	-284.71
合计	26.09	1305.20

■ Structure System

A. connection of wooden structure in three directions
B. joint of diagonal bracing and column
C. joint of wooden pillar and bottom pillar
D. connection of primary and secondary beams

■ Construction Material Analysis

photovoltaic panels   rammed earth   bamboo fine   steel   glass tlie

■ Plant Footprint

ginkgo   reed   Salix babylonica   Arundo donax   Taxodium hybrid

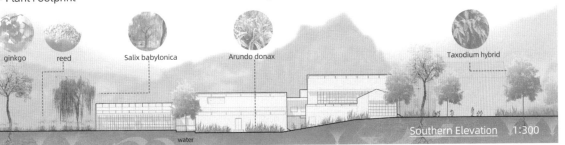

Southern Elevation   1:300

综合奖・优秀奖
General Prize Awarded・
Honorable Mention Prize

注 册 号：100821
项目名称：东风知我欲山行，
吹断檐间积雨声
Knowing that I Want to Go Hiking, the East Wind Blows off the Accumulated Rain between Eaves
作　　者：郝泽厚、潘静涵、郑子豪、姚黄城
参赛单位：福州大学
指导教师：王　炜

东风知我欲山行，吹断檐间积雨声   02
KNOWING THAT I WANT TO GO HIKING, THE EAST WIND BLOWS OFF THE ACCUMULATED RAIN BETWEEN EAVES

SITE ANALYSIS

VALLEY SITUATION

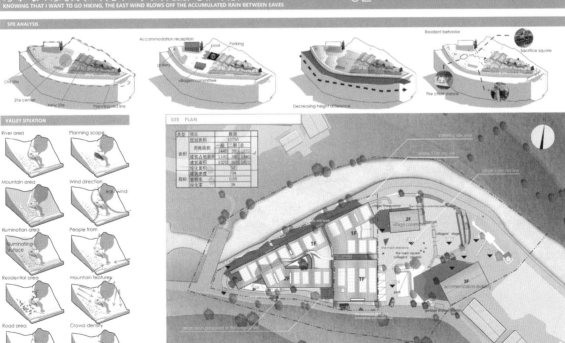

总层平面图 1:500

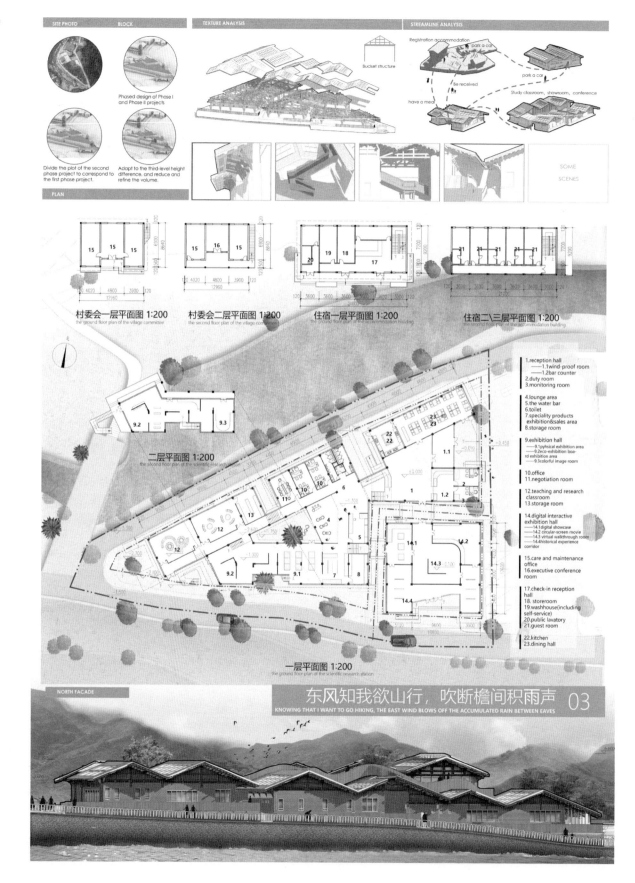

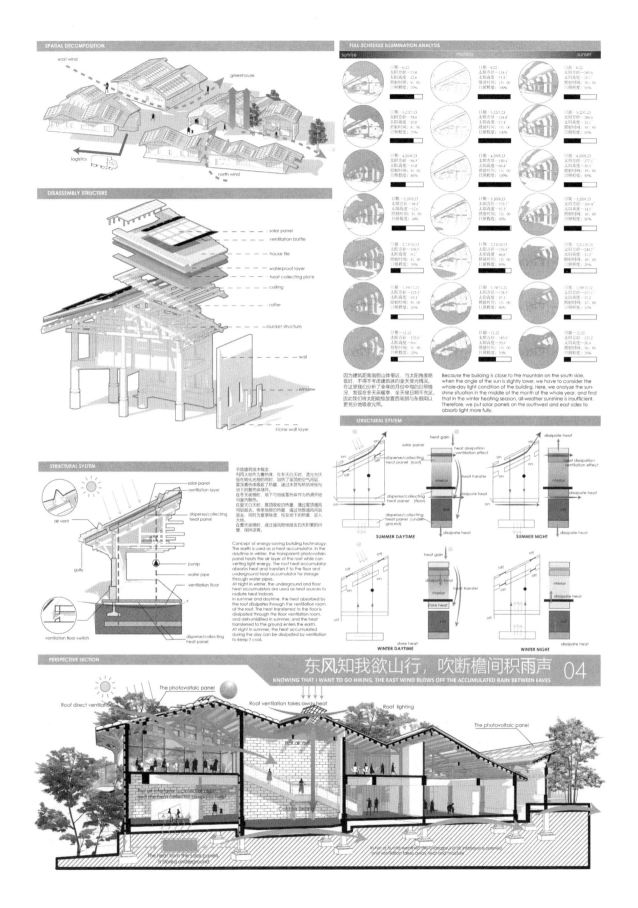

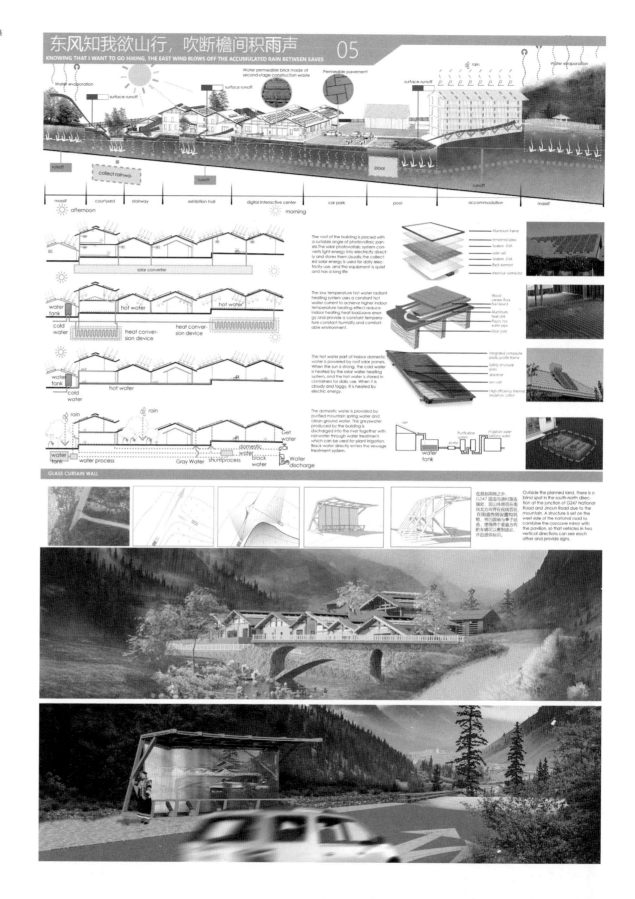

东风知我欲山行，吹断檐间积雨声   06
KNOWING THAT I WANT TO GO HIKING, THE EAST WIND BLOWS OFF THE ACCUMULATED RAIN BETWEEN EAVES

GLASS CURTAIN WALL

中空玻璃采用LOW-E玻璃，双层玻璃的构造，在冬天保证建筑采光的同时保温。在夏天，玻璃间设作为通风层可以带走阳光直射时产生的热量。

The atrium is made of LOW-E glass and double-layer glass, which can ensure the lighting of the building and keep warm in winter. In summer, the glass interlayer as a ventilation layer can take away the heat generated by direct sunlight.

HOUSE TILE

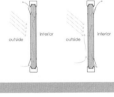
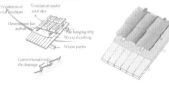

WINDPROOF ROOM

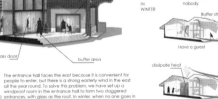

入口的门厅因为方便人员进入而面向东面，但东面常年有很强的东风，为了解决这个问题，我们在门厅中设置了一道窗风室，形成两个交错的入口，并以玻璃作为屋顶。在冬天，当没人进出时，第一道门很好的阻隔了东风的进入，并且屋顶的阳光会加热防风室，防止热量散失。当有人进出时，第二道门会阻隔东风。在夏天，没人进出时，打开第一道门，东风会带走屋顶阳光照射产生的热量。有人进出时，打开双道门，东风会进入建筑内，促进通风。

The entrance hall faces the east because it is convenient for people to enter, but there is a strong easterly wind in the east all the year round. To solve this problem, we have set up a windproof room in the entrance hall to form two staggered entrances, with glass as the roof. In winter, when no one goes in or out, the first door blocks the entrance of the east wind, and the sunlight on the roof will heat the windproof room to prevent heat loss. When someone comes in and out, the second door will block the east wind. In summer, when no one goes in or out, open the first door, and the east wind will take away the heat generated by the sunlight on the roof. When someone comes in and out, open the double doors, and the east wind will enter the building to promote ventilation.

SECTION

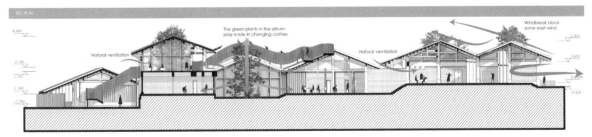

综合奖·优秀奖
General Prize Awarded · Honorable Mention Prize

注 册 号：100836
项目名称：山·望
　　　　　Ridge·Telescopic
作　　者：樊泽宇、蒲丹妮、单司辰
参赛单位：西安科技大学
指导教师：李雪平、孙倩倩

山·望 RIDGE·TELESCOPIC
DESIGN OF SCIENTIFIC RESEARCH STATION IN NATURE CONSERVATION DISTRICT

• Vegetation Analysis

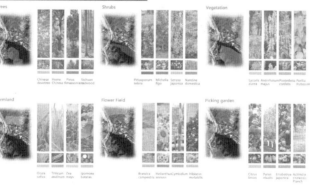

• Element Extraction

山·望 RIDGE·TELESCOPIC
DESIGN OF SCIENTIFIC RESEARCH STATION IN NATURE CONSERVATION DISTRICT

- **Functional Partition**
- **Block Generated**

- **The Total Floor Plan 1:500**

DESCRIPTION OF DESIGN:

本设计一方面在形态上呼应传统四川民居群落形态，并加以现代化设计与规整。丰富的室内外空间期待着良好的观景视野。另一方面在技术上，采用主被动式技术互相结合的设计策略，选用南方建筑特有的被动式空间形式、当地原材、太阳能技术，在最大经济化的条件下给人以最舒适的体验。另外，在室内外流线上，采用古典园林式组织手法，在空间体验和视线组织上均有丰富的变化。

On the one hand, this design echoes the traditional sichuan folk house community form in form, and is modernized and orderly. The rich indoor and outdoor space has a good view. On the other hand, in terms of technology, the design strategy of combining active and passive technologies is adopted. The passive space form unique to southern architecture, local raw materials and solar energy technology are selected to give people the most comfortable experience under the maximum economic conditions. In addition, in the indoor outflow line, the use of classical garden-style organization, in the space experience and visual organization are rich changes.

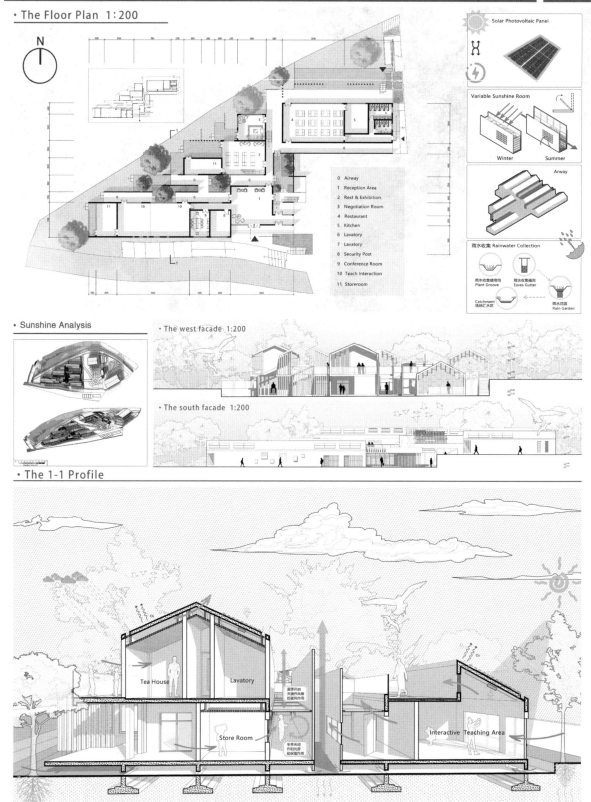

山·望 RIDGE · TELESCOPIC
DESIGN OF SCIENTIFIC RESEARCH STATION IN NATURE CONSERVATION DISTRICT

IV

• The Explosion Figure

Its operating mechanism is mainly based on solar energy as a heat source to induce ventilation, heat is transferred to the ventilation pipe in the way of heat conduction and heat convection, and the air continues to rise under the action of solar radiant heat, forming thermal pressure difference and rising to the air outlet, and forming flow circulation with the indoor air

Before Dismantling

Trombe Wall was proposed by French professor Felix Trombe and his collaborators in 1967 and was the first to combine passive solar technology with architecture

1. Brick and concrete heat storage wall
2. Hydrophobic rock wool insulation pavement
3. Translucent ceramic tile slice cover plate
4. Rainwater collecting tank
5. Ventilation skylight
6. Sichuan traditional small green tiles
7. Gutters hide cover
8. Roofing with composite solar photovoltaic panels
9. Ventilated roof

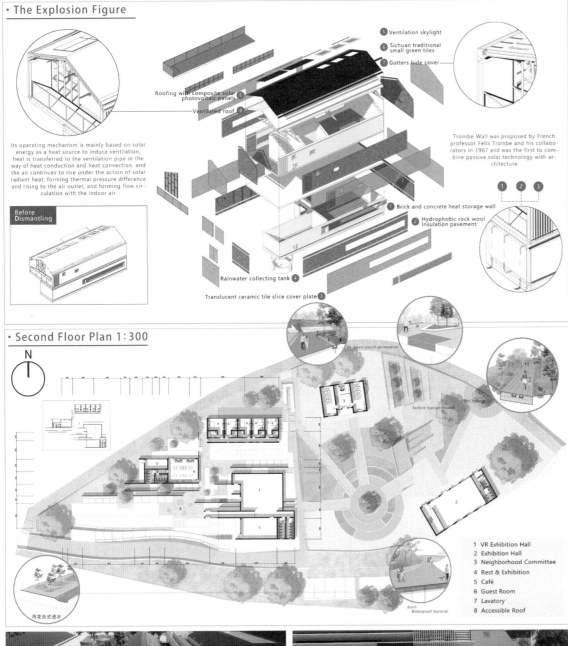

• Second Floor Plan 1:300

1. VR Exhibition Hall
2. Exhibition Hall
3. Neighborhood Committee
4. Rest & Exhibition
5. Café
6. Guest Room
7. Lavatory
8. Accessible Roof

综合奖·优秀奖
General Prize Awarded · Honorable Mention Prize

注 册 号：100845
项目名称：栖木·归尘
　　　　　Live under Woods · Back to Dust
作　　者：曹裕龙、阮　聪、刘伊琪、沈书睿
参赛单位：南京工业大学
指导教师：薛春霖

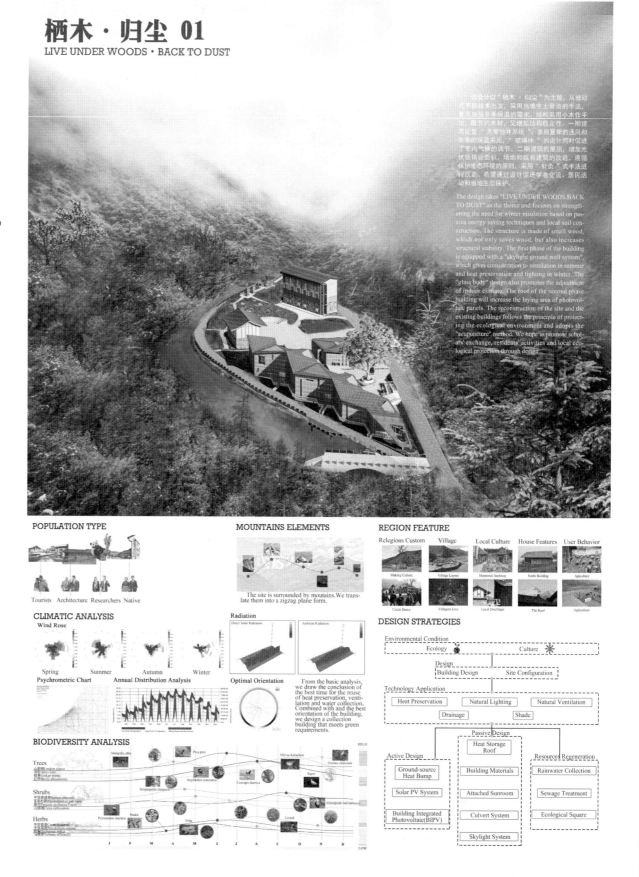

栖木·归尘 02
LIVE UNDER WOODS · BACK TO DUST

MASTER PLANE 1:400

ECONOMIC AND TECHNICAL INDICATORS:
BUILDING LAND AREA: 2097 m²
FLOOR AREA: 2861.5 m²
FLOOR AREA(FIRST PHASE OF CONSTRUCTION): 997 m²
FLOOR AREA(SECOND PHASE OF CONSTRUCTION): 303 m²
PLANNED LAND: 10709 m²
BUILDING HEIGHT: 9.8M
PLOT RATIO: 0.19
BUILDING DENSITY: 0.34

ENTRANCE SIGN

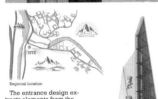

Regional location

The entrance design extracts elements from the nearby Tibetan headdress, combines with the local sacred tree culture, and uses wooden triangular panels to design, forming landscape markers at the entrance of Guanba Village. The setting of the structure can guide visitors from outside into the site, and at the same time can be used as a symbol of spiritual culture.

FUNCTIONS & STRATEGIES

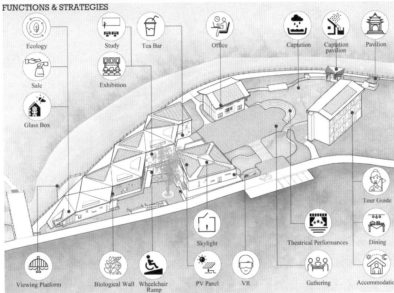

SITE GENERATION

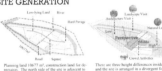

Planning land 10677 m², construction land for depression. The north side of the site is adjacent to the lake, and the south side is a motor vehicle lane.

There are three height differences inside the site, and the site is arranged in a divergent form in response to the river.

Connect three height differences with a unified streamline to activate the square. Diverges the water flow line inside the site to form more ecological space.

Put into the building to form a complete site space streamline. Interspersed with the ecological system, highlighting the ecological theme.

BLOCK GENERATED

① A Long Bock
② Three Fnctional Pieces
③ The Top Dressing
④ South Facing Plywood Roof

According to the site characteristics, put rectangular blocks.

Divide the whole into three volumes, echoing another volume.

Echoing the surrounding site, the shape of the building is reversed.

Reorienting the roof, placing solar panels and skylights.

SECOND FLOOR PLANE 1:200

FIRST FLOOR PLANE 1:200

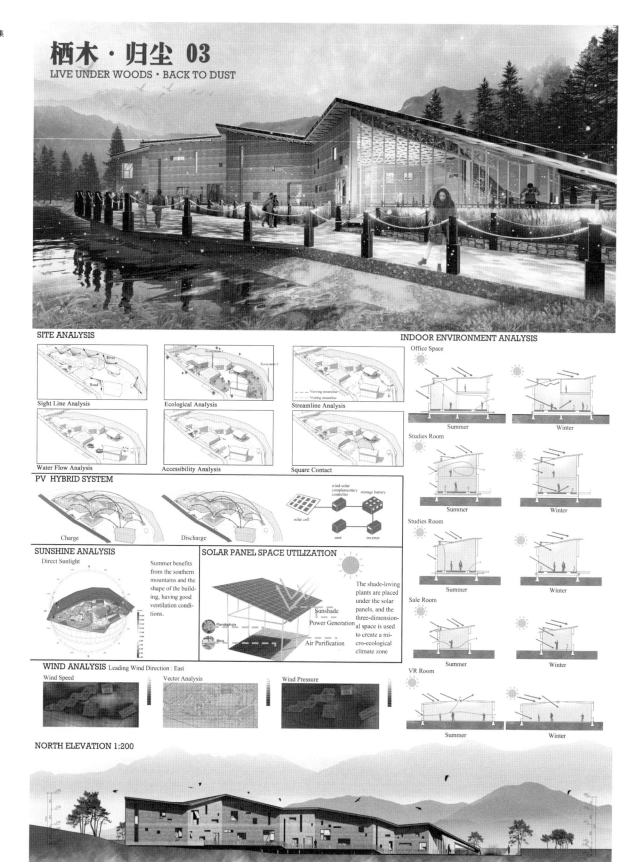

栖木・归尘 04
LIVE UNDER WOODS · BACK TO DUST

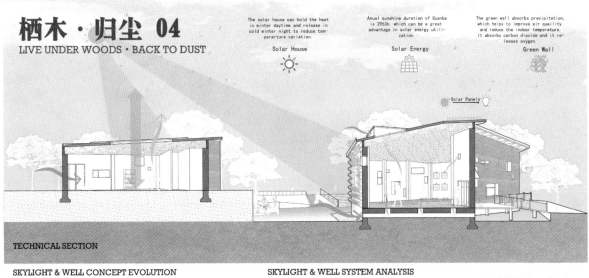

TECHNICAL SECTION

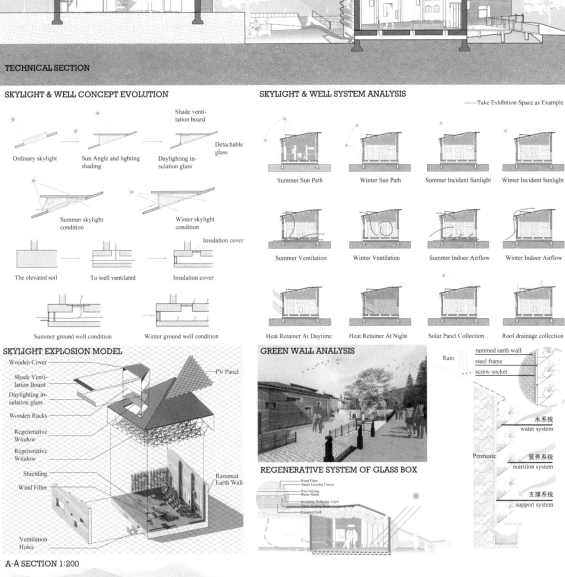

A-A SECTION 1:200

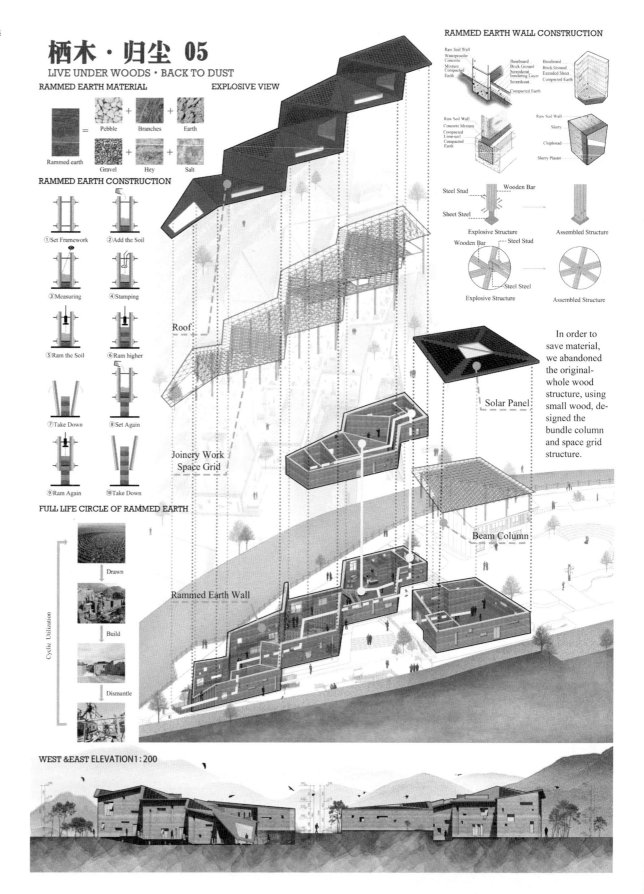

栖木·归尘 06
LIVE UNDER WOODS · BACK TO DUST

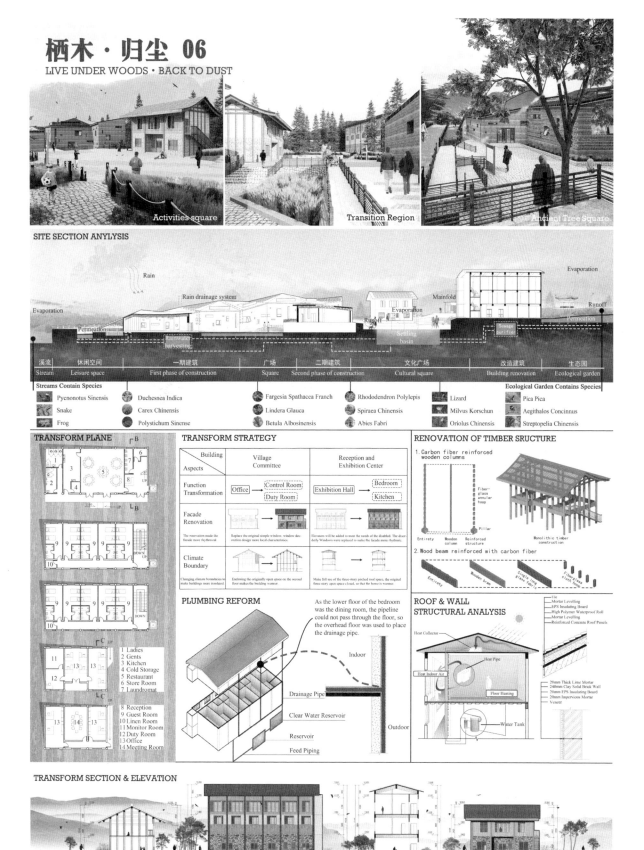

综合奖 · 优秀奖
General Prize Awarded ·
Honorable Mention Prize

注 册 号：100931
项目名称：林遇·镜缘
　　　　　Palpitating in the Forest ·
　　　　　Forest in the Water
作　者：宁馨儿、孙晴波、倪云杰、
　　　　王　彤
参赛单位：南京工业大学
指导教师：罗　靖、刘晓光

林遇·镜缘

Palpitating in the Forest, Forest in the Water

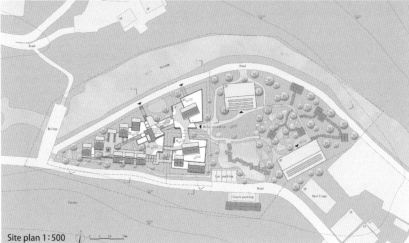

Site plan 1:500

South elevation 1:200

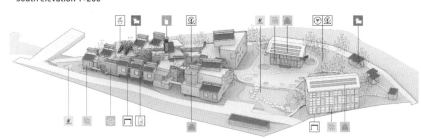

Techno-economic indicater

Planning land area: 10709 ㎡
GAR: 82.4%

New building:
Site Area: 2853 ㎡
GBA: 530.3 ㎡
GFA: 2176.33 ㎡
BD: 0.69
FAR: 0.76
Storey Height: 17.3m

Renovation building No. 1:
FAR: 110.94 ㎡
GFA: 221.88 ㎡
Storey Height: 8.9m

Renovation building No. 2:
FAR: 190.75 ㎡
GFA: 572.25 ㎡
Storey Height: 13.7m

Inspiration source

Cabin in the woods   Puddles after rain

Form-creation analysis

1. Site topographic analysis: The middle is a depression.
2. According to the wind direction, determine the main dehumidification wind road of the site —— divide the venue into three pieces.

3. In order to prevent moisture and protect biodiversity, the main part of the building is lifted.
4. The shape is partially hollowed out to form the atrium and entrance space.

5. Adjust building mass relationship.
6. Modular building units are placed.

7. Strengthen the connection of each building block, and achieve a certain degree of wind obstruction while dehumidifying.
8. Add solar chimneys to modular blocks.

Analysis of green technology

Based on the local climate, we form the biodiverse and ecological strategy to solve the existing problems like wetness by reconstructing water treatment and rainwater collection, and improve the heating and ventilation conditions of buildings through the solar building design strategy.

Solar energy utilization

Tromble wall   Solar chimney   Attached sunspace
Solar panels   Solar thin films

Environmentally friendly measures

The elevated ground floor   Ecological microclimate zone
Bird habitat   Planting test chamber
Site ecological construction method transformation
Rain water collection and preliminary filtration

2/6

1st Floor Plan 1:200

1 Rest reception area
2 Exhibition hall
3 Folklore exhibition hall
4 Climate exhibition hall
5 Botanical Exhibition hall
6 Animal exhieebition hall
7 Village development exhibition hall
8 Toilet
9 Specialty exhibition and sales room
10 Storeroom
11 Hotel reception
12 Leisure room
13 Standard room
14 Loft room
15 Public Living room
16 Research room
17 Laundry
18 Auxiliary room
19 Terrace
20 Courtyard

2nd Floor plan 1:200

1st floor plan 1:100   2nd floor plan 1:100
Loft room plan

B1 Floor plan 1:200

North elevation 1:200

林遇·镜缘

Palpitating in the Forest, Forest in the Water

Wind simulation

summer solstice   winter solstice

Sun light calculation

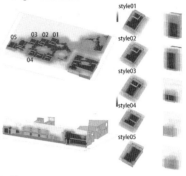

style01
style02
style03
style04
style05

Building sewage treatment system

Solar panel workflow

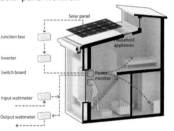

Equipped with power monitors in the room, it popularizes solar power generation to people and reminds people to save electricity.

Building construction process analysis

1. Factory prefabricated components

CLT beams and pillars | CLT shear wall | Prefabricated solar panels | Rubber bearing

2. Transportation of building materials and disposal of construction waste

The project is expected to use seven heavy-duty trucks to transport the components. | It is planned to pile up construction waste on the square within the site and sort and remove it in a timely manner. | The sorted construction waste will be recycled and reused.

3. On-site hoisting

Phase II construction
Phase I construction

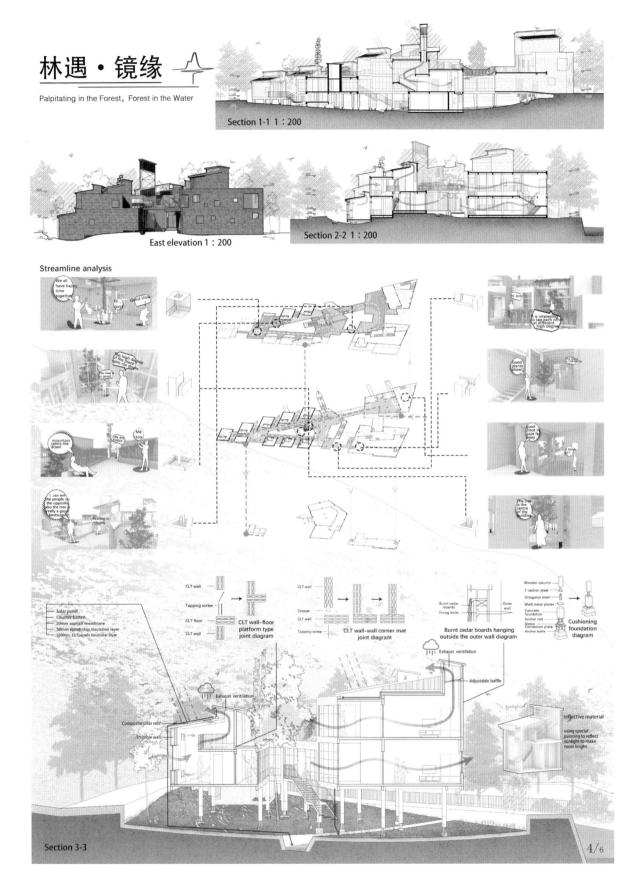

林遇·镜缘

Palpitating in the Forest, Forest in the Water

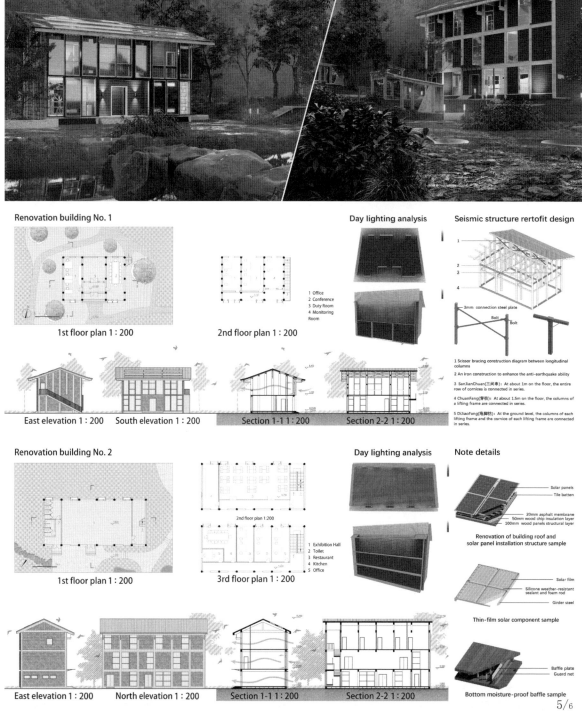

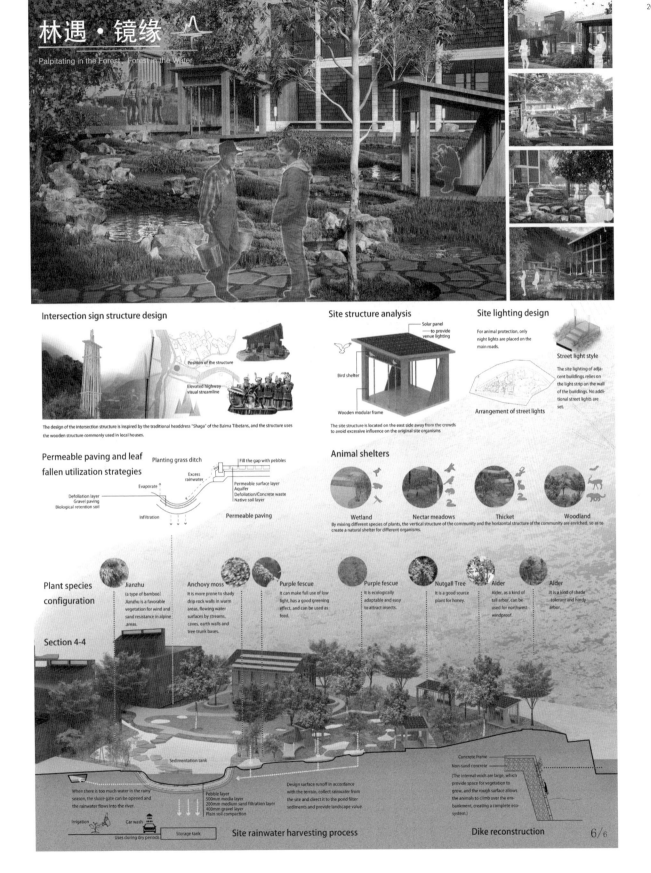

揽风·入垣 1
TOUCH THE WIND

综合奖·优秀奖
General Prize Awarded·
Honorable Mention Prize

注 册 号：101012
项目名称：揽风·入垣
　　　　　Touch the Wind
作　　者：支叶青、赵春燕、郑雅娴
参赛单位：北京建筑大学
指导教师：孟璠磊、刘 烨

Climate Analysis

最佳方向/Optimum Orientation　　盛行风/Prevailing Winds　　湿度图/Temperature Relative Humidity

太阳位置/Solar Position　　每月盛行风/Per Month Prevailing Winds　　气温分析/Temperature Analysis

气温、相对湿度/Air Temperature、Relative Humidity　　干、湿球温度/Dry/Wet Bulb

Traditional Elements Analysis

水文化/Water Culture

天井/Patio

竹钢/Wooden Bamboo

河道石/Local Stone

竹编夹泥墙/Bamboo Woven Mud Wall

Site Analysis

River, mountains and existing buildings　　Ambient echo effects　　Renovation of existing buildings on site

The wind direction of the site and the influence of the valley wind　　Mountain form and skyline　　Landscape vitality of the site

设计说明

方案从平武县山谷风由东向西吹出发，对场地进行风模拟。场地位于山谷中，引入河水，利用水的蒸腾作用与通风墙的拔风作用去除场地的湿热。营造舒适凉爽的工作环境。顺应地势，通过架空下沉的做法为人群提供更多的休闲聚集空间，底层架空更少的破坏场地土地，为野生动物提供更加自由的行走行径。利用场地现存建筑进行改造，使场地联系更加紧密，更加符合人群需求。建筑采用主动与被动技术相结合的形式。利用地源热泵技术、引入海绵城市概念，减少地表径流，更加符合场地本身的需求。

Design Description

The plan starts from the wind blowing from east to west in the valley of Pingwu County, and conducts wind simulation on the site. The site is located in a valley, and river water is introduced to remove the damp and heat of the site by the transpiration of water and the pulling effect of the ventilation wall. Create a comfortable and cool working environment. In accordance with the terrain, the method of overhead sinking provides more leisure gathering space for the crowd, and the ground floor is overhead with less damage to the site and land, and provides more free walking for wild animals. The building takes the form of a combination of active and passive technologies. Using ground source heat pump technology and introducing the concept of sponge city to reduce surface runoff, it is more in line with the needs of the site itself.

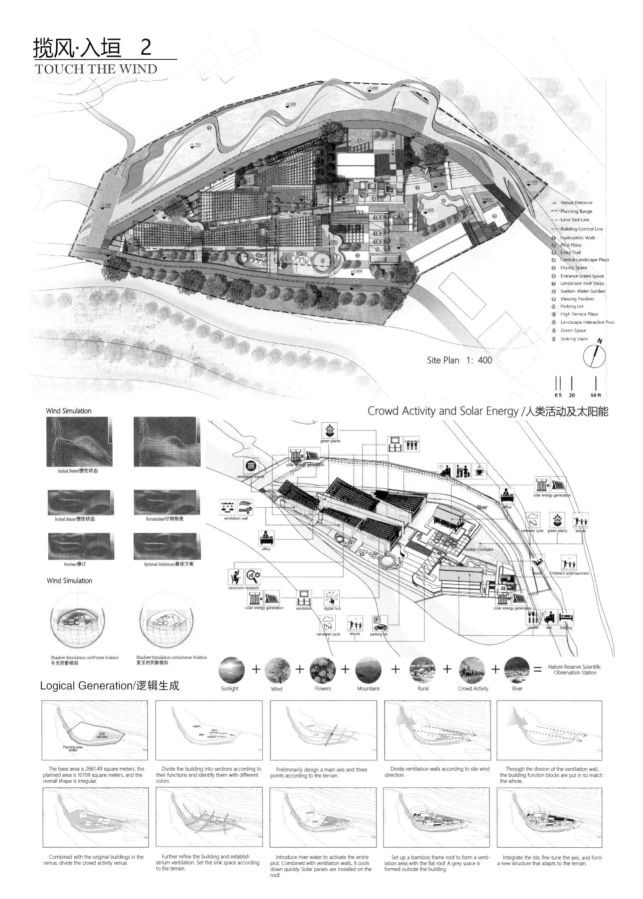

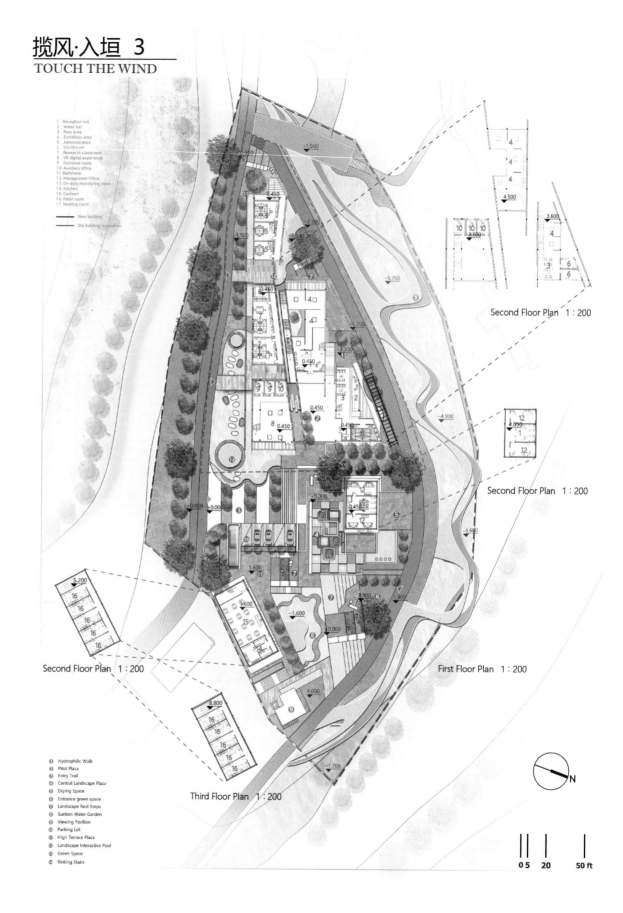

揽风·入垣 4
TOUCH THE WIND

Ventilation Wall Analysis

Traditional-style patio

The patio of traditional buildings is used for ventilation and heat dissipation, while new buildings are built on the basis of the traditional patio and the bottom floor is elevated to add pools of water to bring water into the air circulation.

New construction patio

- Patio plucking wind
- Double wall ventilation
- Double roof ventilation
- Avoid direct sunlight on the roof
- Blocking heat entry
- Solar panels collect energy
- Indoor lighting
- Indoor heat dissipation through windows
- Insulation in winter

Sensory

- th side ventilation wall
- Sunken plaza
- tilation wall at the display
- Exhibition space ground floor
- e ventilation wall
- Exhibition space, first floor

Section A-A

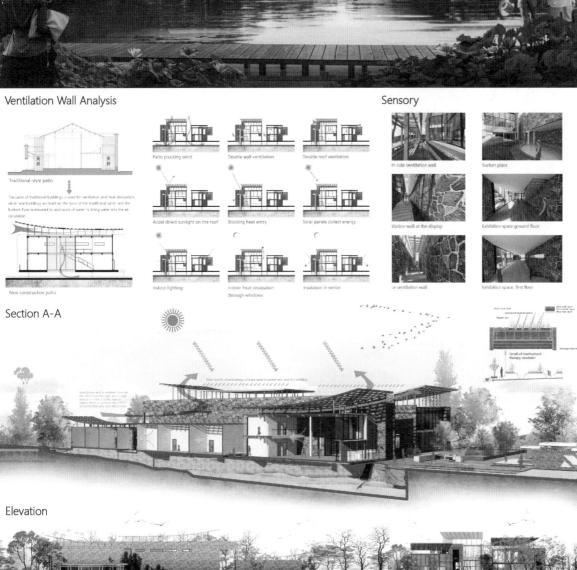

Elevation

North Elevation 1:200

East Elevation 1:200

揽风·入垣 5
TOUCH THE WIND

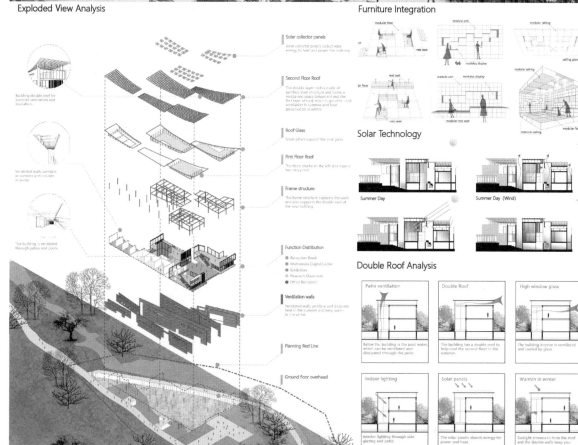

揽风·入垣 6
TOUCH THE WIND

Trombe Wall

Structural Analysis

Active Architectural Technology Analysis

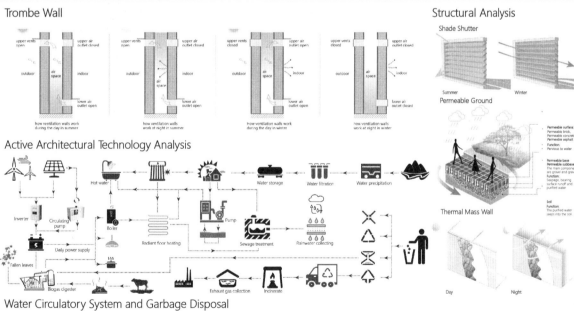

Water Circulatory System and Garbage Disposal

综合奖 · 优秀奖
General Prize Awarded · Honorable Mention Prize

注 册 号：101019
项目名称：光栖坤灵
　　　　　Growing in Soil
作　　者：吕宇蓓、吴　爽、陈　茜、
　　　　　陈烁遠、陈迪菲
参赛单位：南京工业大学、同济大学
指导教师：杨亦陵

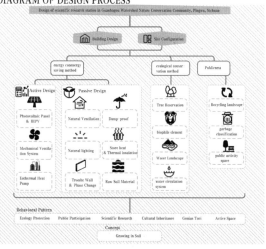

光栖坤灵 II
Growing in Soil

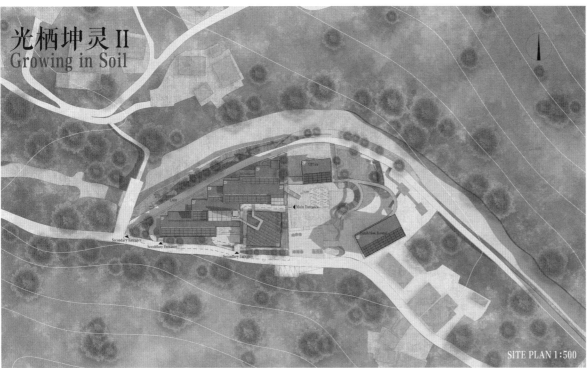

SITE PLAN 1:500

PLAN GENERATION
Spatial Structure

Context — Our site is approximately 1412 square meters for the first phase construction, which is a triangle.

Function Division — We divide the plot into six blocks: research area, exhibition area, VR experience area, reception area, office area and living area.
- research area
- exhibition area
- VR experience area
- reception area
- office area
- living area

Tree Reservation — The shape of the volume is determined according to the size of the area required by each function.

Lane — In order to enrich the spatial moving lines, we made the layout of the building form a horizontal moving line and three vertical moving lines.

Public Space — Next, we planted some outdoor public spaces.

Then we mapped out the landscape space.

Outdoor public activity space
— landscape space

Topographic Adjustment — In order to minimize the damage to the original site, we adjusted the building vertically to better fit the original terrain.

To prevent moisture, we raised the building and built it on a stone foundation.

CROWD ACTIVITY AND SOLAR ENERGY

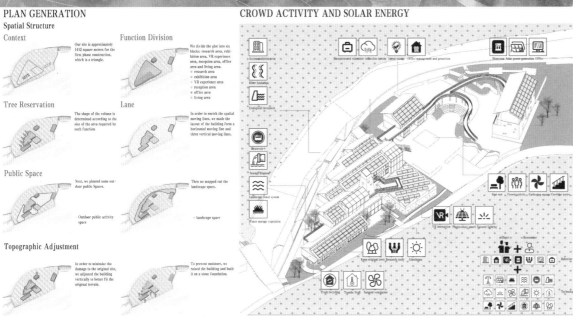

Insolation Effects · Wind Effects

	Direct Sun Hours	Incident Radiation
Winter Solstice		
Summer Solstice		

Wind Speed Simulation | **Wind Direction Simulation**

1.2m/s East Wind Simulation

Based on light and wind simulation results, the south side of each building receives basic light and the wind flows through the entire site. Due to the blocking of the mountain, there is a lack of light inside the site on the winter solstice, so shade-loving plants can be considered.

Details

Sloping Roof — A pitched roof is inserted to allow the building to better drain water and collect solar energy.

Traditional Watchtower — A traditional watchtower is placed to help ventilate and light the building.

Grass Slopes — The next step is to put it into the grassy slope.

Direct Runoff — Direct runoff is also planned by us.

Corridor — The aerial corridor is used to connect the first phase building, the second phase building, the site and the existing building.

Solar House — Finally, put in some solar cells and atrium covered by a shimmering glass roof to collect heat.

光栖坤灵 III
Growing in Soil

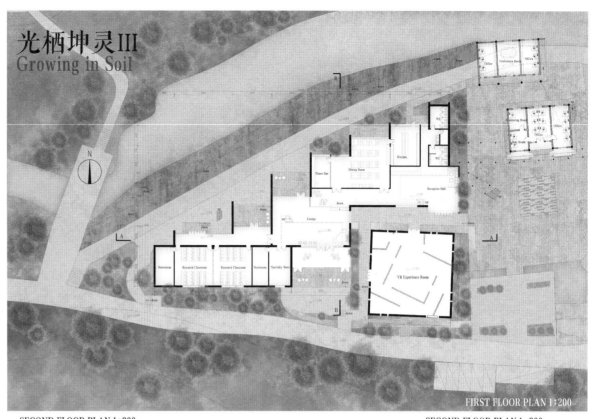

FIRST FLOOR PLAN 1:200

SECOND FLOOR PLAN 1:200

SECOND FLOOR PLAN 1:200

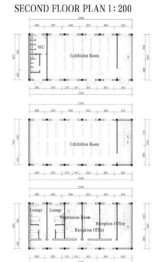

RECONSTRUCTION OF EXISTING BUILDINGS

Existing buildings | Add an openable solar house on the periphery | The solar house is insulated in winter and can be opened in summer for ventilation

Dig out part of the roof slab | Connect the excavated roof slab to the roof slab on the other side to form a herringbone roof | Wrap the facade of the first and second floors of the building with rammed earth to achieve the effect of thermal insulation

ENTRY SIGNAL

The entry signal is in the form of a combination of traditional local elements, a traditional watchtower and a pitched roof, which is made of rammed earth material. There are small windows on the facade, which are in harmony with the building.

光栖坤灵 IV
Growing in Soil

INDOOR ILLUMINATION ANALYSIS

Place floor-to-ceiling Windows on the on the north side.

Skylights are cut in the roof to enhance the interior light.

Day Light Factor
❄ Winter ☀ Summer ❄ Winter ☀ Summer

Indoor light condition is not good, which can not meet the use requirements.

Indoor illuminance is brighter, illumination is more uniform, forming a more uniform light environment.

TROMBE WALL & CAPSULE

The use of normal phase change materials can improve the thermal storage performance of the enclosure structure, but leads to material strength reduction. Therefore, we applied phase change microcapsules to our design, to remedy the limitation and improve the thermal performance of our trombe walls.

The modelling of physical properties of phase change microcapsules proves that this design can improve thermal storage performance both in summer and in winter.

Application of PEDF
Photovoltaics

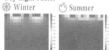

PV panels (placed in the direction which maximize the collection of solar energy)

The total area of PV panels = 500 squaremeters
It is expected to provide 70kW electrical power, and theoretically can provide the electrical power needed for the whole site. (Electrical consumption estimated as 50kW·h per squaremeter.)

Energy Storage

The solar power collected in summer and in daytime can be stored to be used.

Combination of Direct Current and Indirect Current

The application of DC can increase the energy efficiency and the stability.

Flexibility

Electricity curve (per day)

The devices have the property of interruptability and adjustable. By applying the flexibility regulation with the help of energy storage, It can achieve flexible power demand, so that the electricity curve can be smoothened and achieve energy conservation.

SOLAR CHIMENEY ANALYSIS

Summer solar angle | Winter solar angle | Bottom exothermicdampness | Set high and low Windows, using the principle of hot pressing, promote indoor air flow

During the winter days, trombe wall and solar house store heat. | During the winter nights, trombe wall and solar house give off it heat. | During the summer days, the roof is shaded with bamboo, and a valve under the Trombe wall and the door of the solar house are opened to create a ventilation inside. | On summer nights, open the valve under the Trombe wall and the door to the solar house to ventilate the interior.

SOLAR HOUSE ANALYSIS
❄ Winter ☀ Summer

In winter, the sun room collects and stores the sun's heat. | Then, the sun room releases heat into the building. | In the summer, honeycomb curtains shade the sunroom to prevent its overheating. | The sun room can also be opened to facilitate natural ventilation.

ATRIUM AND TRADITIONAL WATCHTOWER ANALYSIS
❄ Winter

Winter sunlight heats the glass atrium, raising the indoor temperature of the glass atrium | The heat heats the surrounding interior of the building. | According to the day light factor, there is plenty of light in the room | The traditional watchtower has the function of lighting, ventilation and viewing

☀ Summer

In summer, the glass atrium is shielded from the sun by a honeycomb curtain. | The glass atrium can be opened for ventilation. | According to the day light factor, indoor light is less, to achieve the effect of insulation | According to the ventilation simulation diagram, Diaolou is helpful for ventilation.

SOUTH FACADE 1:200

光栖坤灵 V
Growing in Soil

B-B SECTION PERSPECTIVE

AXONOMETRIC EXPLODED VIEW

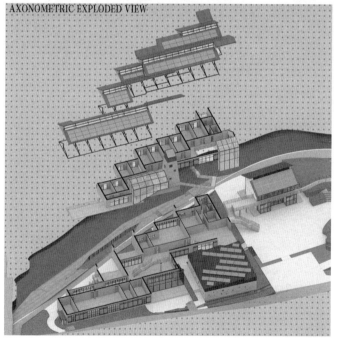

RECYCLE ANALYSIS OF STEEL STRUCTURE

As more cars are sold in China, the number of cars being scrapped is creating more steel waste.

The construction of buildings also produces a lot of steel waste.

Meanwhile, the production of industrial products produces a large amount of steel waste.

The waste is sent to a factory in Mianyang for processing.

RAMMED EARTH WALL ANALYSIS

Advantages

Recycling ability — The rammed earth of a demolished house can be reused. After being dismantled and broken, it can be returned to the land as fertilizer.

Absorption of nitrogen — Rammed earth wall can absorb nitrogen from the air.

Construction convenience — The construction and installation of rammed earth wall is convenient. The materials used can be obtained on site, greatly reducing transportation costs and material processing costs.

Fire safety — Compared with the wooden building, the fire performance is good.

Good thermal inertia — Rammed earth wall has small thermal conductivity, good thermal inertia, strong heat storage capacity, excellent heat insulation effect.

Structure durability — Rammed earth has strong compressive resistance and is suitable for load-bearing structures.

Features

Absorb thermal radiation from outdoor light. The air temperature in the room becomes moderate. Rammed earth walkway out the cold air. Release the heat saved during the day.

Ramming process
1. Erection of first formwork
2. The first wall ramming
3. Removing the first template.
4. Erection of second formwork
5. The second wall ramming
6. Removing the second template.

Detail constitution

Detail drawing of rammed earth foundation structure

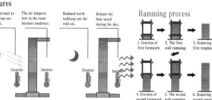

Detail drawing of roof steel structure

Detail drawing of heat storage wall
- Support the keel
- PCM phase change heat storage material
- The first crack resistant and hydrophobic mortar
- Alkali resistant mesh cloth
- The second resistant and hydrophobic mortar
- The putty layer
- Coating

Design of geothermal heat

Ground water thermal heat pump, assisted with the utilization of solar energy.

Our Site
Ground water geothermal heat pump is suitable to be used in small-sized and dispersed architecture group. Our site is also located in the zone suitable to apply ground water geothermal heat pump.

The Evaluation of Suitable Zoning of Geothermal Heat Pump of Sichuan Province

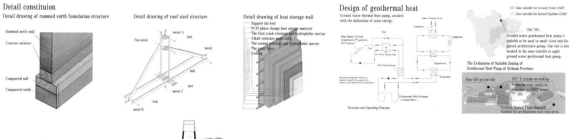

A-A SECTION 1:200

光栖坤灵 VI
Growing in Soil

PURIFICATION OF SEWAGE

INFLUENCE OF CHARACTER BEHAVIOR

TARGET ANALYSIS

Taking all the solar energy utilization, ventilation and geothermal heat pump into consideration, the average indoor temperature in the whole area remain moderate all year round, as the smoothened curve shows.

Electrical Energy Consumption per Year		
Item	Quantity	Notes
Estimation 1		
Indoor Illuminating	14782.5 kW·h	Estimated as 6 hours per day.
Outdoor Illuminating	16425 kW·h	Estimated by the standard of solar assisted outdoor illuminating.
Geothermal Heat Pump System	1272 kW·h	From October 1st to April 30th
Sockets	924 kW·h	
Others	15000 kW·h	
Estimation 2 (Rough)		
Consumption calculated by personal average	49878 kW·h	Estimated by the average consumption data of 2021.
Total	Around 50000 kW·h	

Calculation proves that the quantity of electrical energy consumption is much lower than the quantity that PEDF system can cover. Therefore, it is possible to achieve zero carbon in the whole site theroretically.

The Green Architecture Assessing Standard of China - Chart 1	
Wi	The weight of each index
Land Saving and Outdoor Environment / W1	0.21
Energy Saving and Utilization / W2	0.24
Water Saving / W3	0.20
Material Conservation and Utilization / W4	0.17
Indoor Air Quality / W5	0.18

Calculation: $Q = \Sigma W_i Q_i$

Self-Evaluation According to the Standard of Green Architecture		
Qi	Evaluation	Notes
Land Saving and Outdoor Environment / Q1	75	No excessive contamination source, radiation and Radon-containing soil. Designed based on the daylight standard.
Energy Saving and Utilization / Q2	69	Re≥4.0 ; Rch≥80% ; 10%≤De≤15%
Water Saving / Q3	67	30%≤Rnt≤50% ; Green plants utilized for water purification.
Material Conservation and Utilization / Q4	50	15%≤Rpc≤30% ; Rim≥90%
Indoor Environment / Q5	63	

$Q = \Sigma W_i Q_i = 65.55$

Due to the self-evaluation, the design has reached the standard of China's second level green architecture.

RAIN WATER TREATMENT SYSTEM

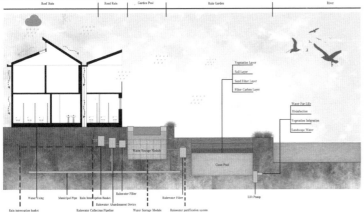

RAINWATER HARVESTING SYSTEM

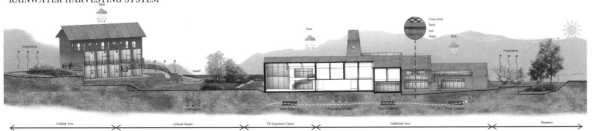

综合奖・优秀奖
General Prize Awarded・
Honorable Mention Prize

注 册 号：101021
项目名称：一川葳蕤
　　　　　The Story between River and Trees
作　　者：吕宇蓓、吴　爽、陈　茜、张佳宁
参赛单位：南京工业大学
指导教师：郭　兰、彭克伟

一川葳蕤 I

The Story between River and Trees

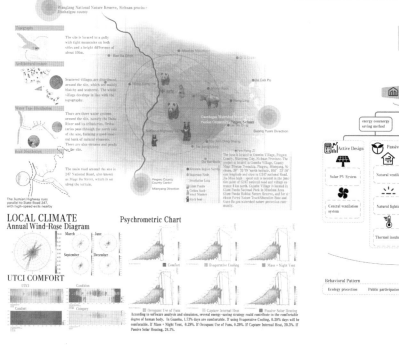

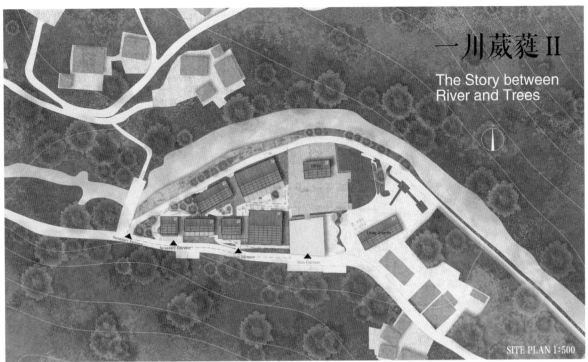

一川葳蕤 II
The Story between River and Trees

SITE PLAN 1:500

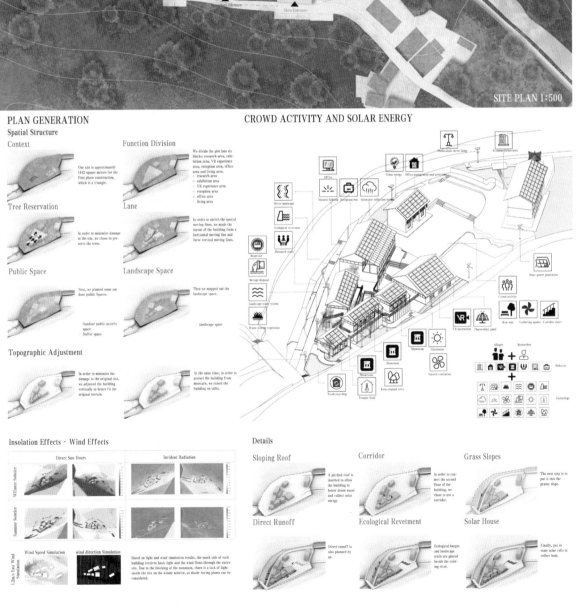

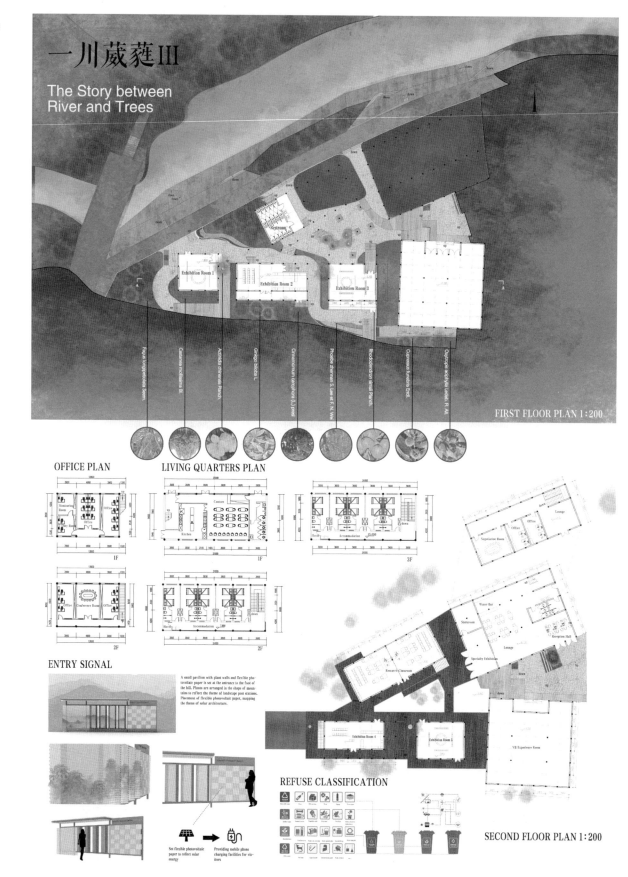

川崴蕤 IV
The Story between River and Trees

INDOOR ILLUMINATION ANALYSIS

Place floor-to-ceiling Windows on the south side and Windows on the north side

Place floor-to-ceiling Windows on the south side, Windows on the north side and High Windows on all sides

 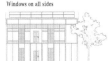

| HDR | Glare Postprocess | HDR | Glare Postprocess |

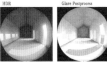

Day Light Factor Day Light Factor

Indoor illuminance is dark, uneven illumination and serious glare.

Indoor illuminance is brighter, illumination is more uniform, forming a more uniform light environment.

SOLAR CHIMENEY ANALYSIS

Summer solar angle   Winter solar angle

Built on stilts, ventilated at the bottom to prevent

Set high and low Windows, using the principle of hot pressing, promote indoor air flow

During the winter days, trombe wall and solar house store heat.

During the winter nights, trombe wall and solar house give off it heat.

During the summer days, the roof is shaded with bamboo, and a valve under the Trombe wall and the door of the solar house are opened to create a ventilation inside.

On summer nights, open the valve under the Tromble wall and the door to the solar house to ventilate the interior.

SOLAR HOUSE ANALYSIS

 ❄ Winter   🔥 Summer

In winter, close the solar house, roll up the bamboo weaving, and let the solar house store heat

In summer, open the solar house and pull down the bamboo weaving to form natural ventilation in the room

PHOTOVOLTAIC PANEL ANALYSIS

Place floor-to-ceiling Windows on the south side, Windows on the north side and High Windows on all sides

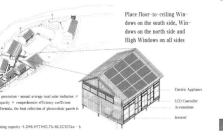

- Electric Appliance
- LCD Controller
- Accumulator
- Inverter

Theoretical power generation = annual average total solar radiation × system installed capacity × comprehensive efficiency coefficient
According to this formula, the heat collection of photovoltaic panels is calculated.

Theoretical generating capacity =1.31*8.1*771*65.7%=66.357857kw·h

UNIT DETAILS

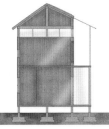 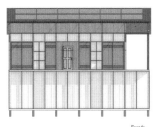

Section   Facade

SOUTH FACADE 1:200

一川葳蕤 V

The Story between River and Trees

SECTION

SHOW RECEPTION AXONOMETRIC BREAKDOWN DIAGRAM

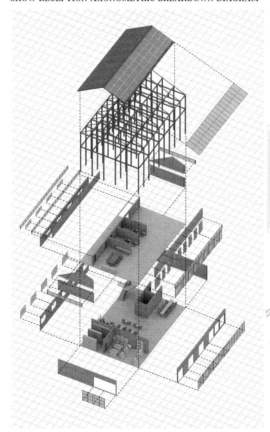

RECYCLE ANALYSIS OF STEEL STRUCTURE

MATERIAL PERFORMANCE ANALYSIS

ANALYSIS OF HEATING EQUIPMENT LAYOUT

ENERGY CONSUMPTION ANALYSIS

A-A SECTION 1:200

一川葳蕤 VI
The Story between River and Trees

PURIFICATION OF SEWAGE

IMPLANTATION OF CULTURAL CUSTOMS

TARGET ANALYSIS

ECOLOGICAL REVETMENT FOR FLOOD RESPONSE

RAIN WATER TREATMENT SYSTEM

ECOLOGICAL REVETMENT FOR FLOOD RESPONSE

RAINWATER HARVESTING SYSTEM

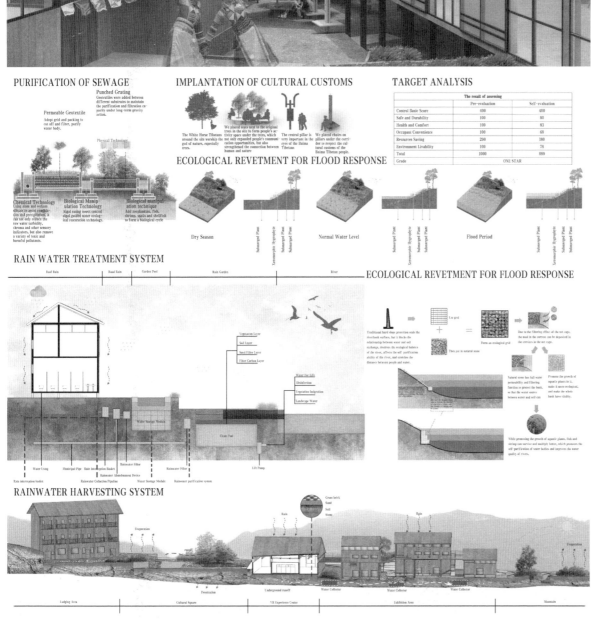

综合奖·优秀奖
General Prize Awarded · Honorable Mention Prize

注 册 号：101027
项目名称：屋檐下
　　　　　Under the Roof
作　　者：张瑞英、祝淑芬、闫　海、
　　　　　张象龙、周禹辰
参赛单位：湖南大学、聊城大学、安徽建
　　　　　筑大学、山东建筑大学、大
　　　　　连理工大学城市学院
指导教师：郑　斐、王月涛

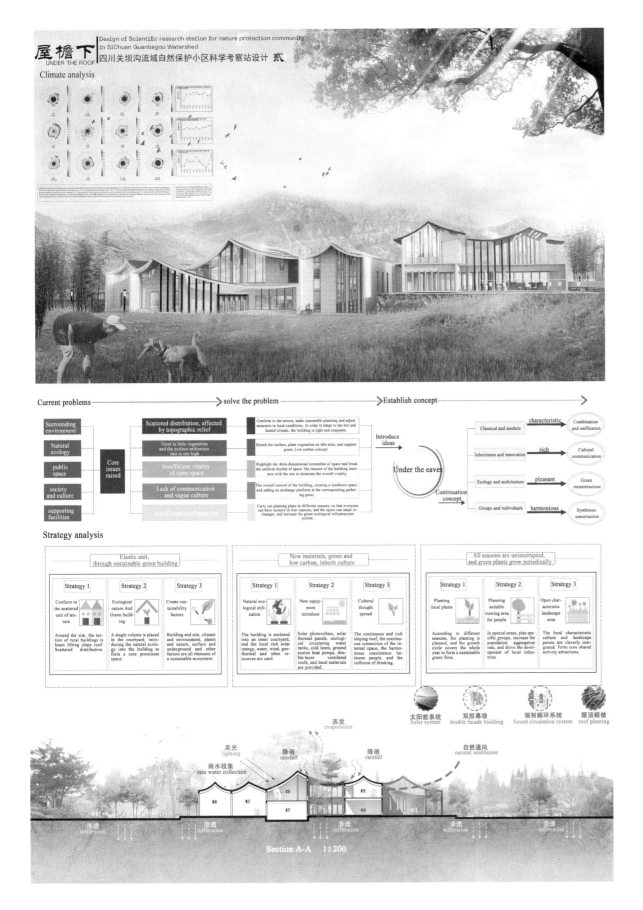

屋檐下 UNDER THE ROOF
Design of Scientific research station for nature protection community in SiChuan Guanbagou Watershed
四川关坝沟流域自然保护小区科学考察站设计 叁

Ethnic Minorities and History and culture

The local villagers are mainly Baima Tibetans. It has a unique history and culture. The main architectural form is close to the wooden structure of western Sichuan folk houses. The White Horse Tibetan people have been living together since ancient times, and dozens of households are a shanzhai

01 Dou yak　02 Suck up wine　03 Jump on the top　04 Rolling felt hat　05 ChuanDouShi　06 Humanoid roof　07 According to the mountain

Population demand analysis

There are four types of users: visitors, researchers, graduate students, and local residents. Each has different requirements for architecture

来访人员 visitor	科考人员 scientific research	研学学生 students	当地居民 local residents
requirement Rest Accommodation Basics of life. The basic life work and communication of the big empty	**requirement** Large space for work and communication. Basic living guarantee with scientific research environment	**requirement** It is necessary to show the local geographical environment, provide basic living materials and learning environment	**requirement** Need square to hold ethnic activities and grain grain and village committee to hold activities space
answer To provide rest for visitors, and a place of communication and material security. And selling local specialties	**answer** Provide a quiet environment with convenience and benefit of service, guarantee the basic life supplies	**answer** To provide graduate students with a place to understand the local culture and geography, and the basic materials of life	**answer** An outdoor square is provided to facilitate the interaction between villagers' ethnic activities and the village committee

Cultural practices
The special geographical location of high mountains and valleys and the closed environment for many years make the living habits and religious beliefs of the Baima people in Pingwu show distinct characteristics.

Architectural features
Houses are generally divided into three floors, the ground floor for livestock, the second floor for people, and the third floor for grain and sundries. The stove is the center of activity for the whole family

Section B-B　1:200

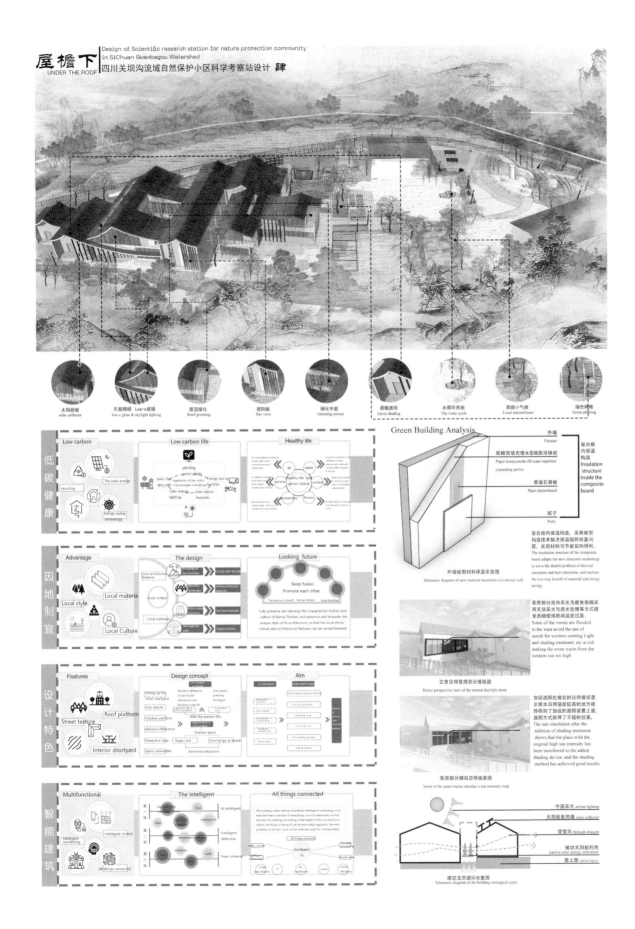

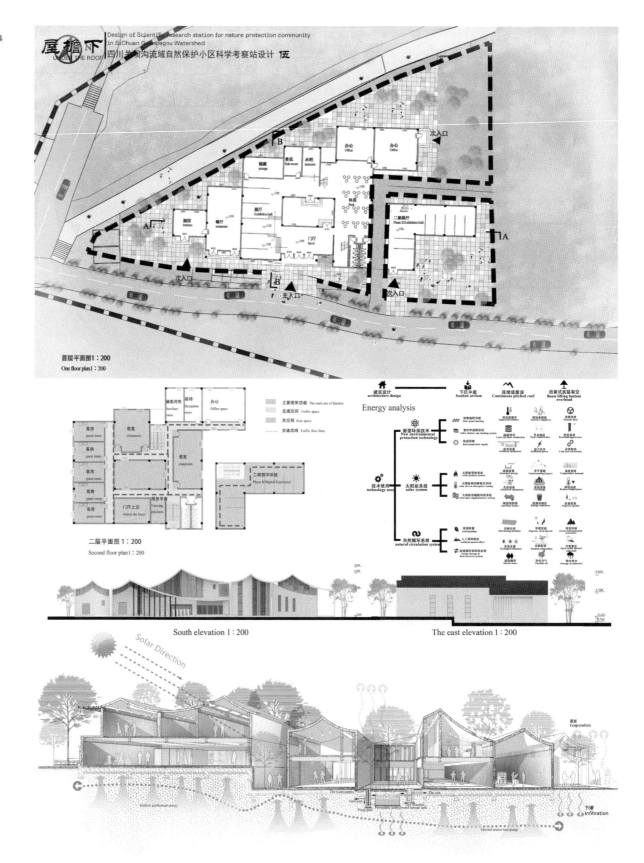

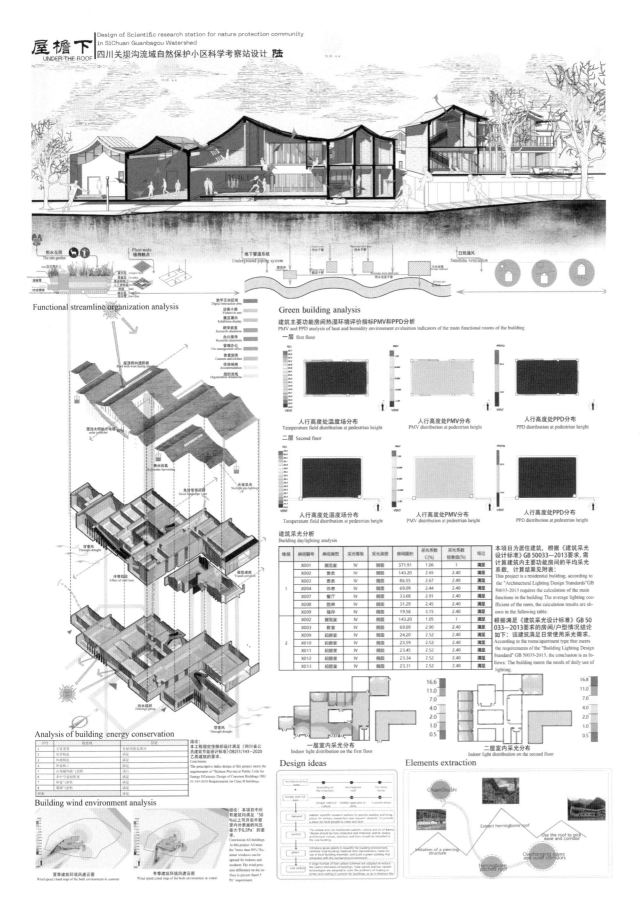

综合奖 · 优秀奖
General Prize Awarded · Honorable Mention Prize

注 册 号：101028
项目名称：细胞呼吸
Spring up Like Mushroom
作　　者：陈漪臻、郑雨婕、陈雨昕、徐雨杭
参赛单位：南京工业大学
指导教师：姜　雷、郭　兰

细胞呼吸-Spring up Like Mushroom Ⅰ

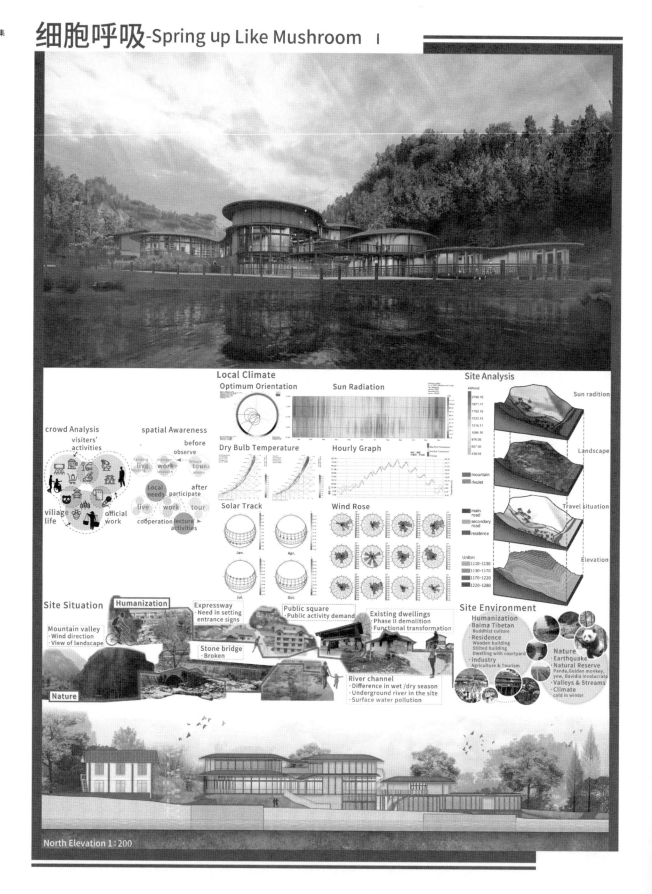

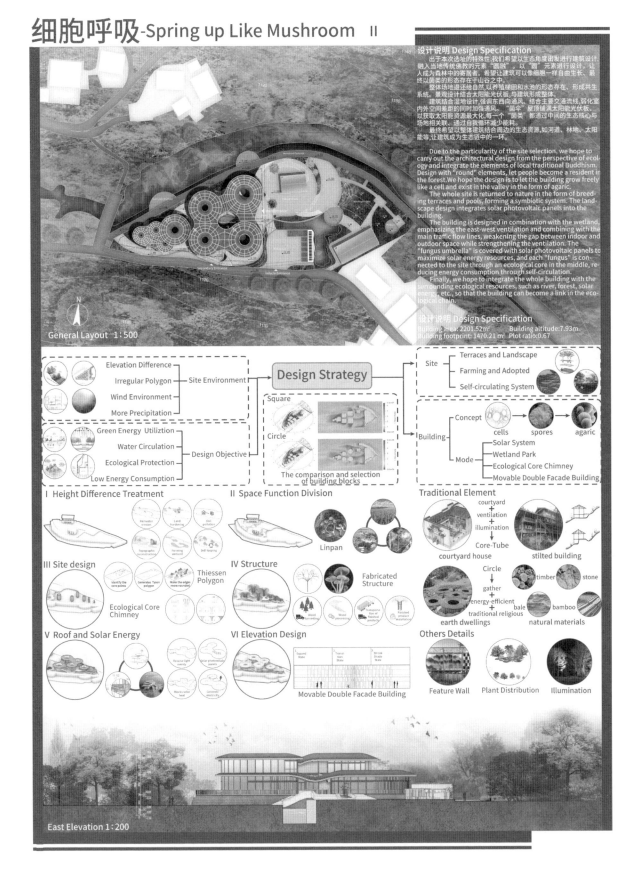

细胞呼吸 -Spring up Like Mushroom III

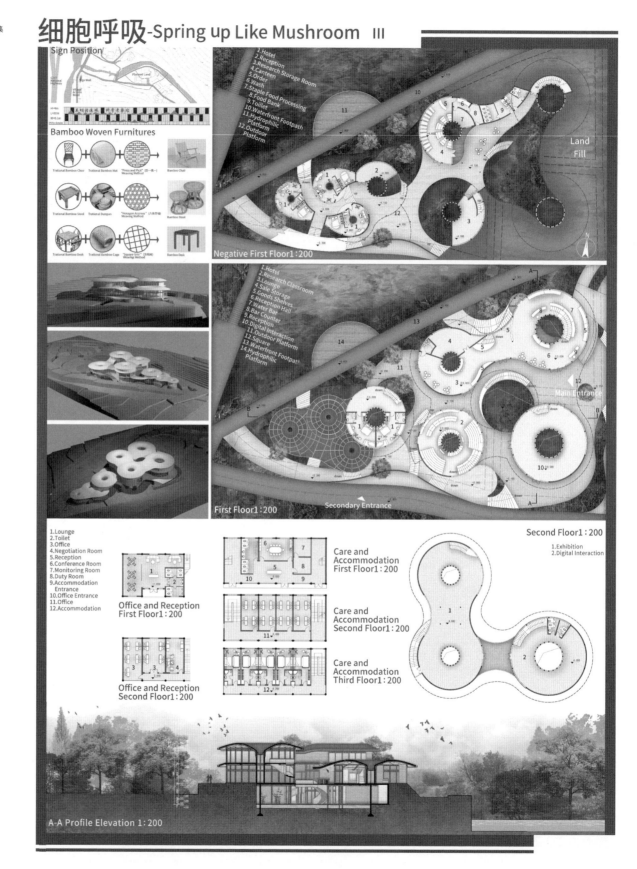

细胞呼吸 - Spring up Like Mushroom IV

Demo of the openable and closable roof

closed / opening / opened / closing

Explosive diagram of single structure

Solar panel / Solar panel frame / Roof / Joist / I-beam / Tree-like structure / Beam-column system / Wood / Bamboo

Roof plus solar panel
Asphalt felt waterproof layer
Horizontal mortar leveling course
Polystyrene organic foam plastic board
Cement carbon slag slope making layer
Reinforced concrete perforated plate

Cornice
Polyurethane board / Roof purlin / Steel beam / Steel gutter

Steel I-beam

Solar Photovoltaic Application

solar energy → PV modules → coupler → controller → DC loads / inverter → AC loads / batteries → power grid (off-grid / DC / AC)

4m 10m 4m 6m 8m 4m
group=20*3*3
Photovoltaic array

Single axis (height angle tracking) solar automatic tracking device
Including sensor, single chip microcomputer, adjusting component and database
1-supporting structure
2-telescopic rod
3-solar panel
4-frame pipeline
5-dc motor
6-fixed shaft

Component schematic diagram side view

Ray offset → Shadow change
Adjust angle → Photoelectric monitoring
For height angle tracking only, the traditional four quadrant device is optimized to three quadrants.

Photodiode / Sensory Edition / Shading plate / Shaded area
Sensor design

Preliminary field trips / Analysis and calculation / On-site installation
Solar energy is converted into electricity / Photovoltaic panels absorb solar energy into electricity / Solar carports supply electric vehicles
Solar wall provides temporary charging / Solar walls use solar energy to heat water / Source of power for site lighting systems

Bamboo weaving generation analysis
Bamboo characteristics → Planes and elevations redesign → Pattern extraction

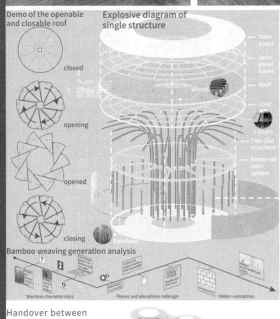

Handover between different core cylinders
Based on the tree structure of the core tube of the foundation monomer, the surrounding corridors are connected with the existing structural foundation to form an overall frame structure system.

Connection of curtain wall and column
Fire protection component / Rivet / Column / Connecting angle steel / Bolt nut / Galvanized steel sheet / Expansion bolt

Tree-structure Sketch Map
Inclosed / Inside & out / half-open

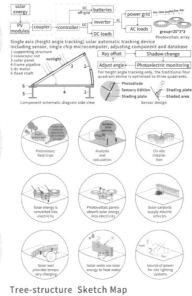

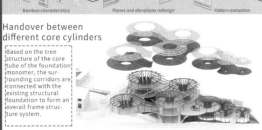

Summer Ventilation
opened
The middle core tube is opened. Use the specific heat capacity of water to reasonably organize ventilation to cope with the high temperature in summer.
Water evaporation cooling / Hot pressing pull-out effect

Winter Insulation
closed
The middle core tube is closed. The heat preservation and the hot water wall system are used to keep warm.

"Four water in one"
Cornice drainage of sloping roof
Wetland water storage / Fabricated reservoir / Landscape water Planting irrigation / Multiple water reuse

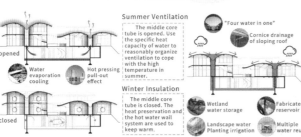

Rainwater collection
The two organized drainage methods, the sloping roof and core pipe, can collect rainwater systematically.

Rainwater storage and reuse
The rainwater is discharged into the lower wetland and reservoir for reclaimed water treatment and purification. Then, it can be reused or discharged without pollution.

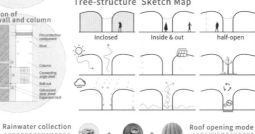

Rotary drive + Ring beam + Aluminum alloy panel

Roof opening mode
The roof can be opened or closed at different times to meet the functional requirements.

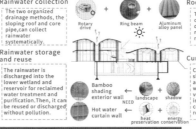

Curtain wall system
Bamboo shading exterior wall / Hot water curtain wall
landscape / shadow / heat preservation / energy conservation

Combining bamboo shading exterior wall and hot water curtain wall. Make building ventilation and insulation more convenient and reasonable.

细胞呼吸 - Spring up Like Mushroom Ⅴ

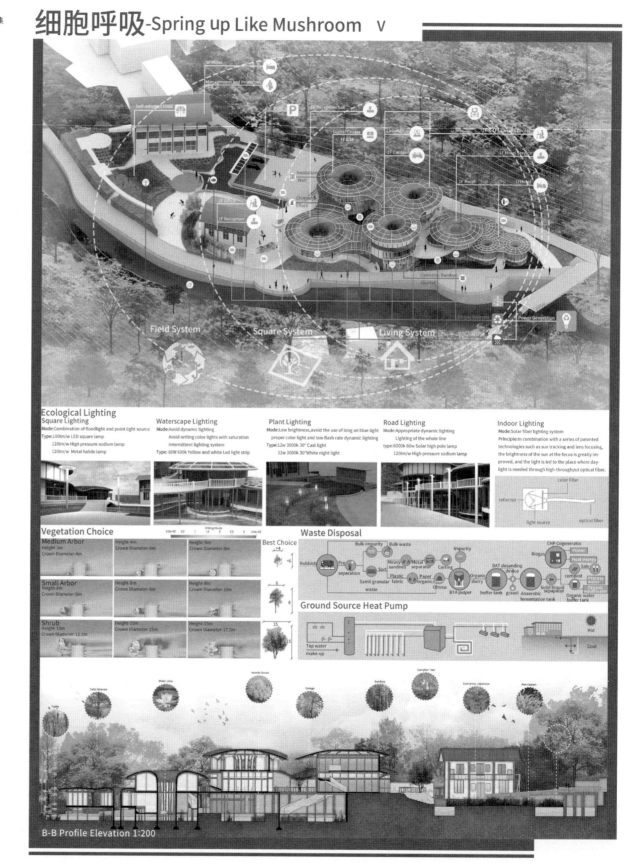

B-B Profile Elevation 1:200

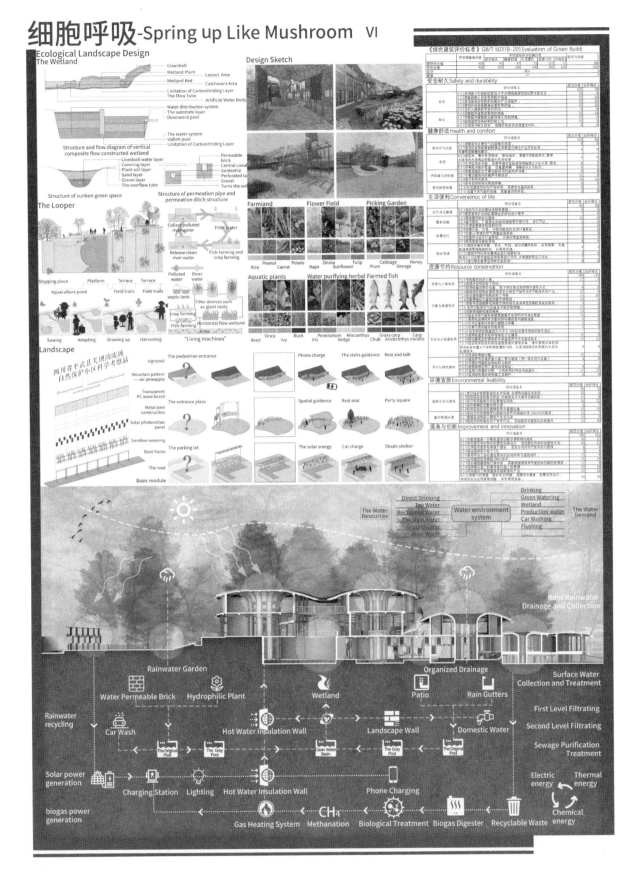

综合奖·优秀奖
General Prize Awarded·
Honorable Mention Prize

注 册 号：101047
项目名称：聚·生
　　　　　Gather and Live
作　　者：胡喆涵、叶雨菲、毛锐杰、
　　　　　杨　政
参赛单位：南京工业大学
指导教师：董　凌、薛春霖

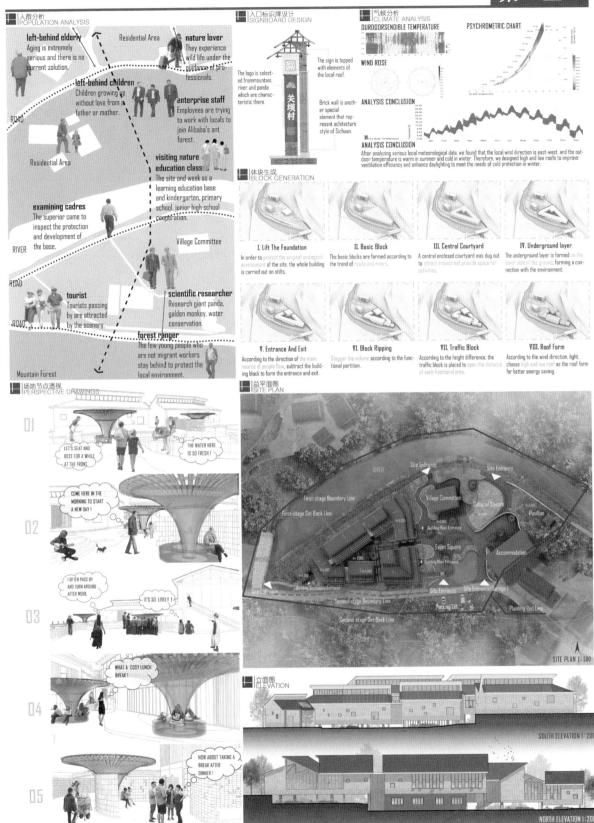

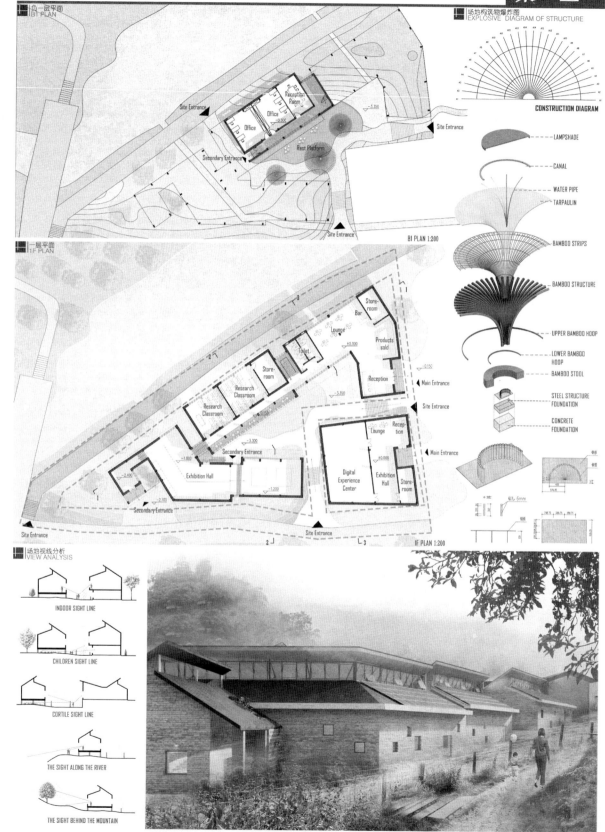

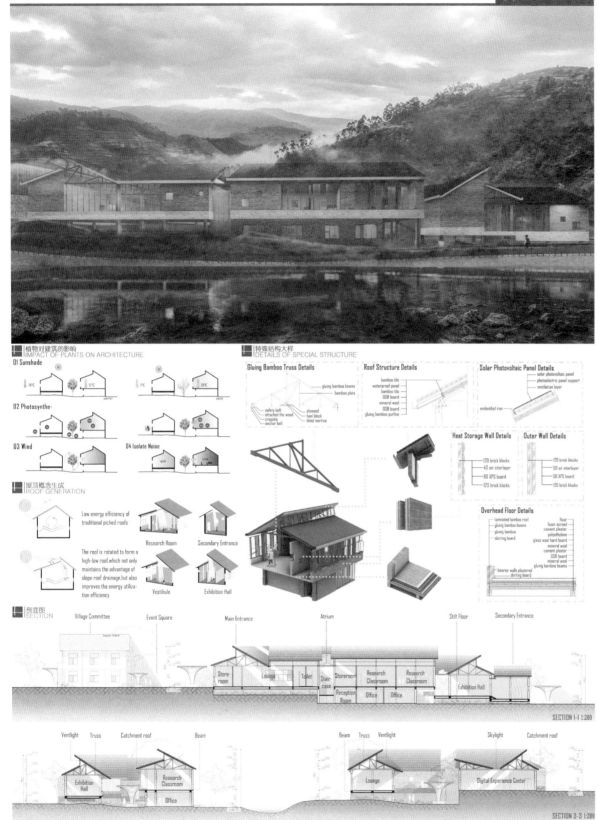

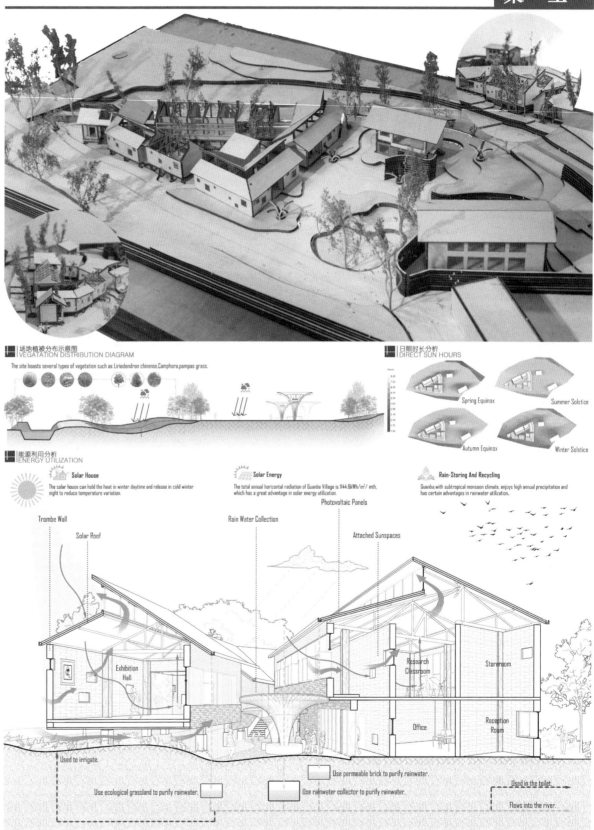

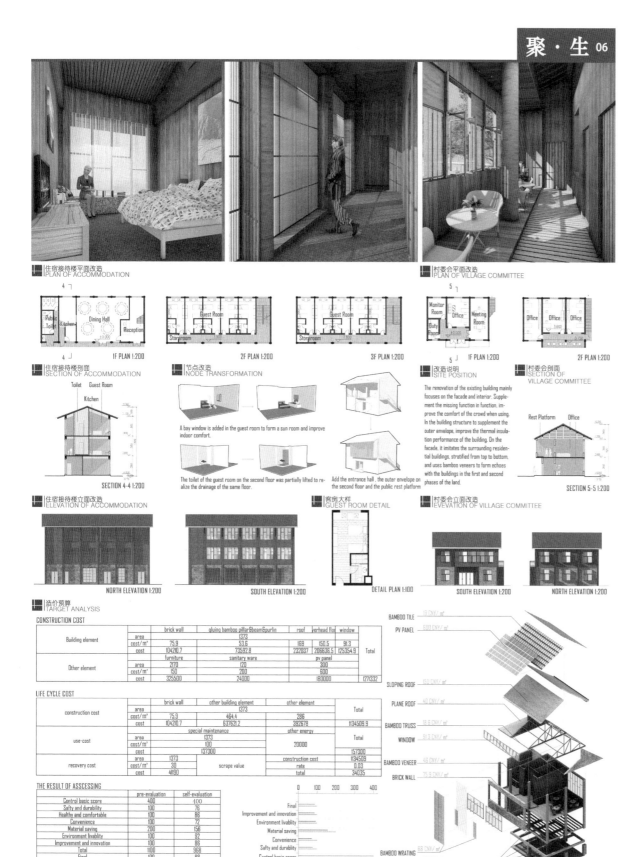

2022 台达杯国际太阳能建筑设计竞赛获奖作品集

综合奖・优秀奖
General Prize Awarded・Honorable Mention Prize

注　册　号：101064
项目名称：溯・曦源
　　　　　Look for the Place Where the Sun Rises
作　　　者：李轶群、刘银露、邱妤菲菲、朱浩博、高 祚、陈雪柯、陈安娴
参赛单位：重庆大学
指导教师：何宝杰

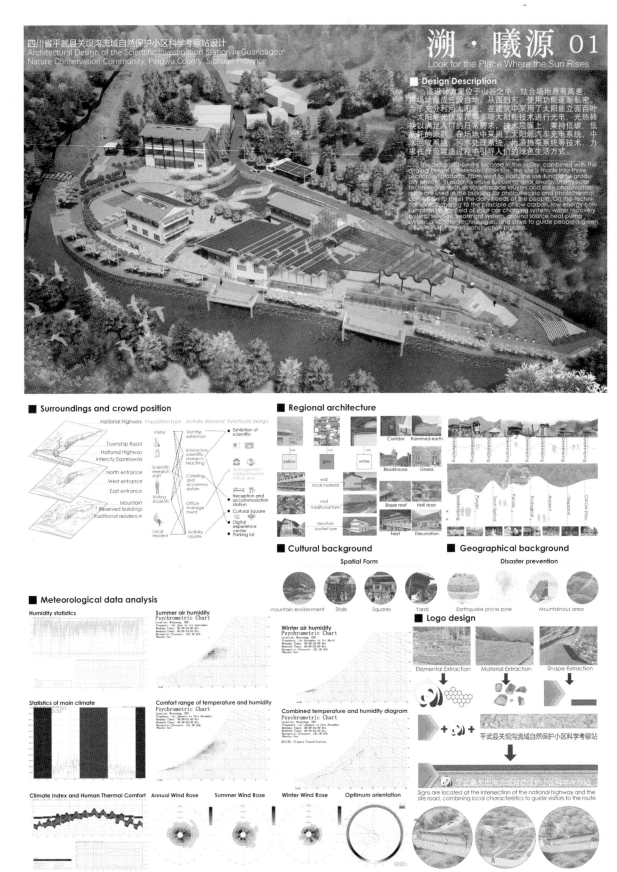

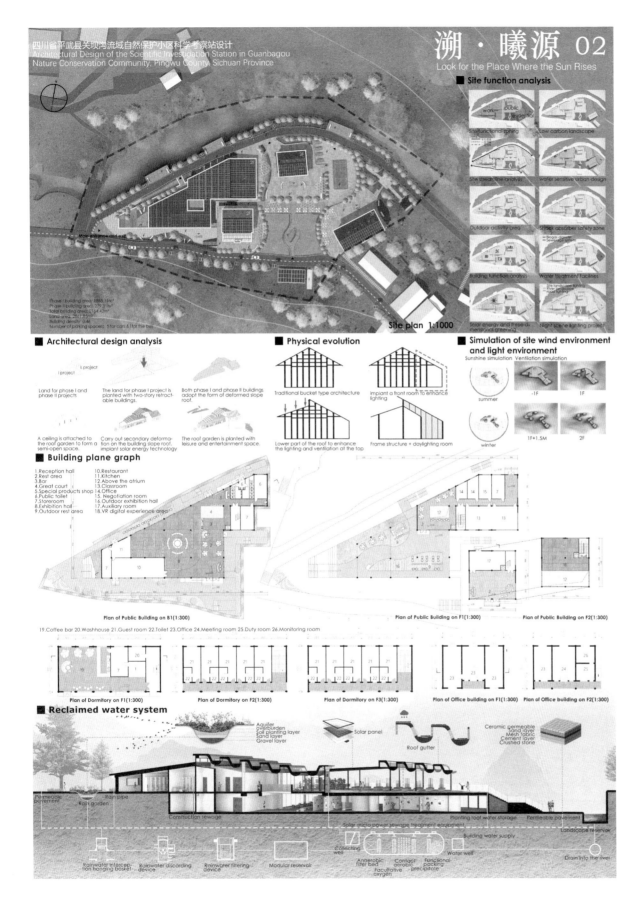

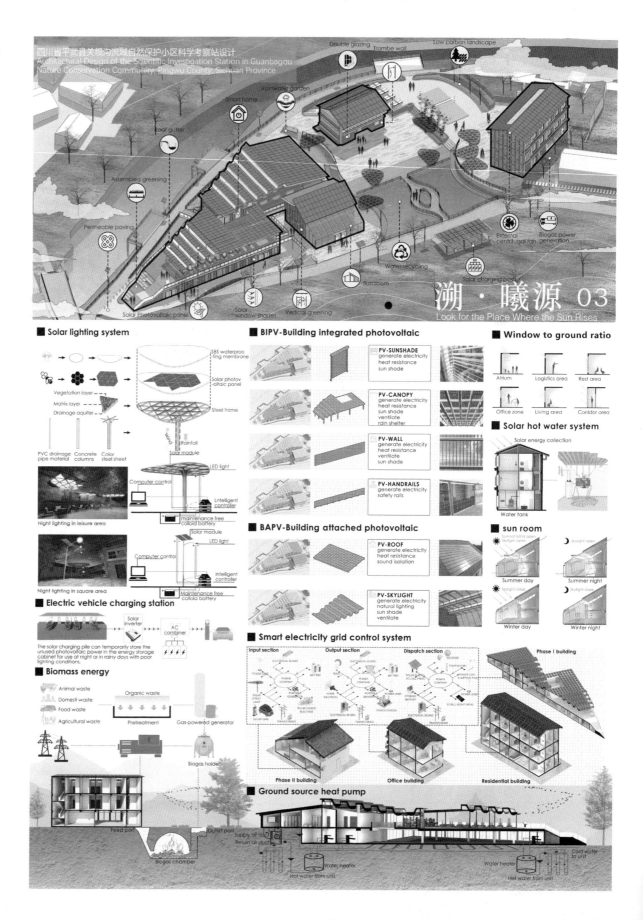

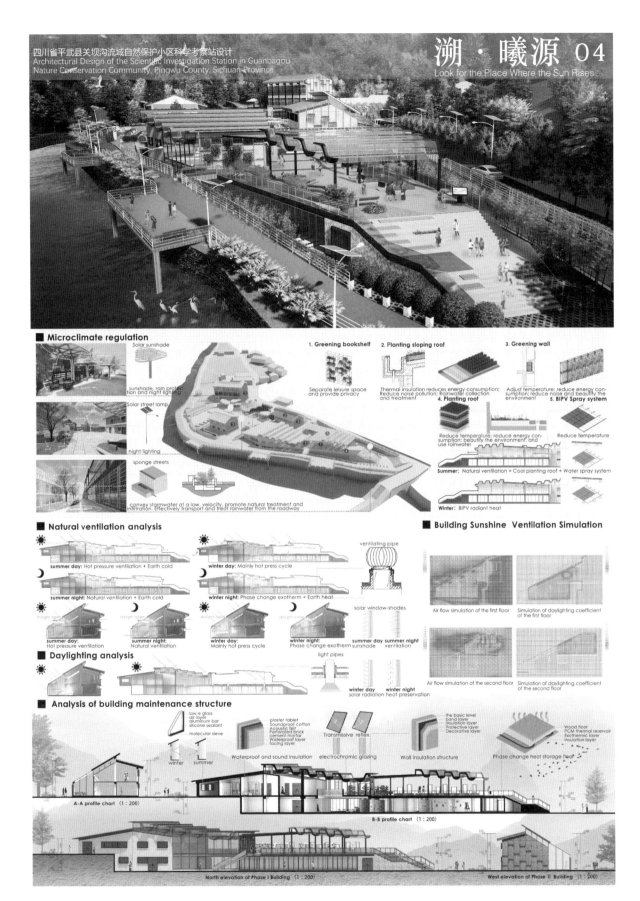

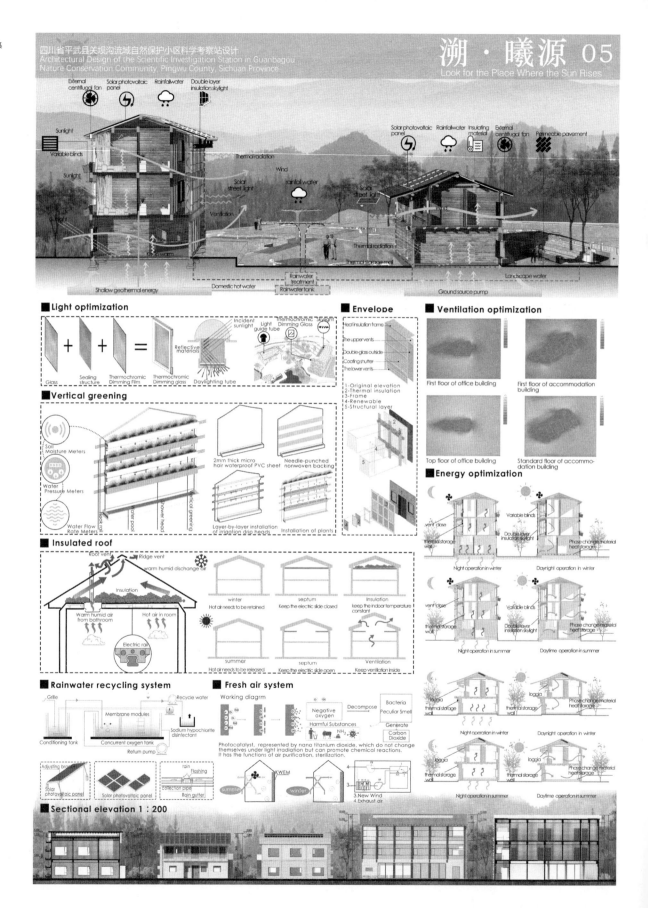

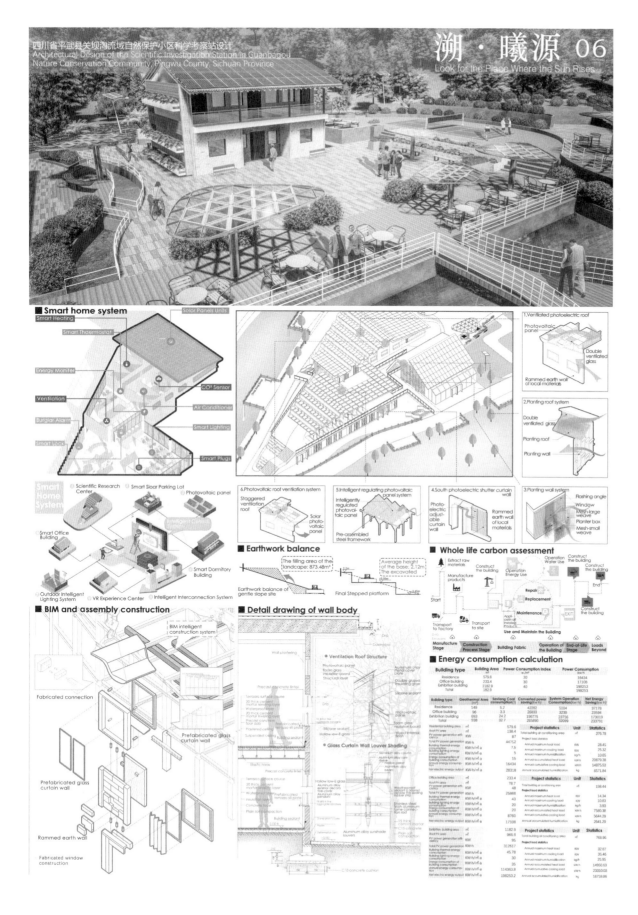

综合奖·优秀奖
General Prize Awarded · Honorable Mention Prize

注 册 号：101099
项目名称："在地""营造"——自然科学考察站设计
　　　　　Design of Scientific Research Station
作　　者：林凯、蒋敬亦、李吕雨阳、刘幸宇
参赛单位：嘉兴学院、昆明理工大学
指导教师：陆莹、毛志睿

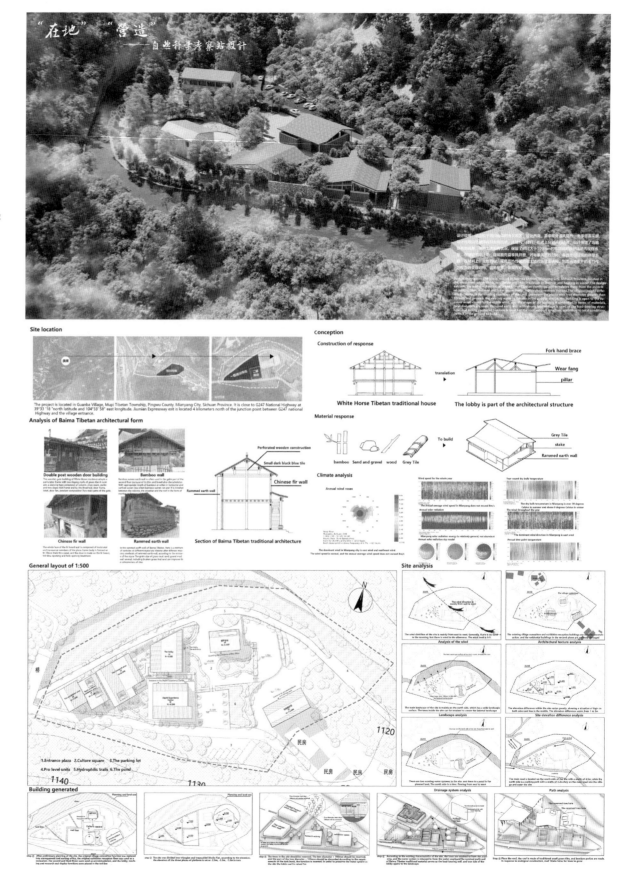

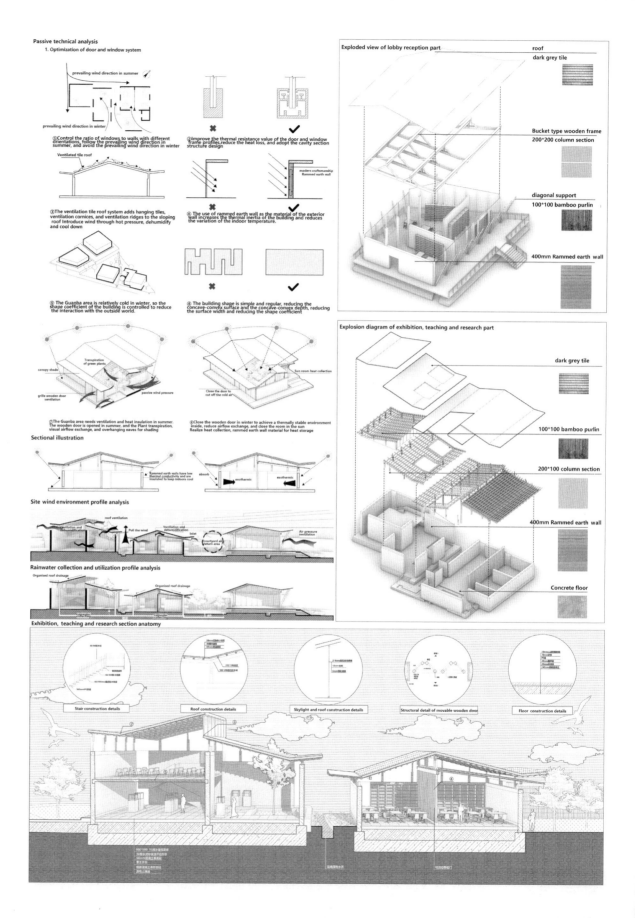

Aerial view of the scene

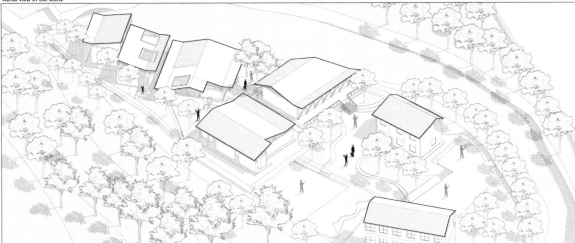

Digital experience center, lobby anatomy

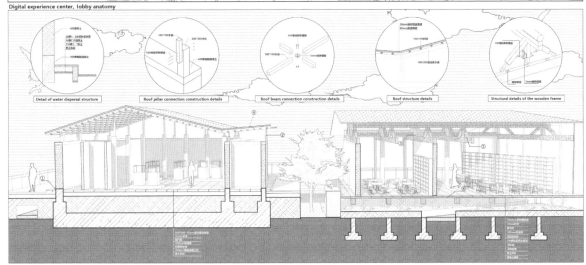

| Detail of water dispersal structure | Roof pillar connection construction details | Roof beam connection construction details | Roof structure details | Structural details of the wooden frame |

综合奖·优秀奖
General Prize Awarded · Honorable Mention Prize

注 册 号：101132
项目名称：檐下·村落
　　　　　Village under Eaves
作　　者：殷嘉蔓、陈忠耀、王思懿、
　　　　　杨　雄、尹家雪
参赛单位：重庆大学
指导教师：黄海静

檐下·村落 Village under Eaves 02

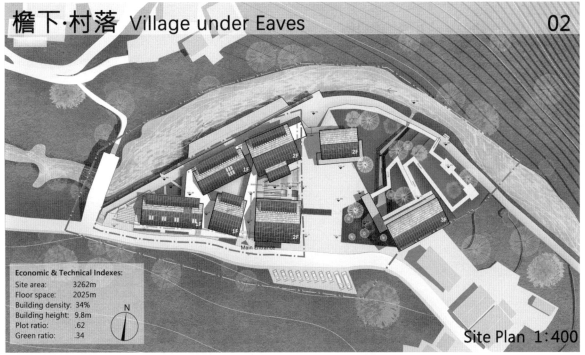

Economic & Technical Indexes:
- Site area: 3262m
- Floor space: 2025m
- Building density: 34%
- Building height: 9.8m
- Plot ratio: .62
- Green ratio: .34

Site Plan 1:400

■ Site Analysis

The base is located in Mupi, Pingwu County, Mianyang CityTibetan township Guan Ba village. The field is green. The terrain is high in the center and low around. Pingwu County is a north subtropical humid mountain. Monsoon climate, mild climate, precipitation. Plenty of sunshine, four distinct seasons, Having more clouds, less fog, and overcast days. The characteristic. The annual average temperature is 14.7℃. The average annual precipitation is 800 mm left on the right. Perennial average sunshine time 1376 hours, annual average frost-free period 252 days.

Surrounding / Road / Transportation / Sight
Sunshine / Wind / Noise / Vegetation

■ Climatic Simulation

Enthalpy diagram

All Year / Spring / Summer / Autumn / Winter

In spring and autumn, a large area is in a comfortable area, which indicates that the climate in spring and autumn in this area is suitable, so that buildings can make the most of the comfort of the environment;
However, in winter and summer, the temperature in summer is much higher than that in comfortable areas, and the humidity is also high. It is considered to reduce the temperature and humidity.
In winter, the temperature is low, and the building needs to consider thermal insulation and heating.

Seasonal wind rose

All Year / Spring / Summer / Autumn / Winter

The wind direction of the site is mainly from east to west. Generally, there is no wind in the morning, and there is wind in the afternoon. The wind force is 3-5. (According to common sense, the wind speed in the canyon will be higher.).
Strategically, avoid the prevailing wind in winter, and actively guide the prevailing wind for ventilation in summer.

Radiation simulation diagram

The annual radiation in the radiation simulation diagram is higher than that in July, which belongs to normal summer radiation in numerical value. There is little direct sunlight in winter.

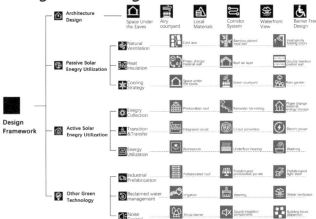

Green hills / Clear waters / The panda and bamboo / Apiary / Pitched roof / Overhanging eaves and space under eaves

■ Diagram of Design Process

Design Framework
- Architecture Design: Space Under the Eaves, Airy courtyard, Local Materials, Corridor System, Waterfront View, Barrier Free Design
- Passive Solar Energy Utilization:
  - Natural Ventilation: Cold lane, Bamboo planted mud wall, Intelligently folding doors
  - Heat Insulation: Phase-change material wall, Roof air layer, Double bamboo plated wall
  - Cooling Strategy: Space under the eaves, Green courtyard, Rain garden
- Active Solar Energy Utilization:
  - Energy Collection: Photovoltaic roof, Rainwater harvesting, Phase change material energy storage
  - Transition & Transfer: Integrated circuit, Circuit converters, Electric power
  - Energy Utilization: Illumination, Underfloor heating, Washing
- Other Green Technology:
  - Industrial Prefabrication: Prefabricated roof, Prefabricated photovoltaic panels, Prefabricated light steel
  - Reclaimed water management: Irrigation, Cleaning, Water landscape
  - Noise Control: Shrub barrier, Sound insulation components, Building block dispersion

Analysis of the midday sun track

January 15th-Midday sun track / April 15th-Midday sun track
July 15th-Midday sun track / October 15th-Midday sun track
Annual sun shadow grayscale

The canyon terrain has an impact on the site sunshine in winter, and there is almost no direct sunlight in winter. The highest annual radiation is in July, which belongs to normal summer radiation.
Therefore, the window opening should avoid east-west direction, and the sun height angle should be considered for shading.

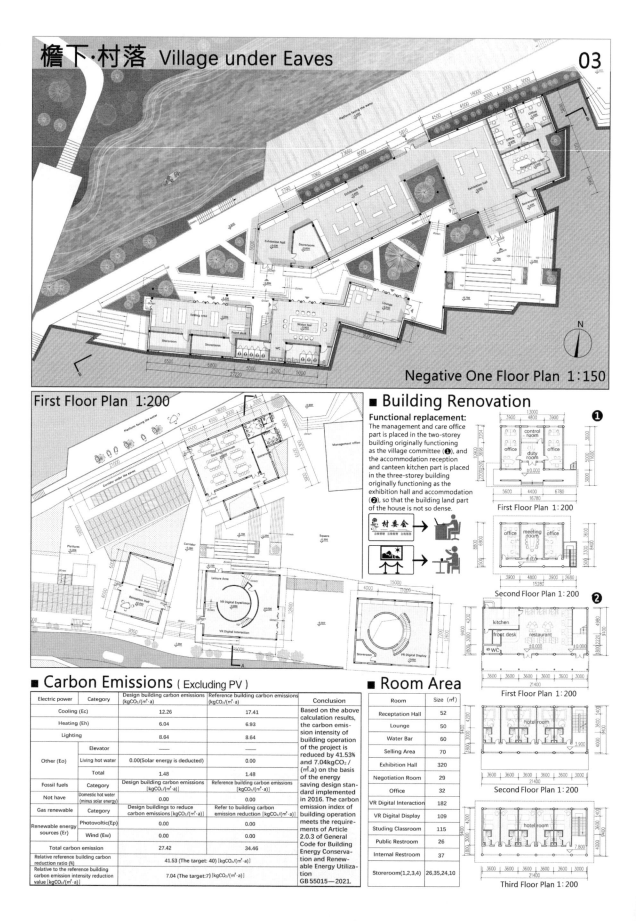

檐下·村落 Village under Eaves

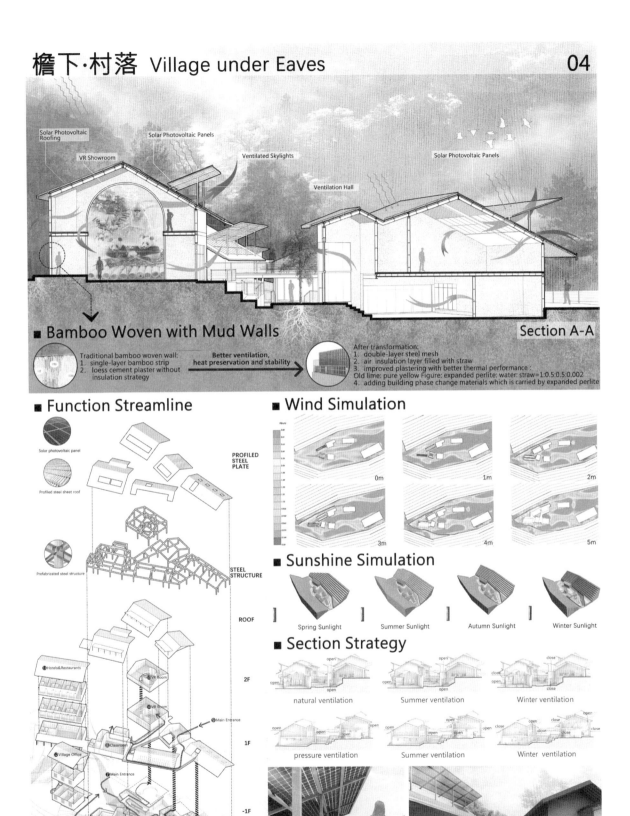

檐下·村落 Village under Eaves

05

■ Active Solar Technology

Ⓐ Photovoltaic roof
Ⓑ Rainwater harvesting
Ⓒ Double bamboo plaited wall
Ⓓ Cold lane
Ⓕ Intelligent control window
Ⓖ Intelligently folding doors
Ⓗ Underfloor heating
Ⓘ Prefabircated light steel

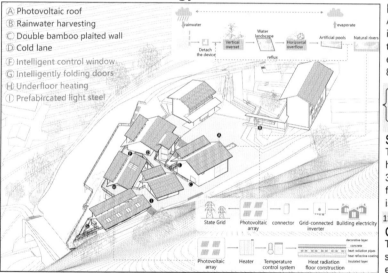

■ Energy Calculation

Building power consumption
According to the electricity consumption index of civil buildings, the office building, the exhibition hall,the classroom's hourly electricity consumption: $70W/(m^2 \cdot h)$, $80W/m^2$, $40W/m^2$ respectively

$$\begin{cases} 200m^2 \times 70W/m^2 \times 8.0h = 112000W/h = 112\,kW/h \\ 600m^2 \times 80W/m^2 \times 6.0h = 288000W/h \end{cases}$$

$$sum = 572\,kW/h$$

Solar panel power generation
The module has effective sunshine for 5.5 hours,considering the 30% loss generate 3.5 electricity per day.The area required for 1kw solar panel to generate electricity is about 7.0m

$$1247m^2 \times 3.5\,kW/h \div 7.0m^2 = 623.5\,kW/h > 572\,kW/h$$

Conclusion:
The current power generation capacity of solar panels can basically meet the electricty demand of the buildings.

■ Plant Footprint

👁 Visual (Colour/Form)
👃 Smell (Scent)
👣 Tour (Atmosphere)
♾ Environment (Friendly)

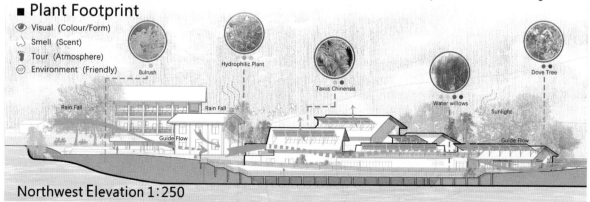

Northwest Elevation 1:250

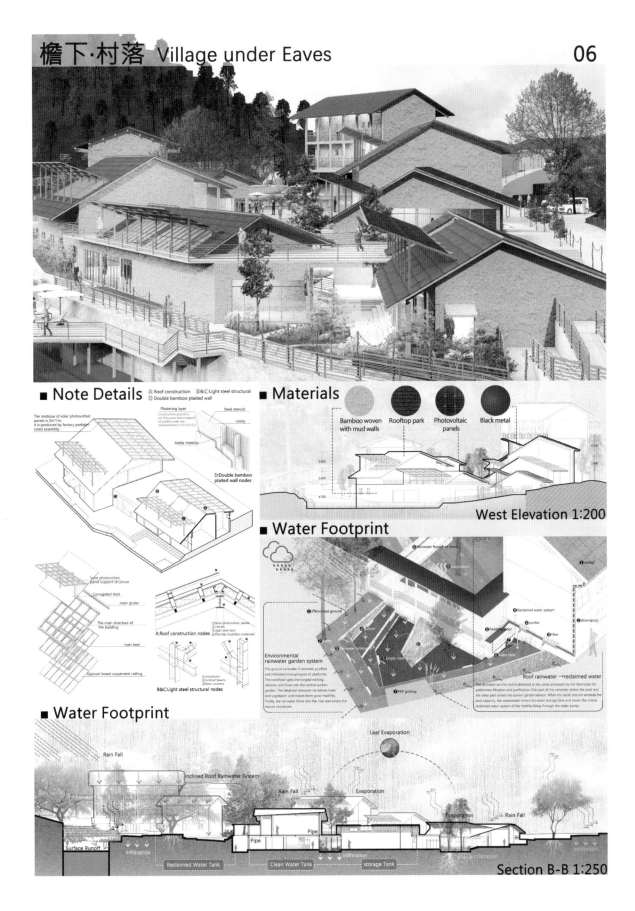

综合奖·优秀奖
General Prize Awarded·
Honorable Mention Prize

注 册 号：101153
项目名称：辉隐驿事
　　　　　Light in the Station
作　　者：陈佳怡、罗　程、刘昕彤、
　　　　　白馨怡
参赛单位：重庆大学
指导教师：周铁军

辉隐驿事 01
LIGHT IN THE STATION

四川平武关坝沟流域自然保护小区科学考察站设计
Research Station in Guanbagou Basin Nature Mini-Reserve, Pingwu County, Sichuan Province

设计说明 Design Description

本次设计融入了"光储直柔"的概念，拟打造集能源生产、消费、调蓄功能于一体的建筑；建筑通过设置集热空腔、风井、架空屋面、太阳能烟囱、吊顶层等构件，形成热风采暖环路，结合相变储热材料、蓄热墙、阳光房、电动车电池资源等实现建筑的储能蓄能；在主动式太阳能技术方面，利用建筑表面设置光伏利用可再生能源进行发电，面向碳中和。技术的运用与传统建筑形象结合，回应当地的气候、自然环境、建筑、文化与材料。

This design incorporates the concept of "photovoltaics, energy storage, direct current and flexibility" (PEDF), and intends to create a building that integrates energy production, consumption, regulation and storage functions: The building forms a hot air heating loop by setting up components such as heat collecting cavities, air shafts, overhead roofs, solar chimneys, and ceiling layer and combines phase change thermal storage materials, thermal storage walls, sun rooms, electric vehicle battery resources, etc. to achieve building energy storage; In terms of active solar technology, renewable energy such as photovoltaics is used to generate electricity on the surface of buildings, aiming at carbon neutrality. The use of technology is combined with the traditional architectural image, responding to the local climate, natural environment, architecture, culture and materials.

Site Analysis

Road　　Surrounding　　Water　　Mountains

The site is at the bottom of the valley | It is high in the southeast and low in the northwest | The valley affects the wind and sunlight of the site | Valleys create unique landscapes

Diagram of Design Process

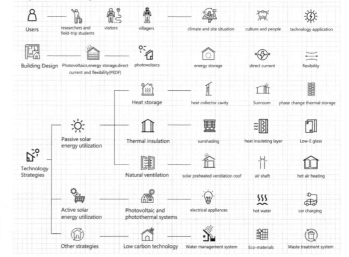

Regional Feature

Ecological resources　　National culture　　Village form　　Traditional buildings

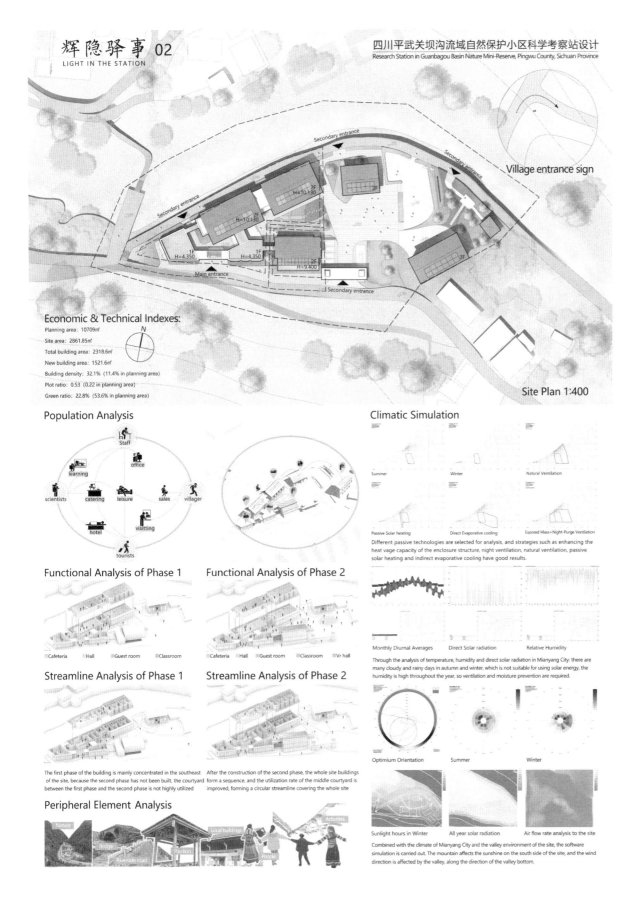

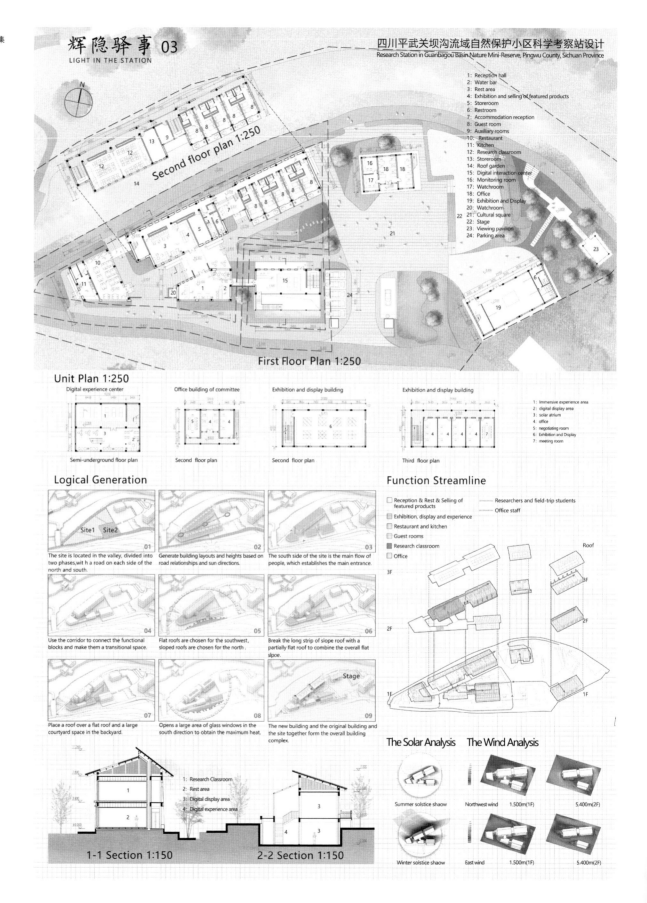

辉隐驿事 04 — Light in the Station

四川平武关坝沟流域自然保护小区科学考察站设计
Research Station in Guanbagou Basin Nature Mini-Reserve, Pingwu County, Sichuan Province

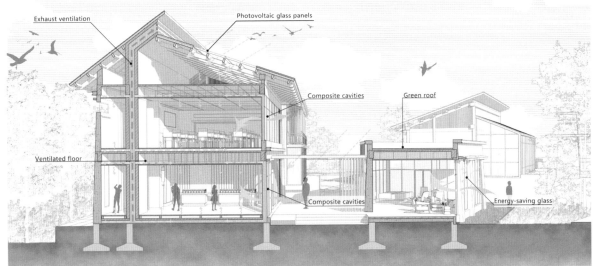

Inspiration

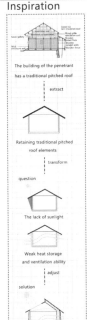

Technology Strategy

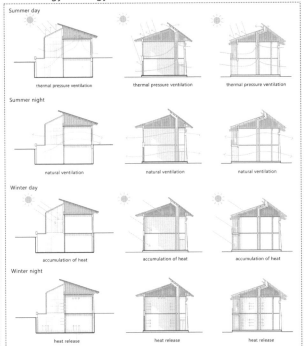

Details

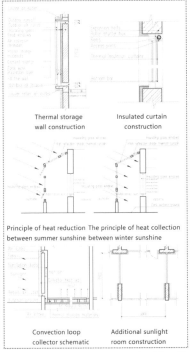

辉隐驿事 05
LIGHT IN THE STATION

四川平武关坝沟流域自然保护小区科学考察站设计
Research Station in Guanbagou Basin Nature Mini-Reserve, Pingwu County, Sichuan Province

Energy Consumption Calculation

Unit Technology Analysis

Reception hall

- Planted roof
- Overhead corridor
- Direct benefit window

Restaurant & Kitchen
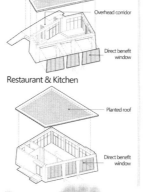
- Planted roof
- Direct benefit window

Rest area & Research classrooms
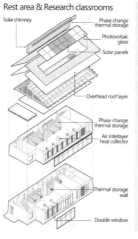
- Solar chimney
- Phase change thermal storage
- Photovoltaic glass
- Solar panels
- Overhead roof layer
- Phase change thermal storage
- Air interlayer heat collector
- Thermal storage wall
- Double window

Guest rooms
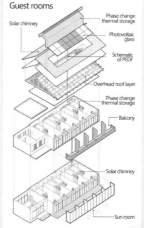
- Phase change thermal storage
- Solar chimney
- Photovoltaic glass
- Schematic of PEDF
- Overhead roof layer
- Phase change thermal storage
- Balcony
- Solar chimney
- Sun room

Digital experience center
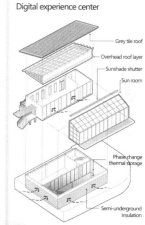
- Grey tile roof
- Overhead roof layer
- Sunshade shutter
- Sun room
- Phase change thermal storage
- Semi-underground insulation

Grey tiles · Concrete · Pine wood · Pebbl · Glass

North Elevation 1:250

辉隐驿事 06
LIGHT IN THE STATION

四川平武关坝沟流域自然保护小区科学考察站设计
Research Station in Guanbagou Basin Nature Mini-Reserve, Pingwu County, Sichuan Province

2022 台达杯国际太阳能建筑设计竞赛获奖作品集

Technology Integration

PEDF Technology Analysis

PEDF - Photovoltaics, Energy storage, Direct current and Flexibility

"光储直柔" 系统构架
PEDF system frame

PEDF refers to the construction of a new building power distribution system (building energy system) that ADAPTS to the carbon neutral target demand through photovoltaic and other renewable energy generation, energy storage, DC distribution and flexible energy use.

Photovoltaics: Solar photovoltaic technology. Distributed photovoltaic installations are installed on the building site to realize the use of renewable resources.

Energy storage: All kinds of devices and facilities with energy storage/storage capacity available in the building can be used as energy storage resources in the PEDF system.

Direct current: Direct current technology. The form of building distribution network has been changed from the traditional AC distribution network to the low-voltage DC distribution network.

Flexibility: Flexible power technology. Building electricity equipment should have the ability to be interruptible and adjustable to provide a degree of flexibility to the power system, so that building electricity demand changes from rigid to flexible.

"光储直柔" 建筑系统示意
Schematic of PEDF building system

Low carbon Integration

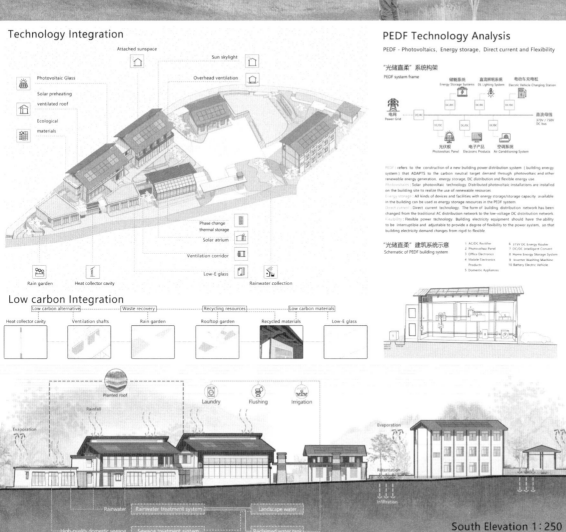

South Elevation 1:250

综合奖·优秀奖
General Prize Awarded·
Honorable Mention Prize

注　册　号：101157
项目名称：沐光山园
　　　　　A Post Station in Guanba Village in the Sun
作　　　者：陈思安、孙上词、钟　锐、吴俊豪
参赛单位：沈阳建筑大学
指导教师：侯　静、武　威

沐光山园

A Post Station in Guanba Village in the Sun
四川关坝村自然保护区科学考察站设计 02

Site Plan 1:500

Site Analysis

Sun Analysis · Wind Analysis · Noise Analysis
Surrounding Buildings · Road Conditions · Surrounding Environment

Climate Simulation

Sunshine Duration

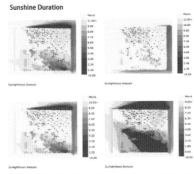

According to the analysis, this is the distribution of sunshine duration in four quarters, and we can determine the sunshine duration according to the distribution. Determine the formal layout and orientation of the building.

Digram of Technology Uses

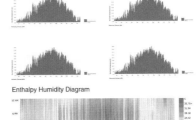

Enthalpy Humidity Diagram

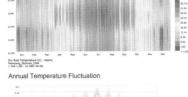

Annual Temperature Fluctuation

Digram of Technology Uses

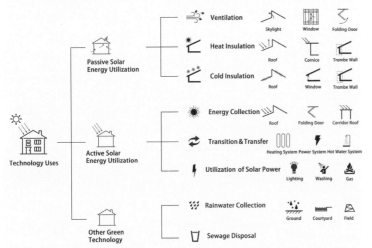

Technology Uses
- Passive Solar Energy Utilization
  - Ventilation (Skylight, Window, Folding Door)
  - Heat Insulation (Roof, Cornice, Trombe Wall)
  - Cold Insulation (Roof, Window, Trombe Wall)
- Active Solar Energy Utilization
  - Energy Collection (Roof, Folding Door, Corridor Roof)
  - Transition & Transfer (Heating System, Power System, Hot Water System)
  - Utilization of Solar Power (Lighting, Washing, Gas)
- Other Green Technology
  - Rainwater Collection (Ground, Courtyard, Field)
  - Sewage Disposal

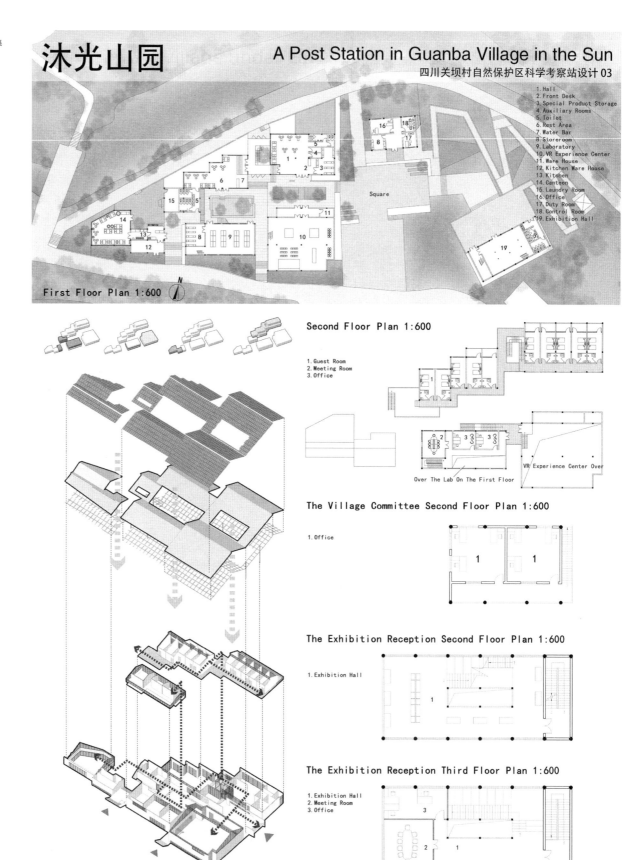

沐光山园

A Post Station in Guanba Village in the Sun
四川关坝村自然保护区科学考察站设计 04

Active solar Technology

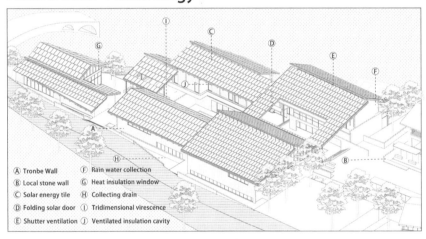

- Ⓐ Tronbe Wall
- Ⓑ Local stone wall
- Ⓒ Solar energy tile
- Ⓓ Folding solar door
- Ⓔ Shutter ventilation
- Ⓕ Rain water collection
- Ⓖ Heat insulation window
- Ⓗ Collecting drain
- Ⓘ Tridimensional virescence
- Ⓙ Ventilated insulation cavity

Node Analysis

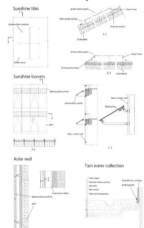

Plant Analysis and Water Cycle

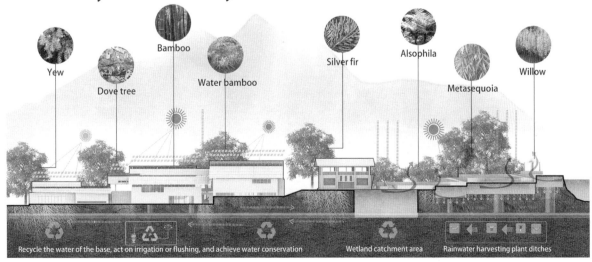

Yew · Dove tree · Bamboo · Water bamboo · Silver fir · Alsophila · Metasequoia · Willow

Recycle the water of the base, act on irrigation or flushing, and achieve water conservation | Wetland catchment area | Rainwater harvesting plant ditches

沐光山园

A Post Station in Guanba Village in the Sun
四川关坝村自然保护区科学考察站设计 05

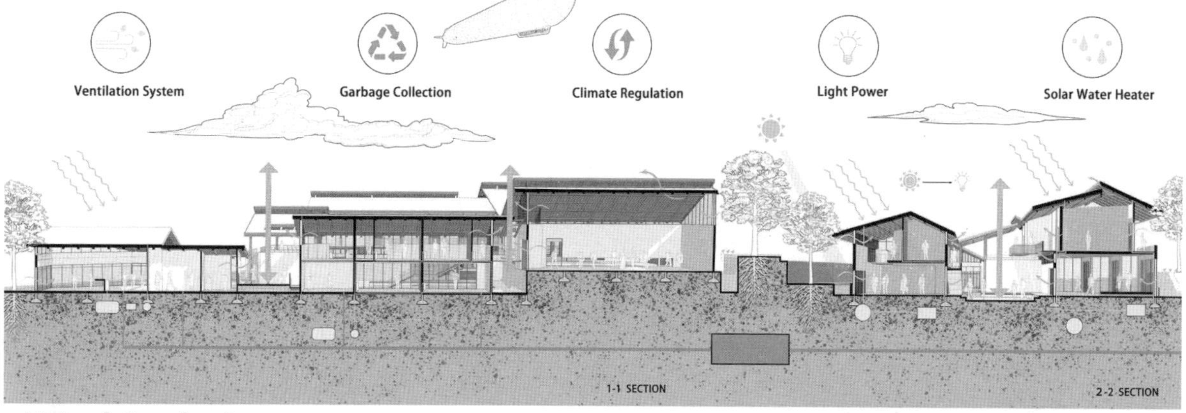

- Ventilation System
- Garbage Collection
- Climate Regulation
- Light Power
- Solar Water Heater

1-1 SECTION 2-2 SECTION

Wind Analysis

Section Strategy

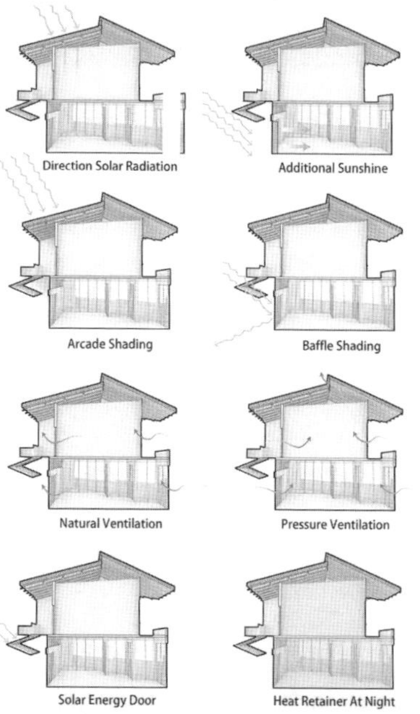

- Direction Solar Radiation
- Additional Sunshine
- Arcade Shading
- Baffle Shading
- Natural Ventilation
- Pressure Ventilation
- Solar Energy Door
- Heat Retainer At Night

Daylight Analysis

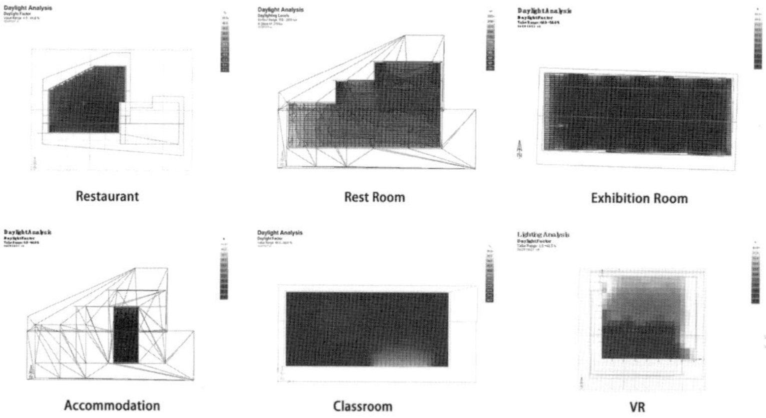

- Restaurant
- Rest Room
- Exhibition Room
- Accommodation
- Classroom
- VR

沐光山园

A Post Station in Guanba Village in the Sun
四川关坝村自然保护区科学考察站设计 06

Spatial Translation

Update Mode

The buildings extends linearly along the river.

Spatial Composition

Terrain
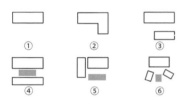

Function
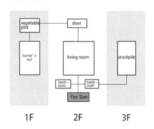
1F　2F　3F

Traffic Analysis
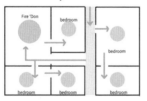
One floor of suites, entering their own bedroom first through other people's bedrooms, to take things also need to cross other people's bedrooms, low privacy.

Water Supply and Drainage

- Water supply pipes
- Drainage pipes
- From the mountain
- Drain
- From the river
- Reservoir

Aqueduct
  Nullah

  Nnderdrain

Material Analysis

 Earth
 Pine wood
 Fir wood
 Masonry
 Bamboo board

North elevation view　1:200

综合奖·优秀奖
General Prize Awarded · Honorable Mention Prize

注 册 号：101188
项目名称：光遇
　　　　　Light Encounters
作　　者：朱雅夫、周天娇、严　帅、
　　　　　刘媛卉
参赛单位：华南理工大学
指导教师：王　静

光遇 I

Description of design 设计说明：

场地位于四川省绵阳市平武县关坝村，结合场地夏季通风隔热、冬季采暖、年均降水量较大的气候特征，考虑平面底图关系和场地地貌，设计分布式布局的起承关系。相互错位有助于保温和阻挡冬季风，连续的折线屋顶将采暖功能空间串联。对当地藏式文化空间的探索和发掘，以"火塘"空间为中心原型，分别转译不同公共功能的单体，包括接待、研学、数字展示和内向性宿空间的被动式空间，结合主动式光伏技术打造出契合当地自然环境和优美景观的保护科考站。

The site is located in Guanba village, Pingwu County, Mianyang City, Sichuan Province. Considering the climate characteristics of the site, such as ventilation and heat insulation in summer, heating in winter, and large average annual precipitation, and considering the relationship between the floor plan and the site terrain, the general layout of the distributed layout is designed. Mutual dislocation is helpful for insulation and blocking the winter wind. The continuous broken line roof connects different functional spaces in series. The exploration and excavation of the local Tibetan cultural space, taking the "Huotang" space as the central prototype, translates the single units with constant public functions, including the passive space for reception, research, digital display and internal accommodation space, and combines the active photovoltaic technology to create a conservation scientific research station that matches the local natural environment and beautiful landscape.

■ Site Analysis

· Site traffic and status analysis

· Landscape analysis

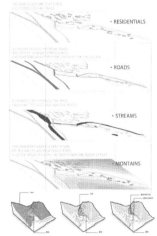

The site is close to the national road, and the current buildings within the planning scope include an office building of the village committee and an exhibition reception office. There are two houses in the second phase of the site, which are scheduled to be demolished. Most of the buildings in the site are sloping roofs and their combinations, with the characteristics of White Horse Tibetan folk houses.

· Architectural analysis

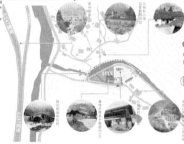

White Horse Tibetan Residence

· Population analysis

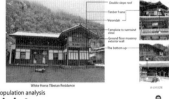

The total forest area of the village is 6499.3 mu, with a forest coverage rate of 96.3%. There are more than 70 species of rare animals and plants under state protection, such as giant panda, golden monkey, taxus chinensis and davidii tree. The area is rich in biological resources and high in ecological environment quality.

Local villagers except for the young workers, most of the village to keep bees, walnut planting. With the protection of ecological environment in recent years, a group of field researchers and local ecological patrol personnel have emerged, and the region also shoulders the function of popular science education.

■ Climate Analysis

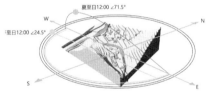

Sun path diagram in the venue

Annual solar radiation

Solar radiation is highest throughout the year from April to August and lowest in February and April. Due to the higher altitude of the site compared to the plain area, the solar radiation will be relatively high.

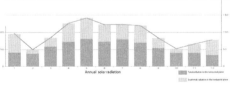

Year-round wind rose | Summer wind rose | Winter wind rose

It can be seen from the wind rose diagram that the dominant wind direction of the site is from east to west throughout the year, and the average wind speed is 3m/s. Therefore, it is necessary to consider the influence of wind in the site design and building layout, to maximize the use of wind for air conditioning, and to consider the wind blocking strategy in winter.

Year-round dry bulb temperature | Annual relative humidity

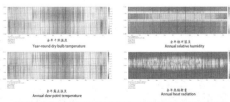

Annual dew point temperature | Annual heat radiation

According to the weather data generated by the software and the basic understanding of the site, the temperature of the site is not high in summer, so it is not necessary to consider cooling in summer, but ventilation is particularly important. In winter, the temperature is low, and the heating of the building needs to be considered.

■ Design Strategy

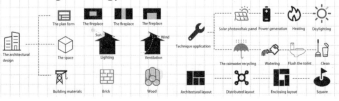

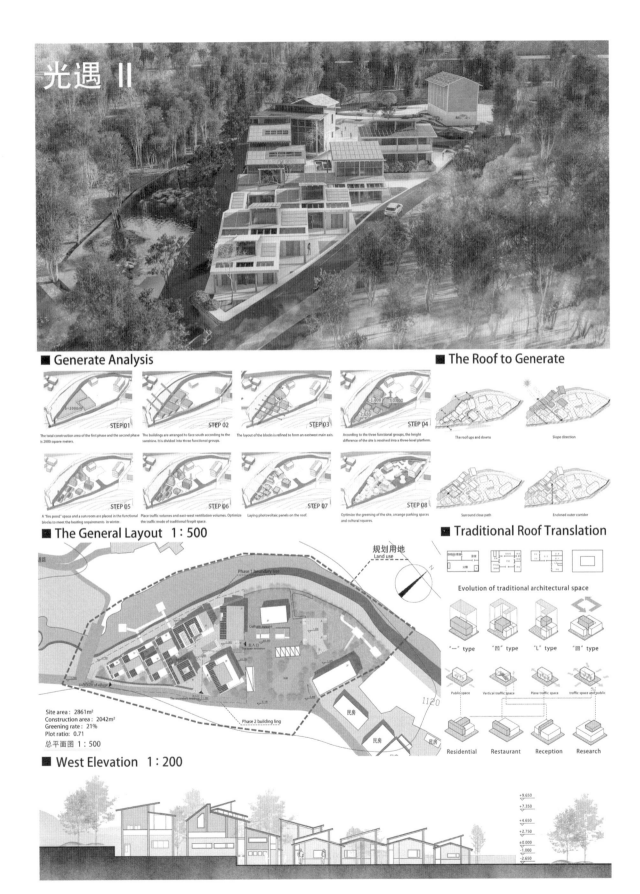

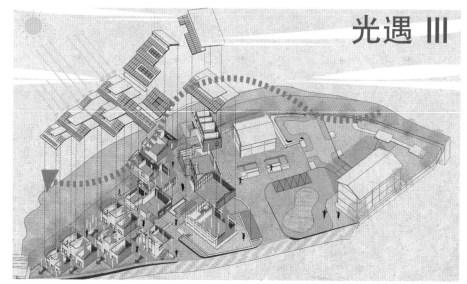

光遇 III

■ First Floor Plan 1:200

■ East Elevation 1:200

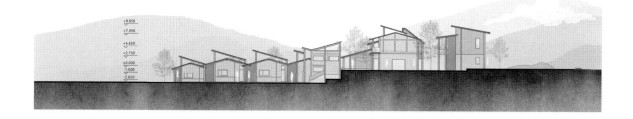

光遇 IV

■ Passive Technique Analysis

Canteen climate buffer space | Ventilated space | Lighting in daytime | PV system

Climate buffer space in reception center | Ventilated space | Lighting in daytime | PV system

Accommodation climate buffer spacer | Ventilated space | Lighting in daytime | PV system

■ Simulate of Technology

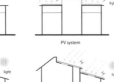

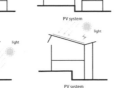

The sun radiation | Winter solstice daylighting | 4.5m wind speed

Ventilation in summer | Ventilation in winter | 1.5m wind speed

■ Ventilated Trumbo Wall

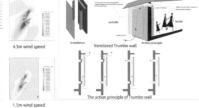

Installation | Ventilated Trumbo wall | Action principle

The action principle of Trumbo wall

■ Exploded Drawing

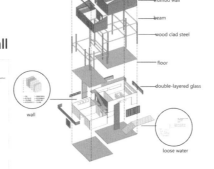

Photovoltaic panels — Green Roof — board — Purlin — rafter — trumbo wall — beam — wood clad steel — floor — double-layered glass — wall — loose water

Explosion analysis of reception center

■ Section A-A 1:200

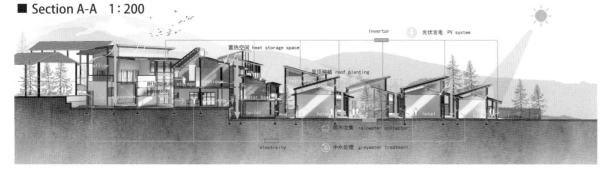

综合奖·优秀奖
General Prize Awarded·
Honorable Mention Prize

注　册　号：101223
项目名称：山谷·漫步
　　　　　Roam in the Valley
作　　者：樊秀君、刘滢、董海音
参赛单位：南京工业大学
指导教师：薛春霖

山谷·漫步 02
Roam in the valley

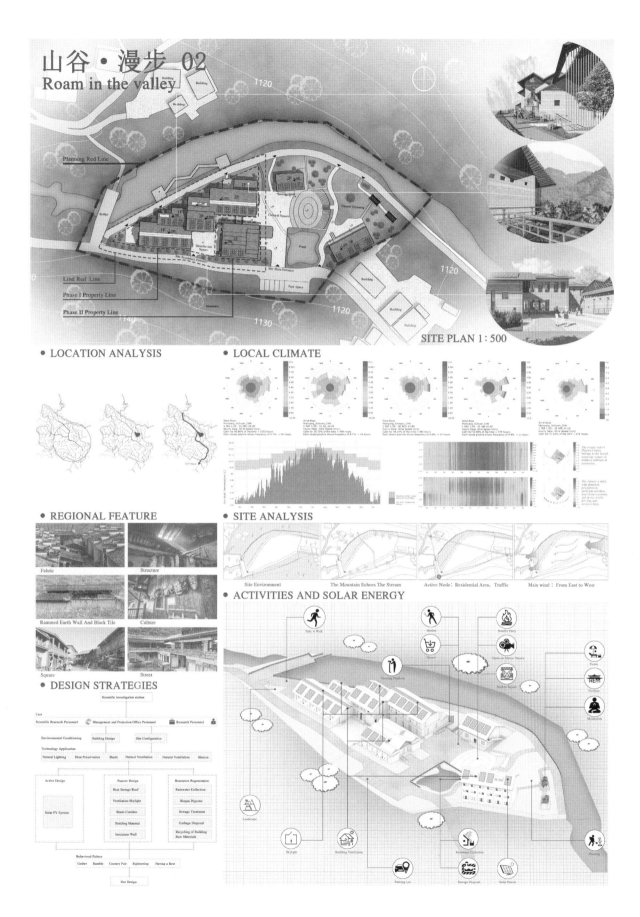

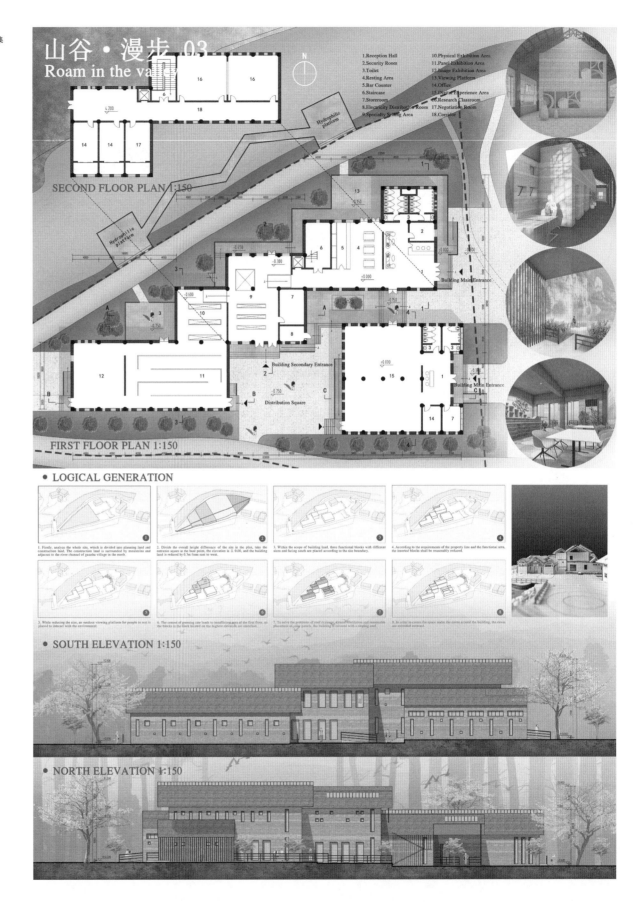

山谷·漫步 04
Roam in the valley

- ECOLOGICAL ANALYSIS
- INSPIRATION
- SOLAR ANALYSIS
- EAST ELEVATION 1:150
- 4-4 SECTION 1:150
- WEST ELEVATION 1:150
- D-D SECTION 1:150

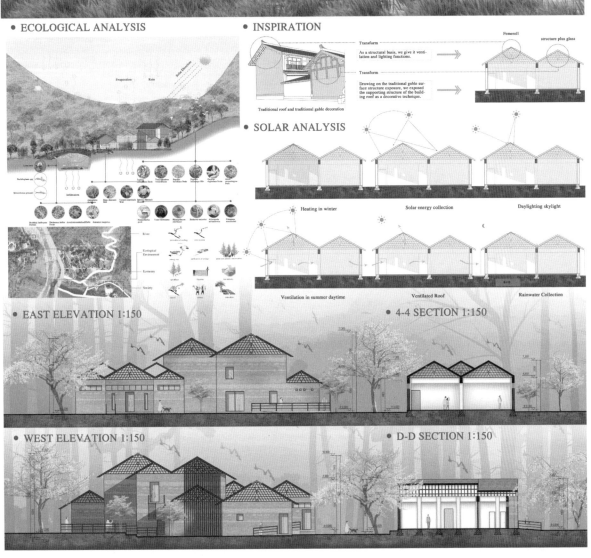

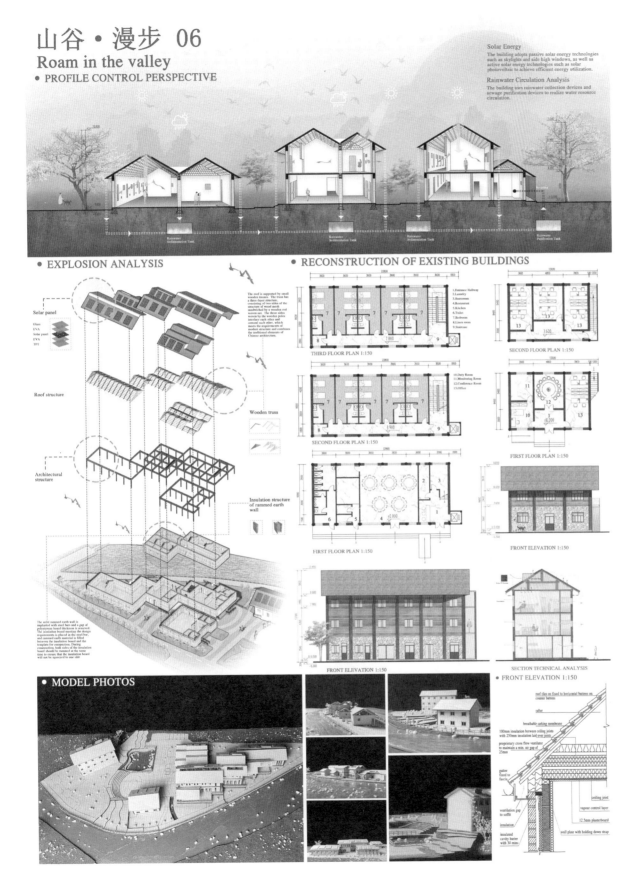

综合奖·优秀奖
General Prize Awarded · Honorable Mention Prize

注 册 号：101231
项 目 名 称：漂浮的森林
　　　　　　Floating Forest
作　　　者：毛　越、孙艺萌、贺晓婷
参 赛 单 位：南京工业大学
指 导 教 师：姜　雷

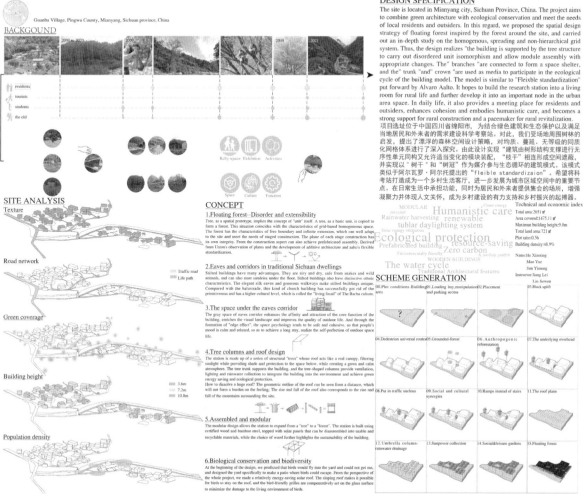

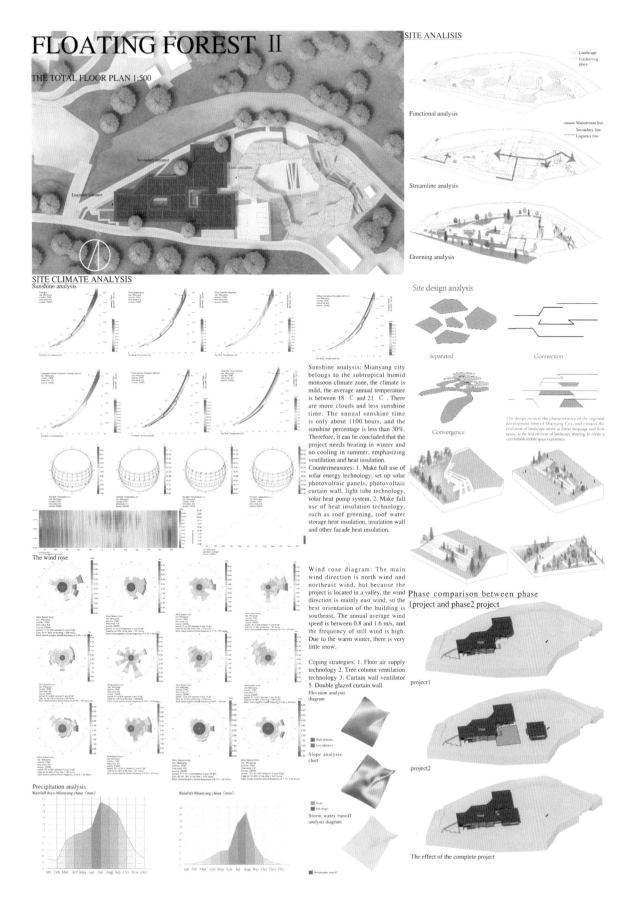

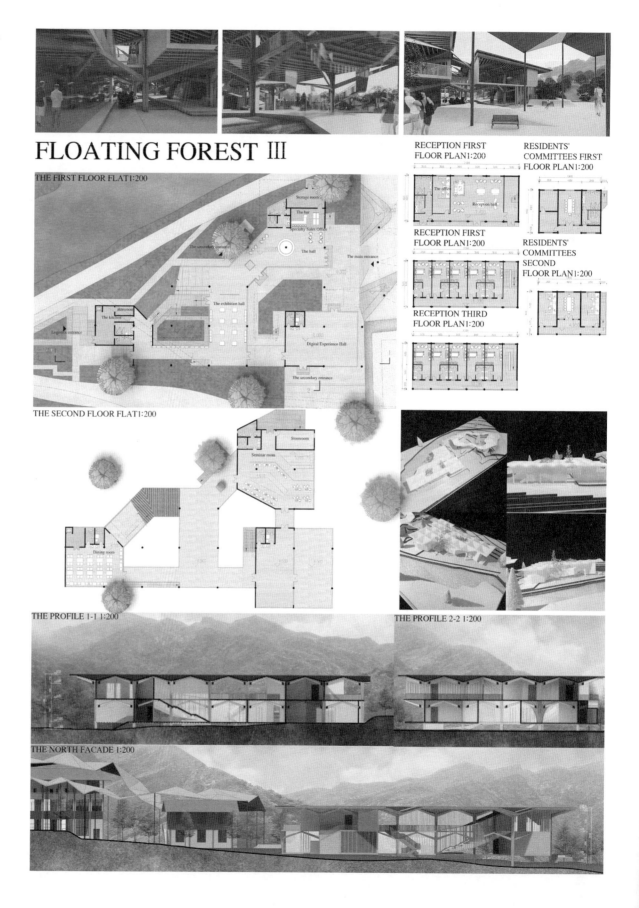

FLOATING FOREST IV

EXPLOSIVE VIEW

DETAIL DRAWING

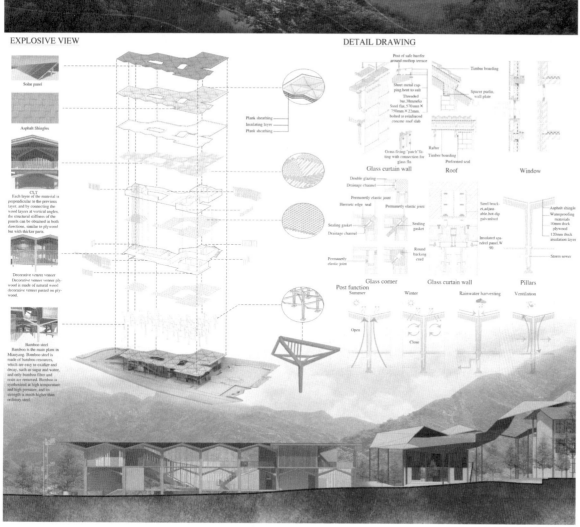

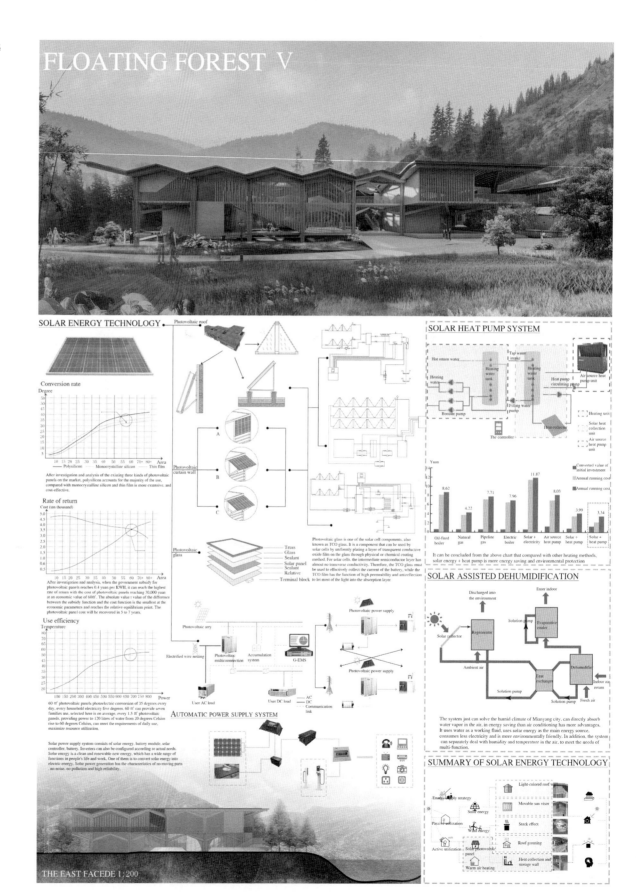

FLOATING FOREST VI

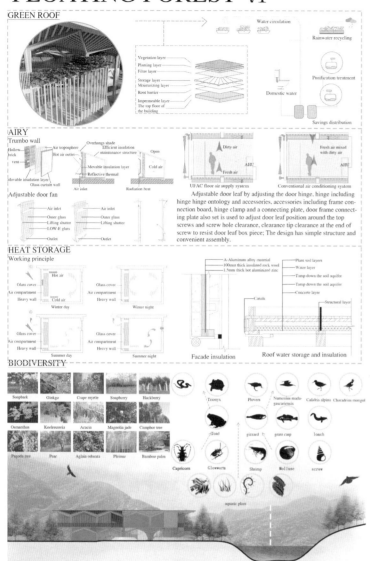

综合奖 · 入围奖
General Prize Awarded ·
Finalist Award

注 册 号：100699
项目名称：光·弋林间
　　　　　Light in the Forest
作　　者：刘杞铭、张军杰、吴辰懿、
　　　　　隋蕴仪
参赛单位：重庆大学
指导教师：黄海静

光·弋林间 Design of Scientific Research Station
自然保护小区科学考察站设计

设计受当地文化启发，形成川西藏族的围合空间；结合场地日照角度，将传统坡屋顶形式进行变形，更好地利用太阳能光伏瓦屋面和屋顶绿化；借鉴川西民居特色抱厅和坝台，形成独特的模数化阳光间和屋顶平台，保证采暖保温和采光且加强了建筑与周边环境的联系；因场地多雨雪，设计湿地生态系统，将雨雪回收，充分利用当地材料，最终达到传承文化、绿色生态、舒适节能和科技创新的主被动式节能的设计目标。

Inspired by local culture, the design aims at forming a traditional western Sichuan's Tibetan enclosure space. In consider of the sun angle of the site, the design's roofs, which are based on the traditional slope roof form, is deformed from its prototype to better utilize solar photovoltaic tile roofs and green roofs. The unique modular sunshine room and roof platform take the characteristic holding hall and dam platform of western Sichuan residential buildings for example, to ensure heating, heat preservation and daylighting. This technique also strengthen the connection between the building and its surrounding environment. Due to the snowy and rainy climate of the site, a wetland ecosystem is designed to recycle the rain and snow. Local materials are massively used in this design as well. Based on the design methods mentioned above, the design innovates in achieving passive-active energy reservation, and realizes its final goal of cultural inheritance, energy saving and a both organic and comfortable living style.

经济技术指标
规划面积 -Planned area : 10674.90m²
红线面积 -Property line : 2861.85m²
建筑密度 -Building density : 10.4%
建筑面积 -Building area : 2100m²
绿地率 -Greening rate : 34.1%
容积率 -Floor area ratio : 19.6%

Site Plan 1:500

光·弋林间

Design of Scientific Research Station
自然保护小区科学考察站设计 2

■ Site Analysis

The project site is located in Guanba Village, Mupi Tibetan Township, Pingwu County, Mianyang City, Sichuan Province, and belongs to the nature reserve. The project site around the traffic is convenient, close to the local Sichuan West residence, because it is located at the intersection of the two mountains, the site is very sunny, there is a small stream on the north side of the water flowing from east to west, there are often valley winds blowing, the environment is beautiful.

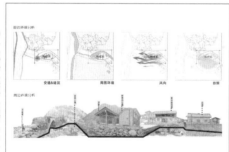

■ Climatic Simulation

Due to the cool and rainy summers and the cold and snowy winters, the building needs to focus on the problems of thermal insulation in winter and ventilation and rain in summer. At the same time, as a scientific research station, the design should emphasize environmental friendliness, environmental demonstration, and rational use of natural resources.

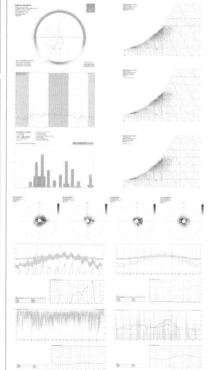

■ Diagram of Design Process

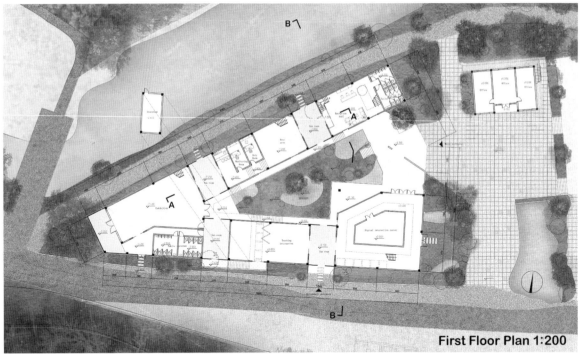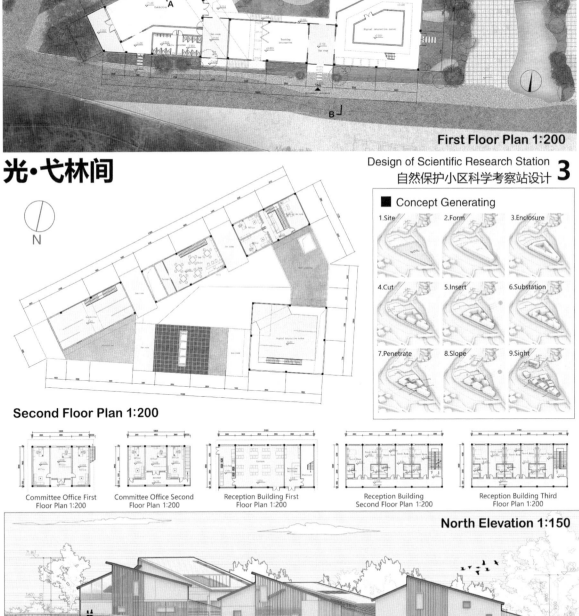

光·弋林间

Design of Scientific Research Station 4
自然保护小区科学考察站设计

■ Active Solar Technology

The design uses a variety of active solar energy technologies, while integrating them into People's Daily life, and diversifies activities within the building through multiple functional switches of sunlight room.

Play | Relax | Eat | Performance | Market | Talk

■ Water Footprint

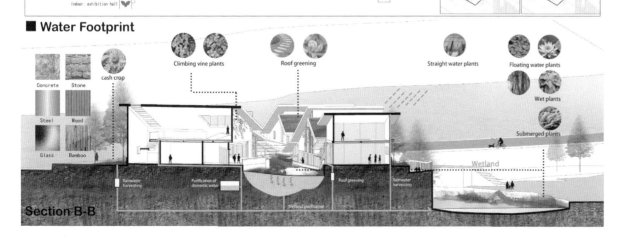

Section B-B

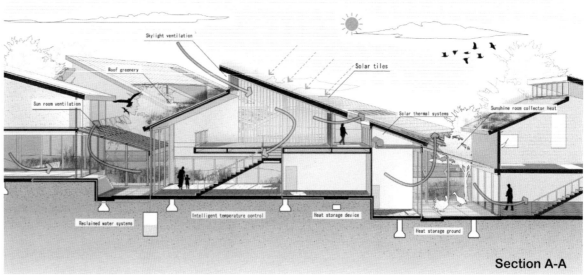

Section A-A

光·弋林间

Design of Scientific Research Station 5
自然保护小区科学考察站设计

■ Wind Simulation   ■ Sunshine Simulation   ■ Note Detail

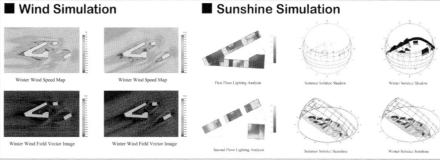

■ Section Strategy

The design emphasizes "passive priority, active optimization", innovative solar building integration technology, and independent adjustment of the physical environment of the building. Solar panels supply energy and hot water needs on the one hand, and absorb heat to improve indoor comfort on the other. The sunroom adopts intelligent opening and closing design, closing the skylight in winter to gather heat to maintain the indoor temperature, and opening the skylight in summer to promote ventilation in cooperation with the high windows of the main house.

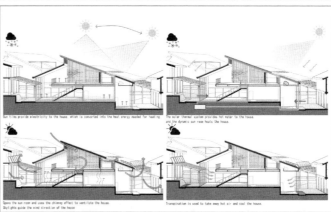

光·弋林间

自然保护小区科学考察站设计 6
Design of Scientific Research Station

■ Energy Consumption Diagram

■ Exploded Drawing

- Solar tiles
- Thermal storage walls
- Sun room
- Ventilated structure
- Roof greening / Green wall
- Double glazing / Adjustable sun shading

■ Target Analysis

Energy Consumption Comparison

Type of energy consumption	Sub type of energy consumption	Designed building (kW·h/m³)	Reference building (kW·h/m³)
Construction load	Cooling consumption	73.53	85.91
	Heat consumption	59.14	73.92
Heat recovery load	Cooling supply	1.01	—
	Heating supply	12.38	—
Cooling energy consumption	Central cold source	0.00	0.00
	Cooling water pum	0.00	0.00
	Frozen pump	0.00	0.00
	Cooling tower	0.00	0.00
	Multi-line / unit air conditioning	21.87	25.78
Heating energy consumption	Central heat source	9.24	37.05
	Heating pump	2.31	0.55
	Multi-line / unit heat pump	0.00	0.00
Air conditioner fan energy consumption	Independent new exhaust	7.78	7.78
	Fan coil	0.00	0.00
	Multi-online indoor machine	1.19	1.40
	All air system	0.00	0.00
Heating and air conditioning energy consumption		42.38	72.56
Lighting energy consumption		13.89	13.89
Comprehensive energy consumption		56.27	86.45
Comprehensive energy saving rate		34.91%	

Renewable Energy Calculation

Solar panel area (m²)	805
Solar water heating (kW·h/m³)	3.78
Photovoltaic power generation (kW·h/m³)	111.60
Wind power generation (kW·h/m³)	0
Renewable Energy (kW·h/m³)	115.51

Building Lifecycle Carbon Emissions

Type of carbon emissions	Annual carbon emissions (tCO₂/a)	Carbon emission (tCO₂/a)
Construction material production	3.080	153.971
Transportation of building materials	0.410	20.593
Building construction	0.158	7.933
Building demolition	0.366	18.289
Building operation	0.000	0.000
Carbon sink	−7.375	−368.750
Totals	0.000	0.000

综合奖 · 入围奖
General Prize Awarded · Finalist Award

注 册 号：100777
项目名称：风隐山居
　　　　　Hide on the Wind
作　　者：何丽诗、黄 李、贺为皓、
　　　　　秦 漓、肖 峰、黄荟瑜、
　　　　　胡诗雨
参赛单位：四川农业大学
指导教师：陈 川、侯超平

设計説明：

本方案结合当地建筑特色，提取远山、廊道等元素，屋檐重叠错落，屋檐下廊道和院落彼此连接，加强冬季保温的同时，尽可能兼顾夏季的通风散热。
考虑到当地石块台基，建筑通过卵石蓄热的被动式设计消化场地高差，并通过太阳能烟囱、阳光房等实现冬季保温。在主动式设计上，将太阳能光伏板和空气集热器结合，部分屋面设计绿化。此外，建筑还加入了雨水收集系统和污水处理系统，最大化地利用雨水，避免人群活动对当地环境造成污染。

This plan combines local architectural features, extracts distant mountains, corridors, and other elements, overlaps eaves, and connects corridors and courtyards under eaves, to strengthen winter insulation and give consideration to summer ventilation and heat dissipation as much as possible. Considering the local stone platform foundation, the building digests the site elevation difference through the passive design of pebble heat storage and achieves winter thermal insulation through the solar chimney, sunlight room, etc. In the active design, solar photovoltaic panels and air collectors are combined, and some roofs are designed for greening. In addition, a rainwater collection system and sewage treatment system are added to the building to maximize the use of rainwater and avoid pollution to the local environment caused by crowd activities.

體塊生成：

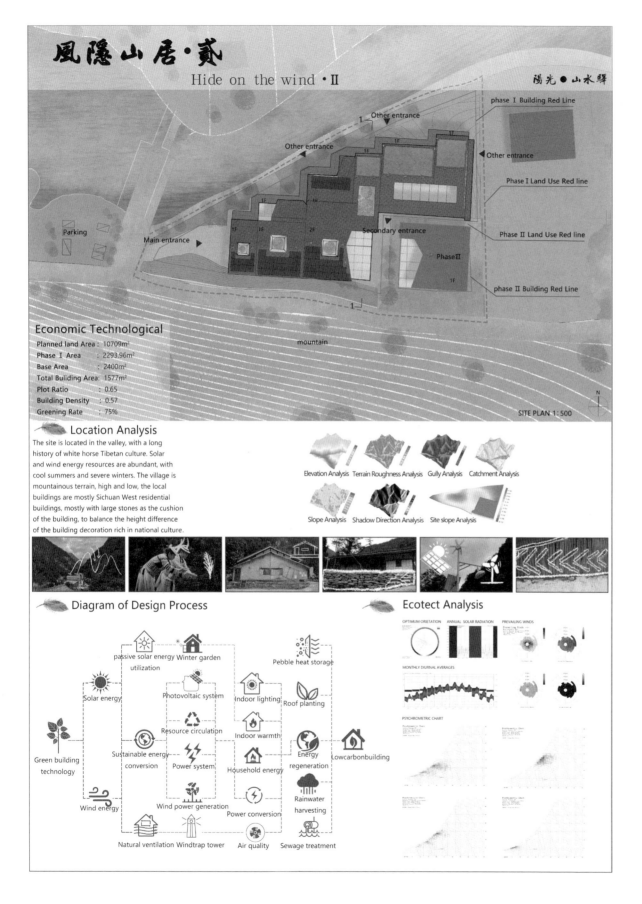

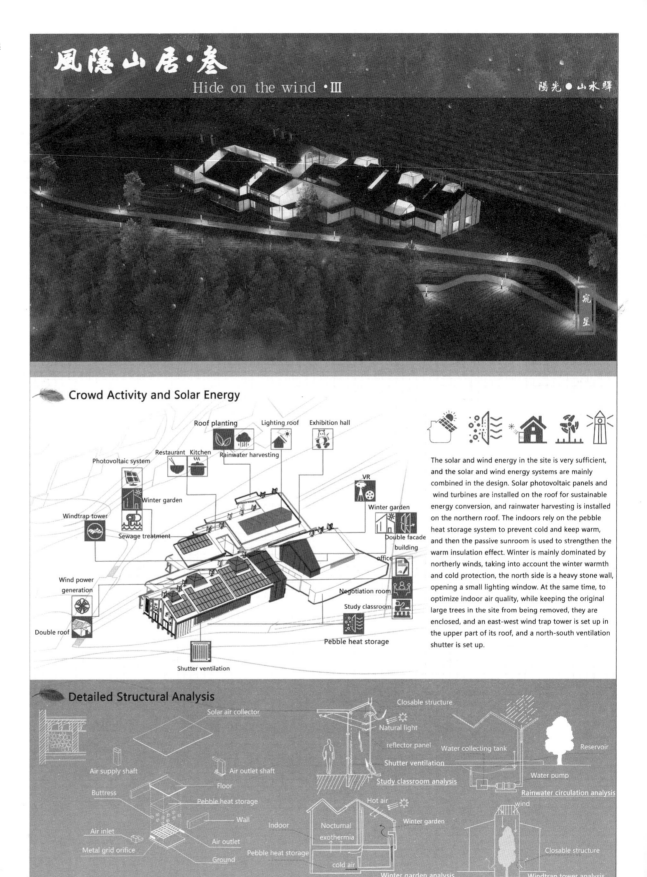

凤隐山居·肆
Hide on the wind · IV

1. Reception
2. Lounge
3. Toilet
4. Winter garden
5. Water bar
6. Exhibition room
7. Storeroom
8. Study classroom
9. Restaurant
10. Kitchen
11. Exhibition hall
12. VR room
13. Auxiliary rooms
14. Courtyard
15. Negotiation room
16. Office
17. Conference room
18. Room for duty
19. Monitoring room
20. Supporting rooms
21. Laundry
22. Accommodation reception
23. Guest room
24. Housekeeping Office

Functional Partition

- Main functional areas
- Management and protection office area
- Accommodation reception area

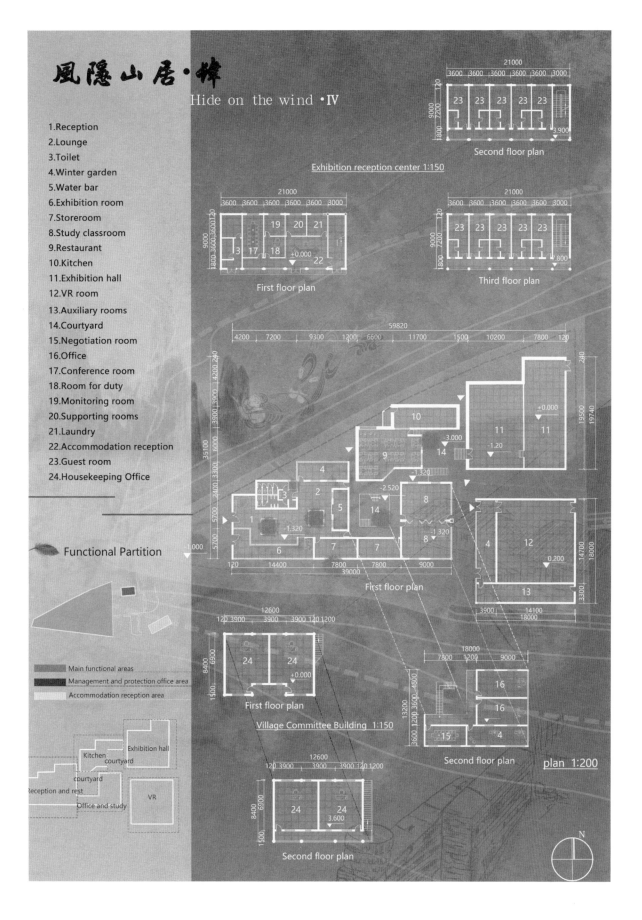

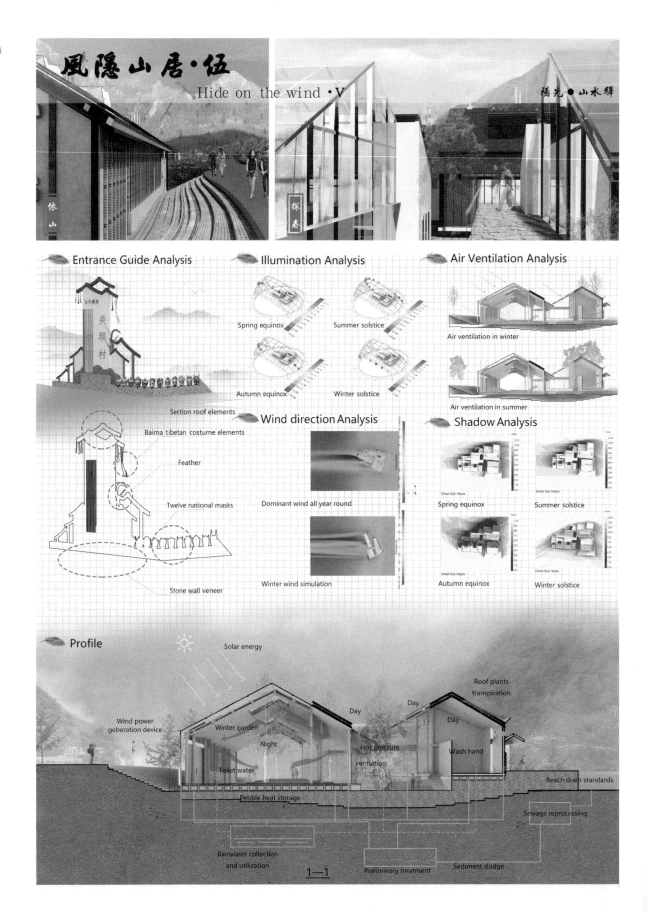

凤隐山居·陆
Hide on the wind · VI

2022 Delta Cup International Solar Building Design Competition

陽光●山水驛

Analysis of Solar Energy Technology

(1) According to the Mianyang Meteorological Bureau, the total annual amount of solar radiation within one square kilometer near the building land is 143,023KJ/m²
(2) Based on the conversion efficiency of photovoltaic panels, the process and the laying form of the cell array, block monocrystalline silicon solar cells are selected
(3) Take the month as the cycle, use the maximum sunshine intensity to determine the battery model, and select photovoltaic panels with a peak power of about 150w/m²

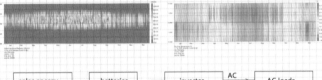

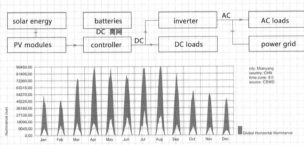

(4) The building uses peak power of 300w, size 1956 * 992 * 45mm photovoltaic panels 115 pieces, an area of about 230m²
(5) According to <Assessment standard for green building in Sichuan Province> the estimated electricity consumption, since the building is composed of multiple types of functions, the calculation is calculated by referring to the corresponding building type to obtain about 360/90% of the building electricity consumption (transmission and distribution loss)=400kWh(days)
(6) From the perspective of user demand and system power generation, solar power generation and wind power generation with an annual cycle can fully meet user needs and achieve zero energy consumption in buildings.

Material Analysis and Roof Structure Analysis

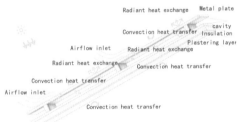

South elevation

West Elevation

综合奖・入围奖
General Prize Awarded · Finalist Award

注 册 号：100783
项目名称：林涧·驿
　　　　　Springdale Station
作　　者：陈曙琪、段于瑄、张　满
参赛单位：河南工业大学
指导教师：张　华、马　静

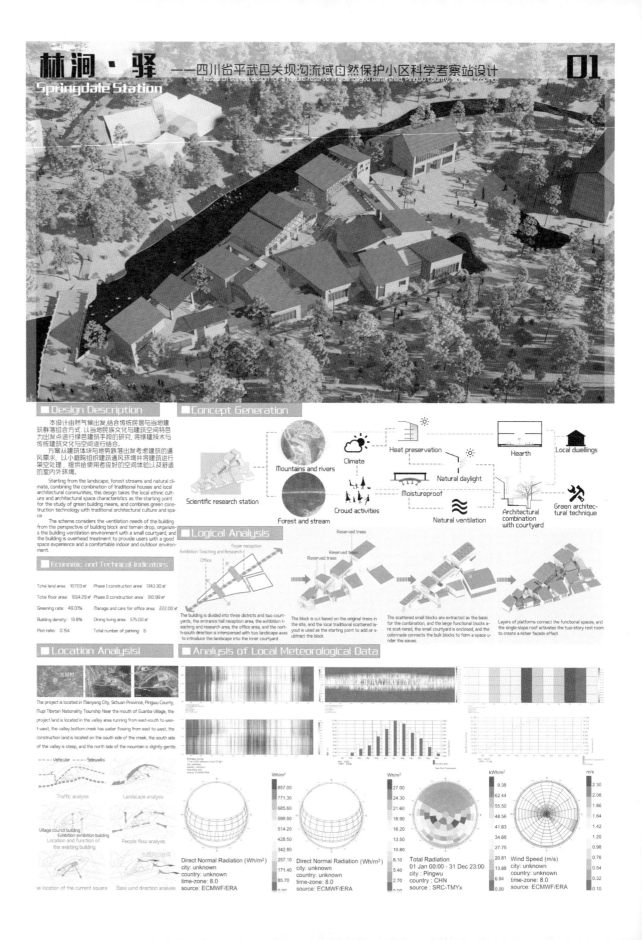

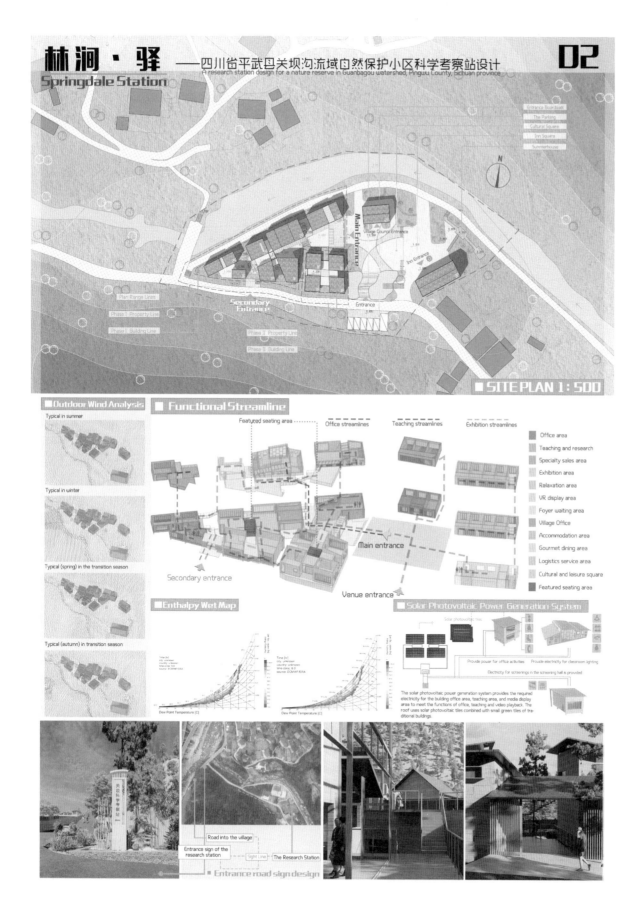

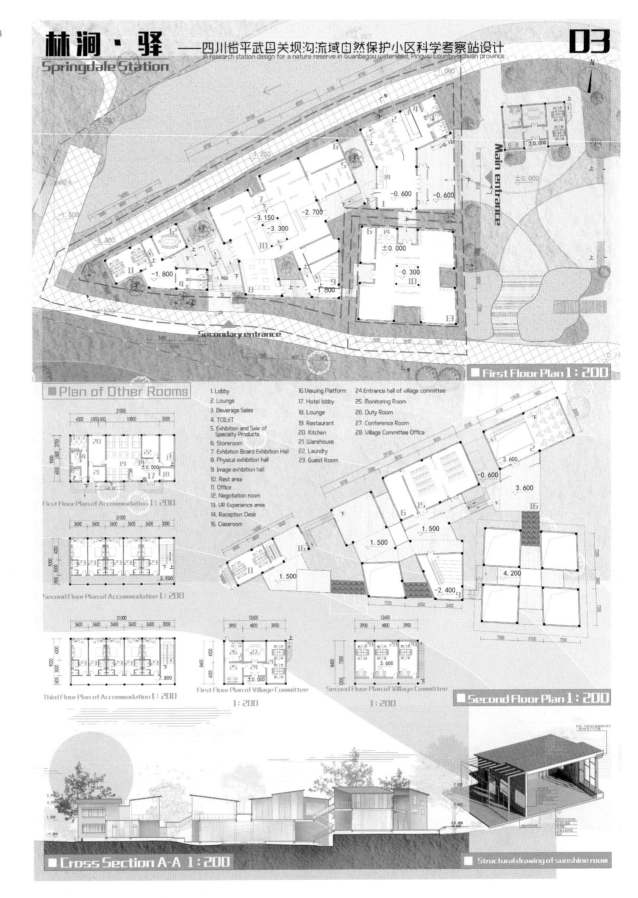

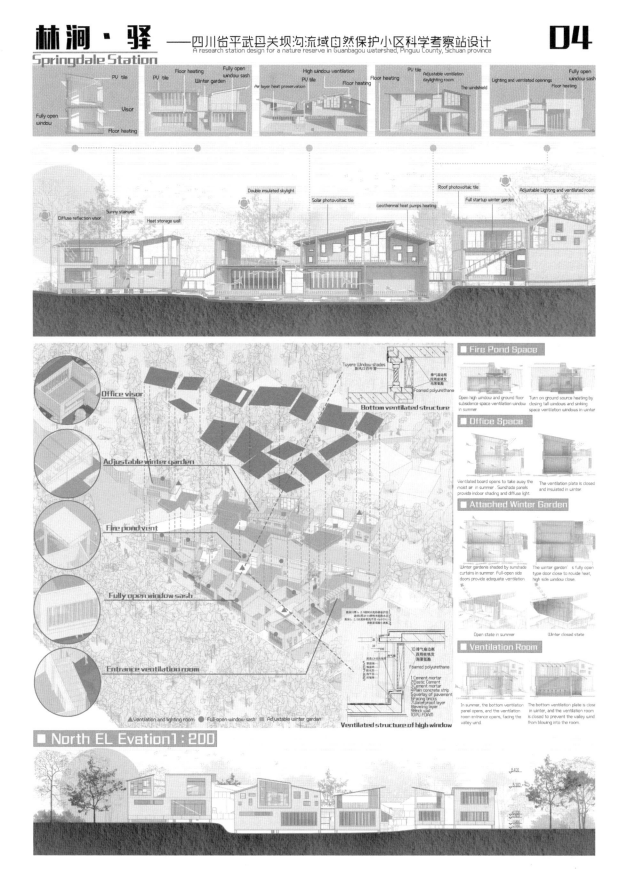

2022 台达杯国际太阳能建筑设计竞赛获奖作品集

土墙板屋
Earth Wall Board House

建筑方案从"体量、材料、构造"三个方面入手，将传统陇南藏族民居"土墙板屋"作为本方案核心。建筑体量的推敲借鉴了《龙安府志》中对白马藏族居所描绘的空间结构，结合场地高差，从而形成了局部二层，顶部架空的建筑空间。在考虑建筑材料的使用时通过对传统建筑材料和现代建筑材料的结合，使其具有现代建筑强度的同时还具备传统建筑之美。通过建筑构造对自然资源加以运用，将绿色建筑通过"风墙"等构造以实现。

The architectural scheme starts from three aspects: volume, material and structure.The traditional Longnan Tibetan dwelling "earth wall board house" as the core of the case.
The construction volume is based on the spatial structure described by the White Horse Tibetan residence in Long'an Fu zhi.Combined with the height difference of the site, a local two-storey building space with an overhead roof is formed.
In considering the use of building materials through the combination of traditional building materials and modern building materials.To give it the strength of modern architecture but also the beauty of traditional architecture.
The use of natural resources through building structures, green buildings through structures such as "wind walls" to achieve.

综合奖 · 入围奖
General Prize Awarded ·
Finalist Award

注 册 号：100784
项目名称：土墙板屋
　　　　　Earth Wall Board House
作　　者：张培俊、金圣煜、薛莹莹、
　　　　　郑　峥
参赛单位：中国美术学院风景建筑设计研
　　　　　究总院有限公司、浙江汇创
　　　　　设计集团有限公司
指导教师：张　军

展板二

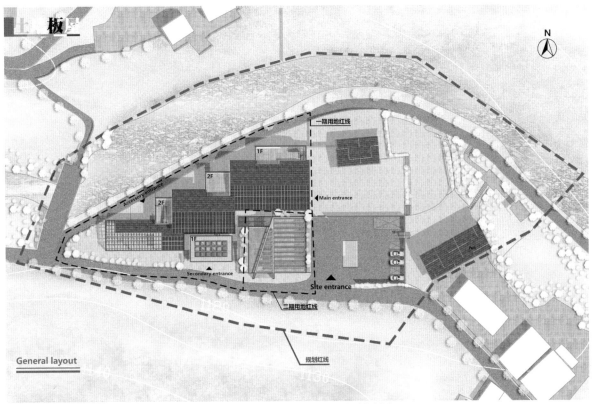

General layout

场地分析
Site analysis

本案场地位于四川省绵阳市平武县木皮藏族乡关坝村
The site of this case is located in Guanba Village, Mupi Tibetan Township, Pingwu County, Mianyang City, Sichuan Province.

功能分析
Functional analysis

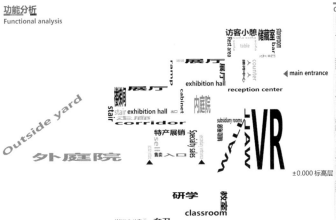

经济技术指标
Economic and technical indicators

名称	数值	单位	备注	名称	数值	单位	备注
一期建设用地面积	2294	m²		建筑占地	1305.45	m²	
二期建设用地面积	567.89	m²		一期占地面积	697.85	m²	
总建筑面积	2112.9	m²		二期占地面积	300.4	m²	
一期建筑面积	1001.85	m²		既有占地面积	307.20	m²	
二期建筑面积	300.40	m²		一期容积率	0.44		
既有建筑面积	810.65	m²		二期容积率	0.53		
一期建筑密度	30%			绿地率	35%		
二期建筑密度	52%			建筑高度	9.95	m	坡屋顶最高处

气候分析
Climate analysis

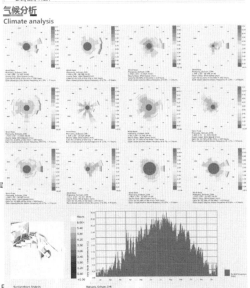

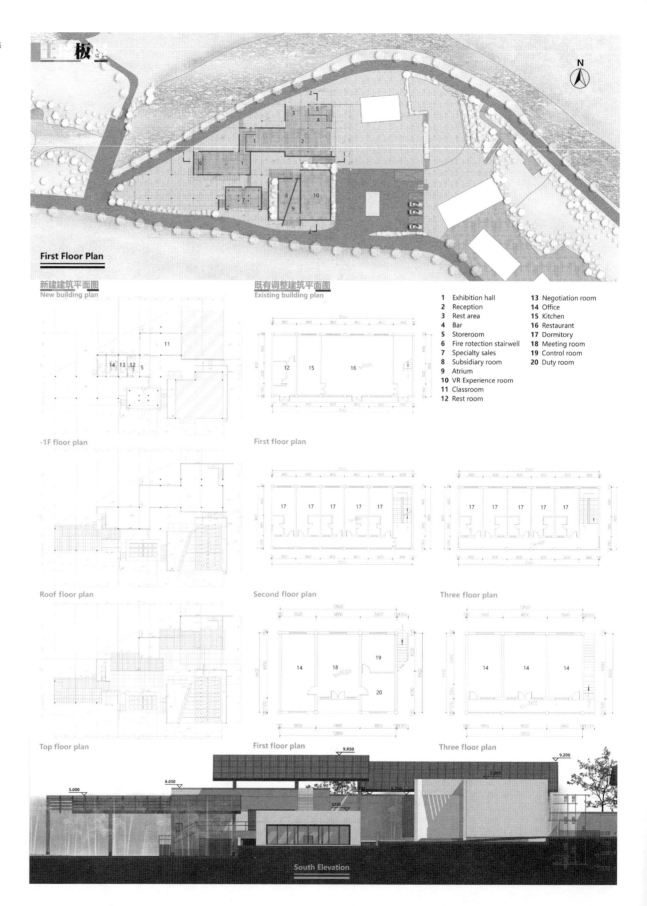

塬板屋

一期建筑体量推导
Phase 1 building derivation

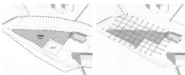

1) 本方案在建筑空间布局上回应场地关系
The layout begins with a response to the site

2) 将 8×8 的柱网置入场地
The layout begins with a response to the site

3) 所有在场地内的交点均为柱子
All intersections in the SITE are post

4) 围合部分柱网，区分出室内外空间和灰空间
Part of the column net is enclosed to distinguis the indoor and outdoor space from the gray space

5) 根据地形局部楼板下降形成局部二层的建筑布局
According to the terrain local floor drops
Forming a local two-story building layout

6) 根据柱网和建筑体量布置连续错层的坡屋面
Part of the column net is enclosed to distinguish the indoor and outdoor space from the gray space

7) 根据场地标高，在最少土方的情况下调整场地高差与建筑相契合
Adjust the site elevation difference, according to the site elevation with minimum earthwork It fits in with the architecture

8) 将建筑"掏"出洞，引入自然光，在南面屋顶增加光伏板
Cut holes in the building to bring in natural light Add photovoltaic panels on the south roof

功能流线图
Function streamline

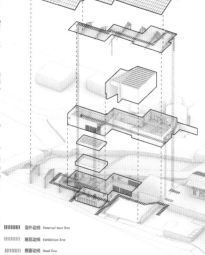

室外动线 External tour line
展览动线 Exhibition line
屋面动线 Roof line

North elevation

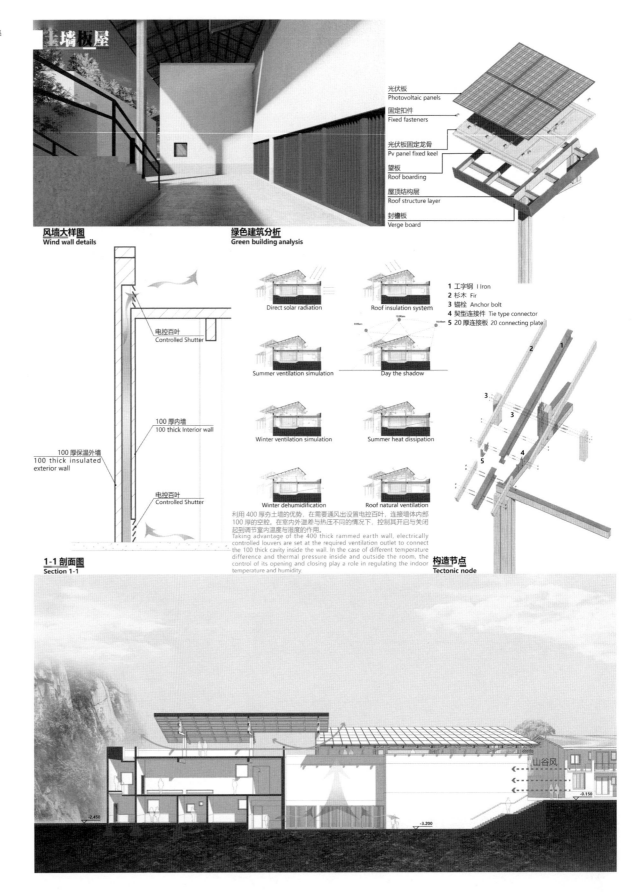

土墙板屋

2-2 剖面图
Section 2-2

1) 二期地块紧邻一期地块
The Phase II is adjacent to the Phase I plot

2) 确定建筑占地轮廓
Define the outline of the building

3) 将建筑分为三个功能分区：VR 数字体验空间、中庭、辅助用房
The building is divided into three functional zones:VR space, Atrium, Auxiliary room

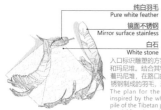

纯白羽毛
Pure white feather

镜面不锈钢
Mirror surface stainless

白石
White stone

入口标识雕塑的方案灵感来源于藏族的白石崇拜和玛尼堆，结合其特殊的功能，白色的羽毛围绕着玛尼堆，在路口的两侧分别放置两个由镜面不锈钢制成的羽毛，起到凸面镜的效果。
The plan for the entrance sign sculpture is inspired by the white stone worship and Mani pile of the Tibetan people. In combination with its special function, white feathers are placed around the mani pile, and two feathers made of mirror stainless steel are placed on both sides of the intersection to play a convex mirror effect.

4) 在中庭立起两面结构墙，起到空间的分割与结构作用
Two structural walls are erected in the atrium to play the role of space division and structure

5) 建筑西面辅助用房空间层高下降将体量进行分割
Lower the height of the auxiliary room space on the west side of the building and divide the volume

6) 将建筑顶板去除，使光线能够最大程度的穿过密肋梁进入室内
The roof plate was removed to allow maximum light to pass through the ribbed beams

2-2 剖面图
Section 2-2

综合奖·入围奖
General Prize Awarded·
Finalist Award

注 册 号：100846
项目名称：于山阆水 环蜂绕竹
　　　　　Guanba Wonderland
作　 者：付嫣然、李玲秀、甘杰文、
　　　　　林逸清
参赛单位：福州大学
指导教师：林志森

Location analysis

Mupi Tibetan township　Guanba village　Site details

The area is influenced and radiated by Chengdu Chongqing urban agglomeration. Its geomorphic types are: eroded plateau and extremely high mountains formed by the humid monsoon climate. Located in the Yangtze River Basin of China, it belongs to the north subtropical humid area.

Social appeal
Policy background

Wild panda　Black bear　Golden Monkey　Takin　Research　Nature Reserve
An important migration channel for wild animals.　An important ecological corridor.

Nationality and religion

Various customs with regional characteristics:
White horse paper cutting, jumping grass cover, white horse custom, white horse "felt hat".

Commercial sale of honey

economic system:
Beekeeping, tourism, walnut, protection community

Meteorological data

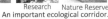

Optimum Orientation
Location: pingwu
Orientation based on average daily incident radiation on a vertical surface.
Underheated Stress:931.1
Overheated Stress:176.9
Compromise:205.0°
Avg.Daily Radiation at -154.0°
Entire Year:076 kWh/㎡
Underheated:0.94kWh/㎡
Overheated:0.55kWh/㎡

Annual Average
Underheated Period
Overheated Period

Best orientation

LEGEND
Comfort:Thermal Neutrality
Temperature
Rel.Humidity
Wind Speed
Direct Solar
Diffuse Solar
Cloud Cover

Wind data chart　　Line chart of climate data

Crowds appeal

the elderly and Young adults:　　Village committee:

Main work:
Beekeeping, walnut planting, ecological protection patrol,home stay reception.

Hopes:
Protect the environment,
develope the local characteristic industries

Tourists　　Researchers

Hopes:
Experience the local customs,
learn the local history, get close to nature

Main work:
Management of ecological reserves,
promote scientific research knowledge

Architecture element analysis

Spatial features:
Stone paving, stone foundation, wood beam column, bamboo veneer and tile roof, blockhouse shape watchtower

Base condition analysis

Village spatial

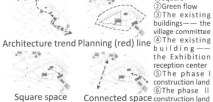
Architecture trend　Planning (red) line
Square space　　Connected space

①Square
②Green flow
③The existing buildings——the village committee
④The existing building——the Exhibition reception center
⑤The phase I construction land
⑥The phase II construction land

 Prensent situation texture
 Traffic analysis

Landscape analysis　　Ventilation analysis

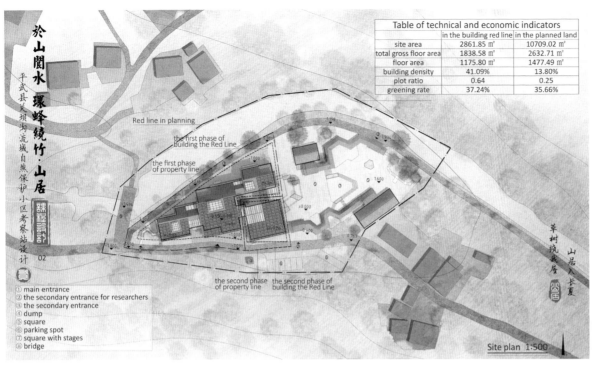

Table of technical and economic indicators		
	in the building red line	in the planned land
site area	2861.85 m²	10709.02 m²
total gross floor area	1838.58 m²	2632.71 m²
floor area	1175.80 m²	1477.49 m²
building density	41.09%	13.80%
plot ratio	0.64	0.25
greening rate	37.24%	35.66%

① main entrance
② the secondary entrance for researchers
③ the secondary entrance
④ dump
⑤ square
⑥ parking spot
⑦ square with stages
⑧ bridge

Site plan 1:500

Inspiration Source

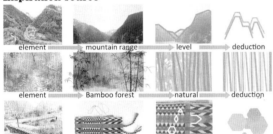

Design Description

本设计建筑提取当地碉楼的形态，设计了吸热通风系统，场地设计上抽取花腰带的六边形元素。整体设计注重对内外空间渗透界壁。综合运用了太阳能光伏系统、太阳能热水系统、底层架空的太阳炕等诸多低碳技术，使节能环保、绿色低碳、宜居舒适的绿色建筑成为现实。设计将彩色太阳能光伏玻璃、生态循环系统、可旋转的太阳能板等与屋顶、立面与架构结合，进行主被动式太阳能及其他低碳技术的利用。

The architectural form extracts the form of the local watchtower to design Heat-absorbing ventilation system. The site design extracts the hexagonal elements of the flower belt . Overall design pay attention to the penetration of internal and external spaces, and comprehensively uses solar photovoltaic system, solar hot water system, overhead Solar Kang at the bottom and many other low-carbon technologies to make the green buildings that are energy-saving, environment-friendly, green, low-carbon, livable and comfortable become a reality. The design combines colored solar photovoltaic glass, ecological recycling system, rotatable solar panels with roofs, facades and structures to make use of passive and active solar energy and other low-carbon technologies.

Site Design Analysis

Phased Construction Process   Streamline Analysis

Block Generation

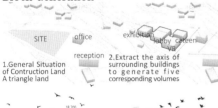

1. General Situation of Contruction Land. A triangle land
2. Extract the axis of surrounding buildings to generate five corresponding volumes
3. Terraces are dug out of the main landscape. Then vertical greening is formed.
4. "Watchtowers" are extracted from traditional watchtowers to form hot pressure ventilation.
5. Sloping roof elements are extracted from surrounding buildings and laid witeh solar panels.

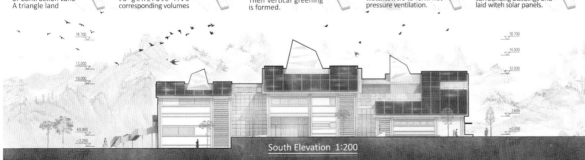

South Elevation 1:200

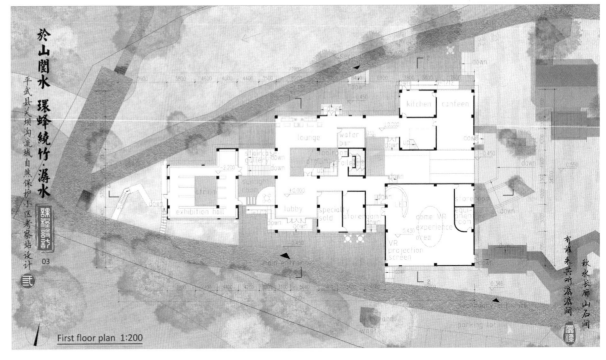

於山間水 環蜂繞竹 瀝水

平武县关坝沟流域自然保护小区考察站设计

First floor plan 1:200

Alley type extraction

    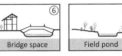 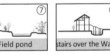

① The pavilion ② Empty platform ③ The public square ④ Mixed space ⑤ Corridor space ⑥ Bridge space ⑦ Field pond ⑧ stairs over the Water

Blind Spot Handling

Convex imitate local watchtowers.

Heat-absorbing Ventilation System

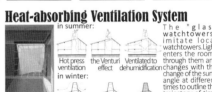

in summer: Hot press ventilation / the Venturi effect / Ventilated to dehumidification

in winter: Cobblestone at the bottom has high specific heat capacity. / Air conditioning isolation / internal circulation of hot gas / Thermal diffusion

The "glass watchtowers" imitate local watchtowers. Light enters the rooms through them and changes with the change of the sun's angle at different times to outline the trajectory of time and to achieve the dialogue between architecture and nature.

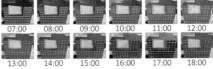
07:00 08:00 09:00 10:00 11:00 12:00
13:00 14:00 15:00 16:00 17:00 18:00

Indoor Perspective

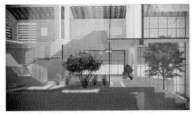

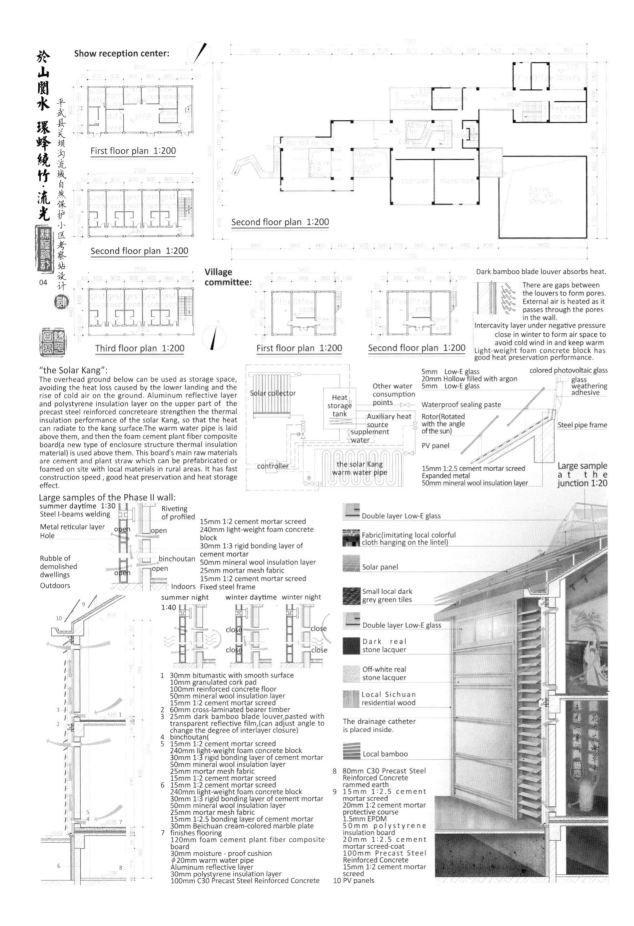

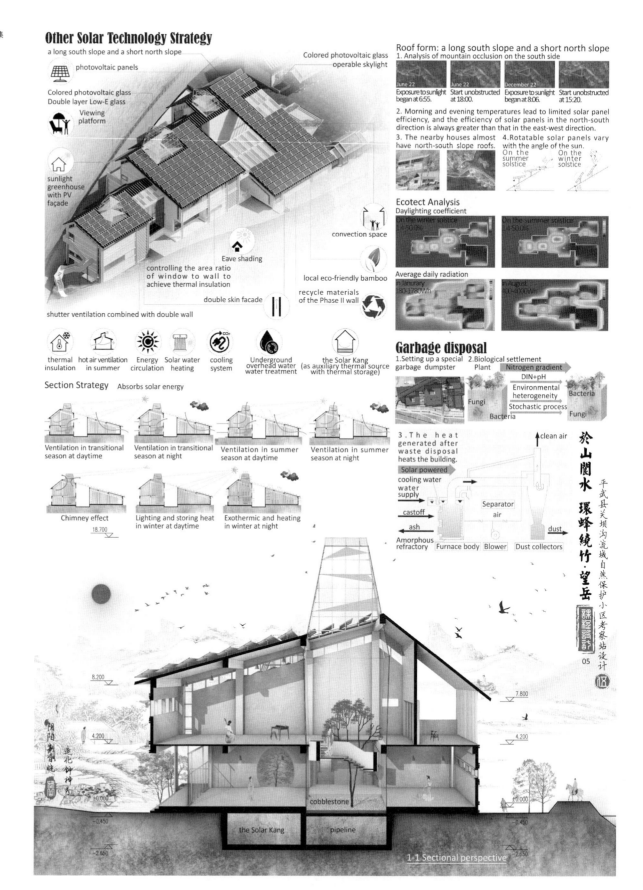

於山間水　環蜂繞竹·聽風

关坝沟流域自然保护小区宣教站设计

别境时听风折竹
断桥闻着水流渐

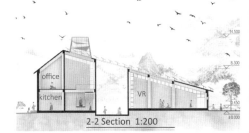

2-2 Section 1:200

Water Treatment System

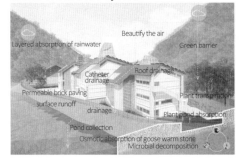

- Beautify the air
- Layered absorption of rainwater
- Green barrier
- Catheter drainage
- Roof drainage
- Permeable brick paving
- surface runoff
- drainage
- Plant transpiration
- Plant pond absorption
- Pond collection
- Osmotic absorption of goose warm stone
- Microbial decomposition

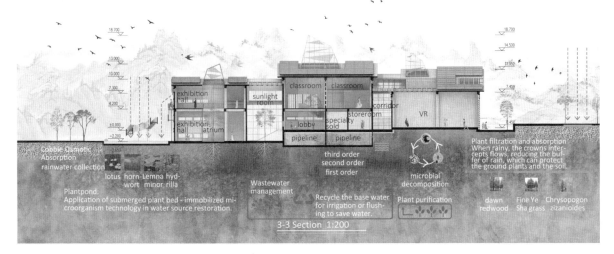

Cobble Osmotic Absorption rainwater collection

lotus　horn-Lemna hyd-wort minor rilla

Plantpond: Application of submerged plant bed - immobilized microorganism technology in water source restoration.

Wastewater management

Recycle the base water for irrigation or flushing to save water.

third order / second order / first order

microbial decomposition

Plant purification

Plant filtration and absorption. When rainy, the crowns intercepts flows, reducing the buffer of rain, which can protect the ground plants and the soil.

dawn redwood　Fine Ye Sha grass　Chrysopogon zizanioides

3-3 Section 1:200

综合奖·入围奖
General Prize Awarded · Finalist Award

注 册 号：100899
项目名称：光之滑梯
　　　　　The Slide of Light
作　　者：王艺霖、许乐萱、刘家锴、
　　　　　赵　方
参赛单位：南京工业大学
指导教师：杨亦陵

光之滑梯 The slide of light

本设计以"光之滑梯"为设计理念，将"滑梯"元素与坡屋顶结合，为在山谷中获得足够的采光与太阳能，本设计将北侧屋顶翘起并延展；考虑到当地常年多雨，将建筑南侧屋顶翘起，收集雨水的同时也扩大了南侧的受光面。
主动式技术方面，在阴雨连绵的情况下，充分利用场地铺设太阳能板，辅以雨能发电；被动技术方面，利用阳光房、蓄热墙体、蓄热屋顶收集辐射热能，设置可调节百叶缓解通风问题。建造方面，将建筑架空并增设夯土层，以抵御霜冻破坏与潮湿侵蚀。
此外，充分考虑当地民俗和生态环境，以减少对周围动植物的影响，同时满足人群生活舒适度。

This design takes the "slide of light" as the design concept, combines the "slide" element with the slope roof, in order to obtain enough lighting and solar energy in the valley, this design will be the north side of the roof warped and extended; Considering the local perennial rain, the roof on the south side of the building is raised to collect rainwater and expand the light surface on the south side.
In terms of active technology, in the case of continuous rain, make full use of the site to lay solar panels, supplemented by rain energy power generation; In terms of passive technology, the sun room, heat storage wall and heat storage roof are used to collect radiant heat energy, and adjustable louvers are set to alleviate ventilation problems. In terms of construction, the building is raised and rammed earth layer is added to resist frost damage and moisture erosion.
In addition, the local folk customs and ecological environment should be fully considered to reduce the impact on the surrounding flora and fauna, while meeting the living comfort of the crowd.

· Regional Feature

· Climate Analysis

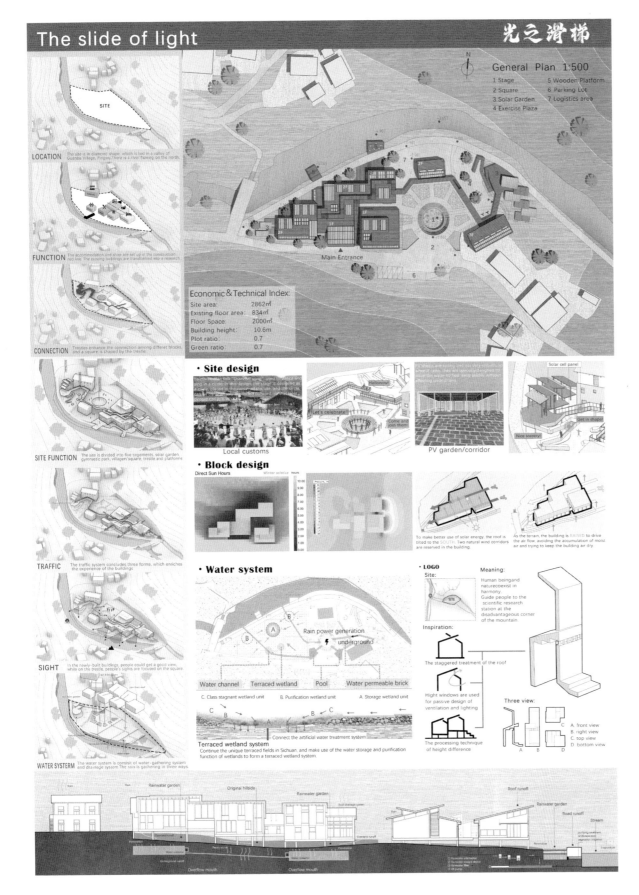

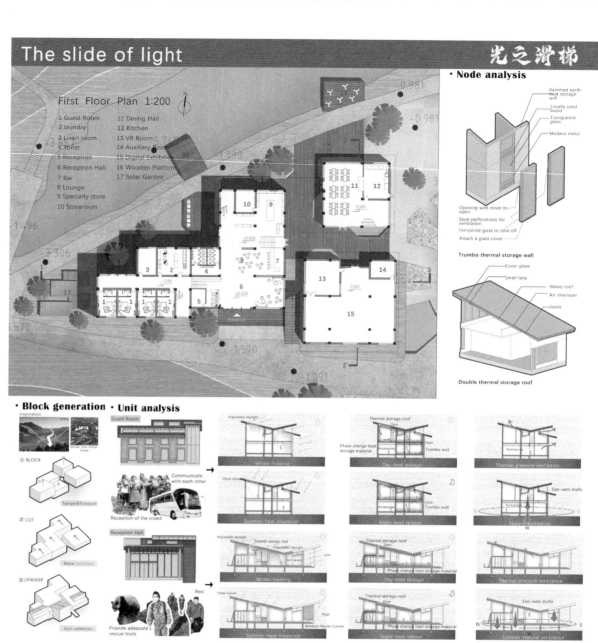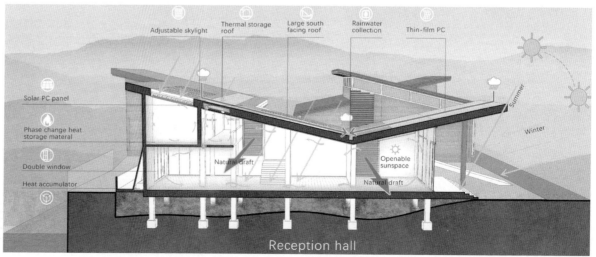

The slide of light

光之滑梯

2022台达杯国际太阳能建筑设计竞赛获奖作品集

Ecological design

Intelligent control | Power monitor | Real time monitor

panda

Through the intelligent system, various nodes can be adjusted, the power generation of the system under different weather can be monitored, and the real-time picture of the hidden camera on the mountain can be realized.

At night, close the blinds on the smart system to reduce the impact of indoor lighting on the flora and fauna in the environment.

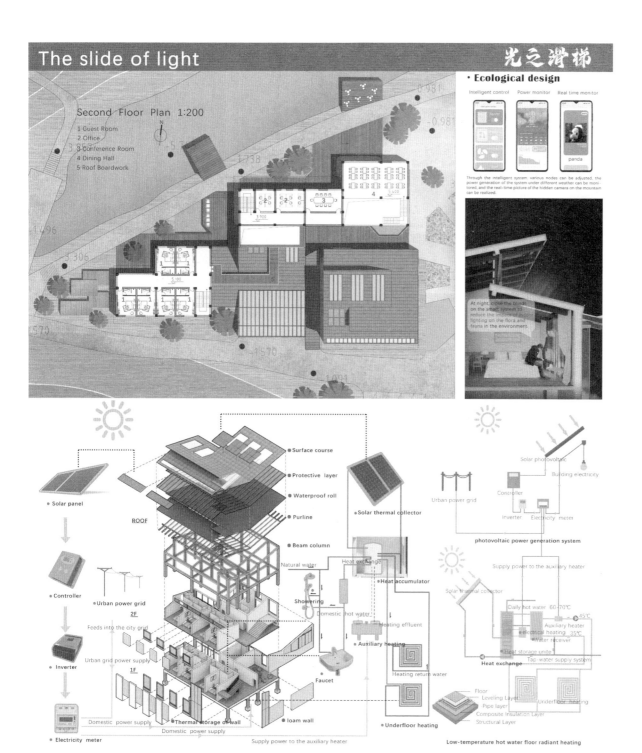

Second Floor Plan 1:200
1 Guest Room
2 Office
3 Conference Room
4 Dining Hall
5 Roof Boardwork

South Elevation 1:200

The slide of light

光之滑梯

Existing building I
Exhibition and Research Center

Third Floor Plan 1:200
1. Seminar Room
2. Storage Room

Second Floor Plan 1:200
1. Theme exhibition hall

First Floor Plan 1:200
1. Reception
2. Auxiliary Room
3. Exhibition Room
4. Lavatory

Existing building II
Village Comittee

First Floor Plan 1:200
1. Control Room
2. Duty Room
3. Office

Second Floor Plan 1:200
1. Office
2. Conference Room

• Office
Heat Storing Wall, Solar Pannel, Solar House

Direct benefit type — Days / Nights

Trombe wall

• Exhibition & Laboratory
solar pannel, rammed earth wall

Additional double glazing / Air space / Heat storage wall / Opened vent

Summer / Winter Days / Winter Nights

Structure Type	Carbon Emission at Different Stages /kg			Total Carbon Emissions/kg	50A Building Carbon Emission per Functional Unit(kg/㎡·A)
	Materialization	Operation	Dismantling		
Timberwork	124724	2917179	1199	3043102	80.7
Lightweight Steel Construction	200877	3190545	1444	3392867	88.26
Reinforced Concrete Structure	244615	3197904	2272	3447910	88.46

• Material analysis

Overhanging eaves — Protect the wall

Wood — Wood structures have low carbon footprint throughout their life cycle and are easy to source.

Bamboo — Local residents use bamboo weaving for shade.

Rammed earth wall
- It has good heat storage effectively regulate the indoor temperature.
- It can absorb and evaporate water, and regulate indoor humidity.
- The earth is widely distributed and can be used for construction in situ; it has practical regression, can be reused and be friendly to the environment.
- Rammed earth houses with less than two floors can resist at least 8.5 earthquakes with seismic measures.

Rammed earth: ROCK 20mm / STONE 2mm / GRAVEL 0.2mm / SAND 0.02mm / CLAY 0.002mm

Cycle: 1.Pour 2.Flat 3.Measure 4.Step 5.Ram 6.Fill 7.Repeat

1-1 Section 1:200

The slide of light

· Design strategies

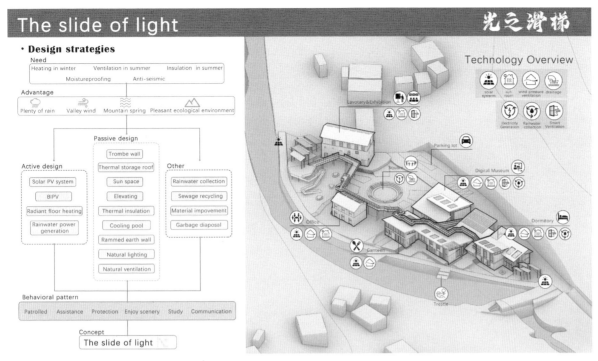

Technology Overview

· Combination of solar energy and water energy

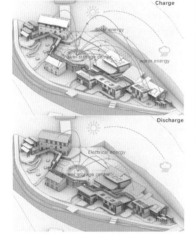

· Summary of heating technology

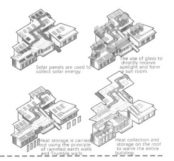

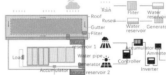

· Life time cycle cost

		Rammed Earth	Timberwork
Construction Cost	Area	228	324
	Cost/㎡/kg	320	50
	Cost	72960	16200
Use Cost		Heating	Maintenance
	Area	228	324
	Cost/㎡/kg	10	80
	Cost/	2280	25920
Recovery Cost	Area	228	
	Cost/㎡	30	Scrap Value
	Cost	6840	

· Unit ventilation analysis in winter
wind pressure   wind speed

综合奖·入围奖
General Prize Awarded · Finalist Award

注 册 号：100942
项目名称：夏时冬语
　　　　　Summer Blooming Winter Warming
作　　者：鲁俊逸、王少潜、韩　旭
参赛单位：中国矿业大学
指导教师：马全明、邵泽彪

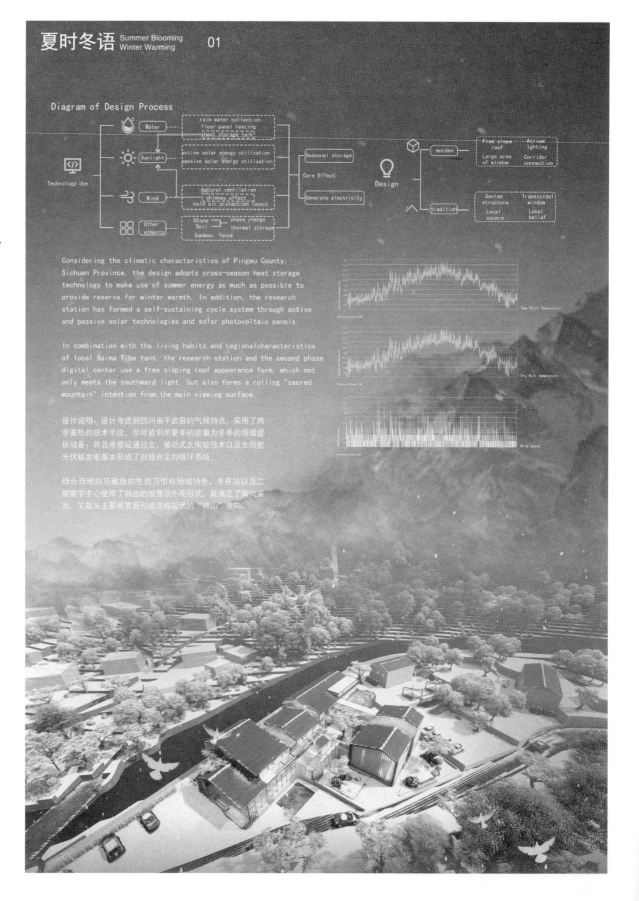

夏时冬语 Summer Blooming Winter Warming 02

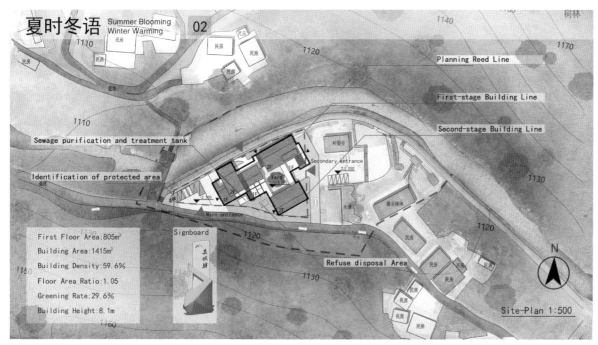

- First Floor Area: 805m²
- Building Area: 1415m²
- Building Density: 59.6%
- Floor Area Ratio: 1.05
- Greening Rate: 29.6%
- Building Height: 8.1m

Site-Plan 1:500

Schemr Generation

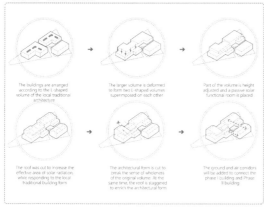

Local Climatic Simulation

Wind Rose (Jan — Dec)

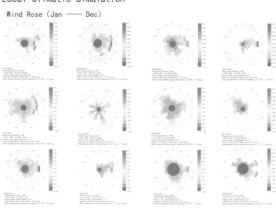

Temprature (Dry/Wet)

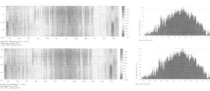

The average annual temperature is 14.7~17.3℃, the extreme maximum temperature is 36.1~39.5℃, and the extreme minimum temperature is -4.5~-7.3℃. Precipitation is relatively abundant, the annual variation of precipitation is great, the annual average precipitation 825.5~1417 mm.

Regional Feature

Tibetan clothing + Decorative Mask + Tun opera + Sacred Mout = "L" settlement / Traditional roof / Enclose Square

Solar Trajectory Analysis

The site is cool in summer and has a large amount of solar radiation. In winter, the temperature is cold and the amount of solar radiation is low. The heating lasts for up to five months of the year, so saving summer's excess solar energy for winter is a better way to resolve the supply-demand mismatch. The site is cool in summer and has a large amount of solar radiation. In winter, the temperature is cold and the amount of solar radiation is low. The heating lasts for up to five months of the year, so saving summer's excess solar energy for winter is a better way to resolve the supply-demand mismatch.

夏时冬语 Summer Blooming Winter Warming 04

Active Solar Energy

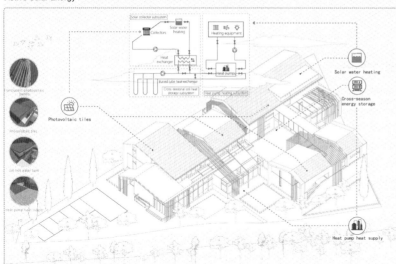

Building Sunshine Analysis

Sun intensity from 9 a.m. to 3 p.m., January to March

Sun intensity from 9 a.m. to 3 p.m., July to September

Sun intensity from 9 a.m. to 3 p.m., April to June

Sun intensity from 9 a.m. to 3 p.m., October to December

Cross Seasonal Heat Cycle

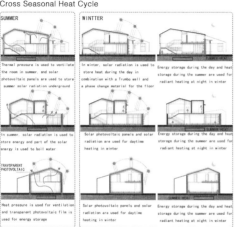

Variable Energy Saving Strategy

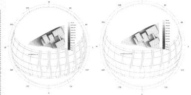
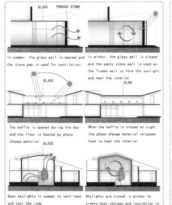
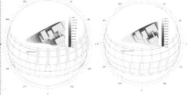

Water-cycling System

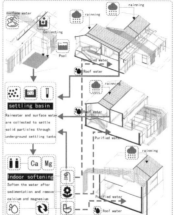

夏时冬语 Summer Blooming Winter Warming 05

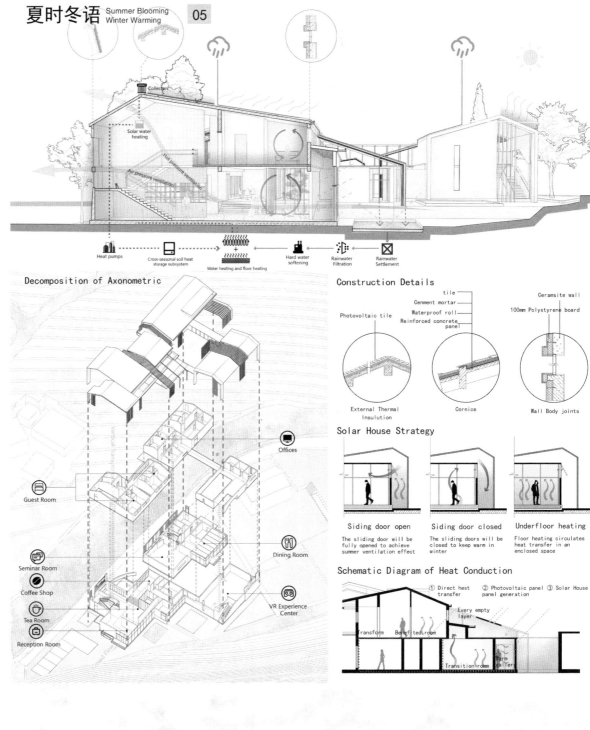

North Elevation 1:150

夏时冬语 Summer Blooming Winter Warming 06

Shematic of Active Solar Energy

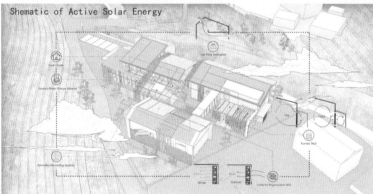

Energy Flow

Function	Power Consumption	Power Available
Office Area (639.61m²)	30-70W/m²	639.61×30=19.2kW
Hostel Area (555.01m²)	40-70W/m²	555.01×40=22.2kW
Total Area (1194.02m²)	30-40W/m²	41.4kW(phase Ⅰ Ⅱ)

Notion: Calculate according to the minimum standard (A day)

Solar Energy Calculation:
South sloping roof area: 72.84+136.36+165.98+58.7=433.88m²
The power of 1m² photovoltaic tile is 110-120W:
433.88×4(h)×110-120(W)×0.7=121.49kW—145.78kW

Area Ratio of Window

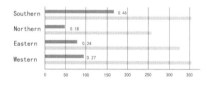

Southern	0.46	
Northern	0.18	
Eastern	0.24	
Western	0.27	

Thermal Properties of Materials:

Category	Area (m²)	HCC W/(m²·K)	HSC W/(m²·K)
Oak board	37.76	0.17	4.66
Rammed earth		1.16	13.05
Limestone	10.3	2.04	18.1
Aerated concrete block	265.6	0.22	3.59

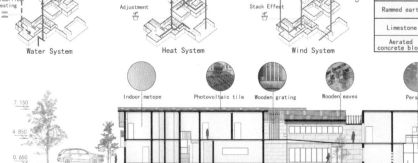

Cross-section 1:150

综合奖·入围奖
General Prize Awarded · Finalist Award

注 册 号：100946
项目名称：流泉·聚旭
　　　　　Flowing Spring · Warm Storage
作　　者：上官玉麒、赵雨婷、曹雯欣、覃文欢、陆澳晨
参赛单位：昆明理工大学
指导教师：陆　莹、毛志睿

Flowing Spring · Warm Storage
流泉·聚旭

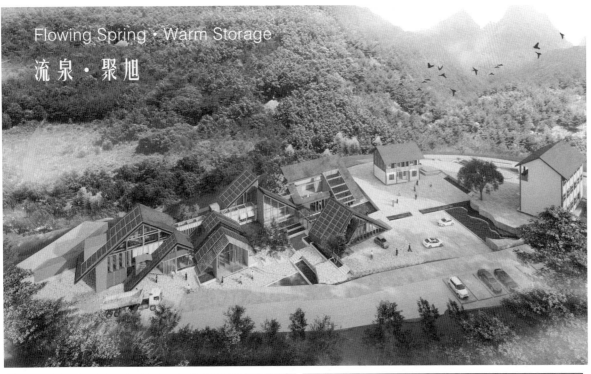

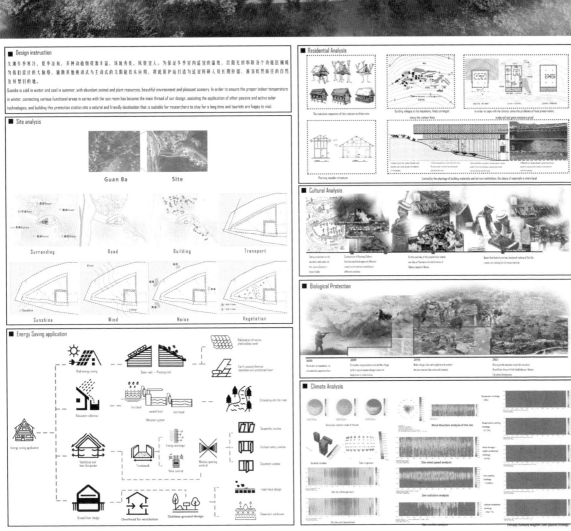

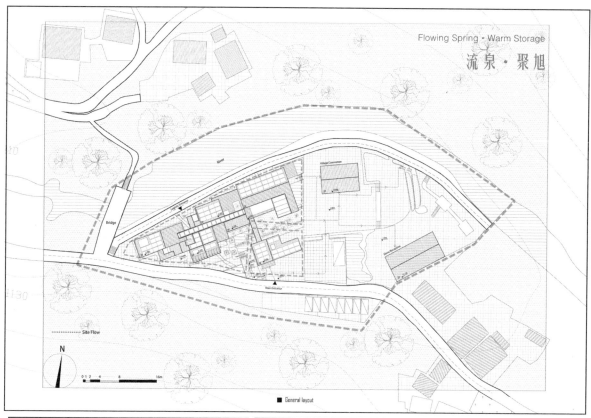
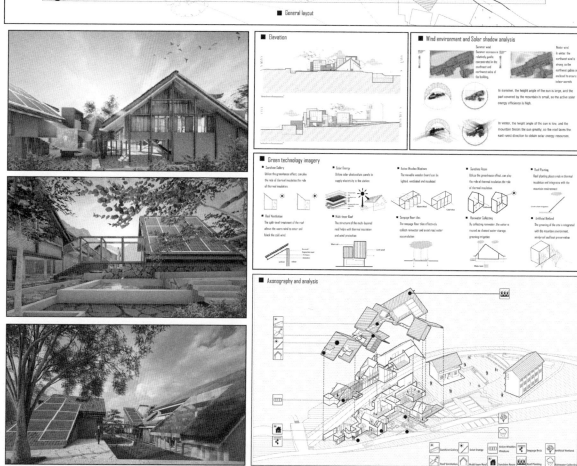

流泉·聚旭

Flowing Spring · Warm Storage

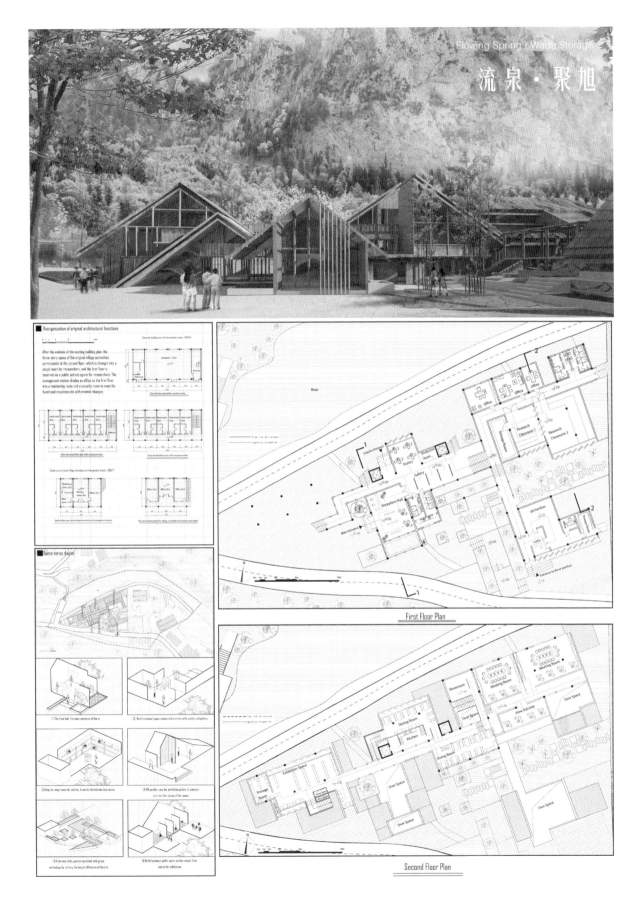

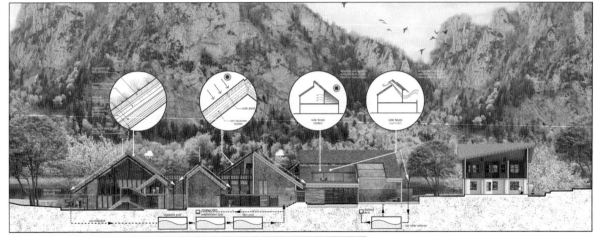
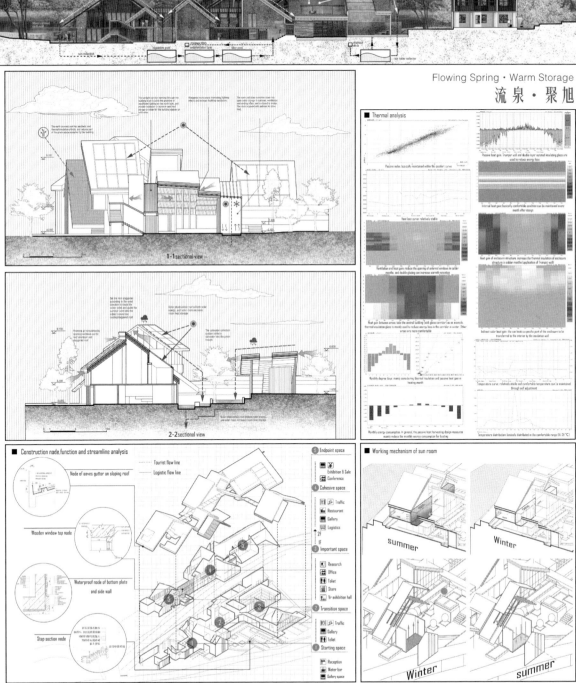

Flowing Spring · Warm Storage
流泉·聚旭

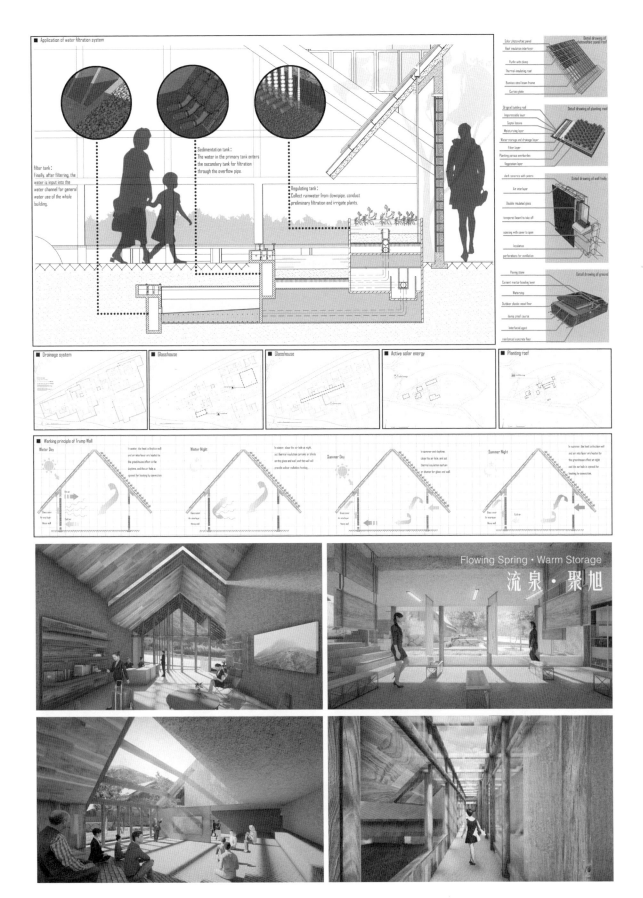

综合奖·入围奖
General Prize Awarded · Finalist Award

注 册 号：100952
项目名称：共生·折驿
　　　　　Symbiosis Folding Corridors
作　　者：沙扬皓、王子倩、阮志鹏、
　　　　　李啸晗、虞静雯
参赛单位：合肥学院
指导教师：丁　蕾、司大雄

Preliminary Investigation and Analysis

The layout form of Tibetan traditional buildings is subject to the dual constraints of natural environmental conditions and spiritual conditions. The layout forms of basic row type and decentralized type are mainly two, usually showing the layout characteristics of heart. Decentrized design is the main form.

The Tibetan "cat dance" is the main form of dance in the region. The dance is charactized by the line type of the dance team. The team is very long. In combination with the characteristics of local traditional dance, the design vein takes the dance streamline as the fitting form, adapts to the terrain and passive technology, and arranges.

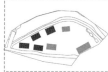

Design Specification

科考站与基地是互利共生的关系，科考站的构建一方面是在寻求人与自然的共生，另一方面是在寻求建筑与物理环境及生态环境的共生。

The scientific research station and the base are in a mutually beneficial relationship. The construction of the scientific research station is to seek the symbiosis of man and nature.on the other hand, it is also to seek the symbiosis of architecture, physical environment and ecological environment.

Digram of Design Process

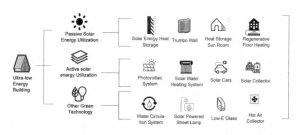

共生·折驿

2022 台达杯国际太阳能建筑设计竞赛获奖作品集

Parti Diagram of the Organisational Idea

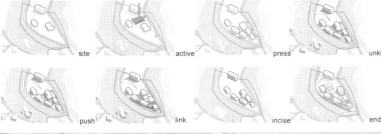

site / active / press / unkint
push / link / incise / end

Site Analysis

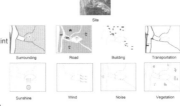

Site
Surrounding / Road / Building / Transportation
Sunshine / Wind / Noise / Vegetation

Climate Analysis

Site plan 1:600

8m 16m 32m

Economic analysis
Base area: 2861.85 m²
Building area: 723.7 m²
Greening aera: 1431 m²
Floor area ratio: 0.25
Greening rate: 50%

共生·折驿

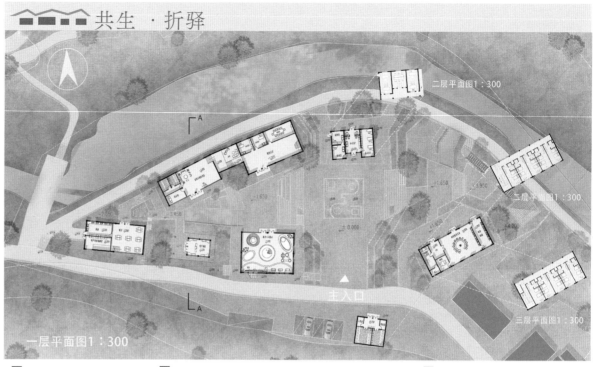

一层平面图 1:300　二层平面图 1:300　三层平面图 1:300

Courtyard Analysis

Bamboo courtyard space brings a certain privacy and interest

The drop space brings more touch experience to the villagers

The stepped space for villagers to rest and watch

Culture Square is an outdoor space for public activities

After Plane

The first and second phase of the new design, the second floor design of a large area of the platform to create a multi-level public space and twists and turns of the interesting courtyard space

Before Plane

The original site space is cramped and copact and the sight is not open. Based on various problems, the later construction is carried out

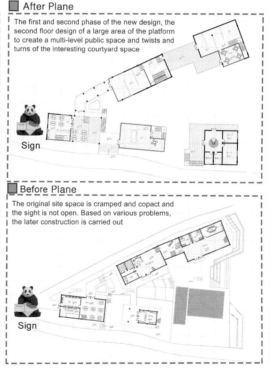

Explode Analysis

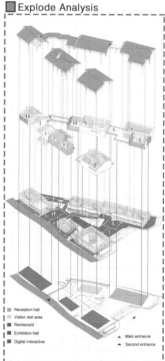

- Reception hall
- Visitor rest area
- Restaurant
- Exhibition hall
- Digital interactive

▲ Main entrance
▲ Second entrance

South Elevation

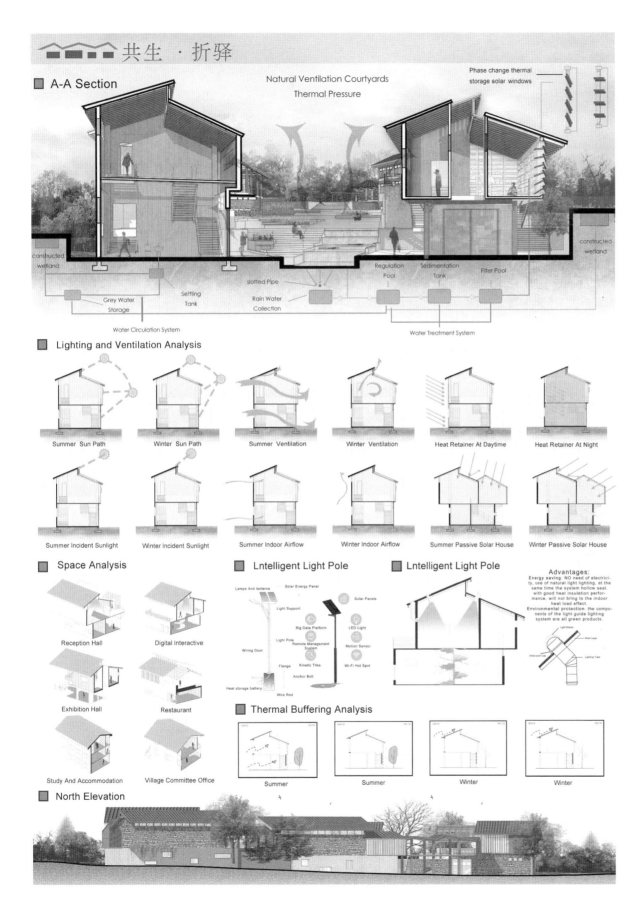

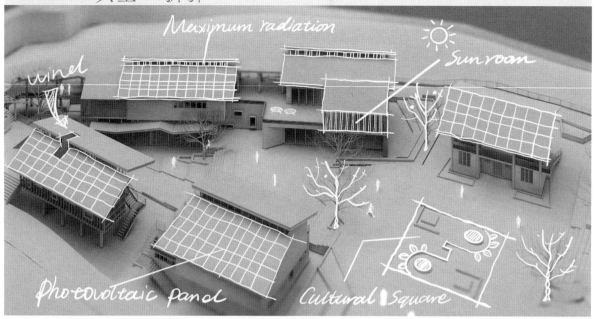

 共生·折驿

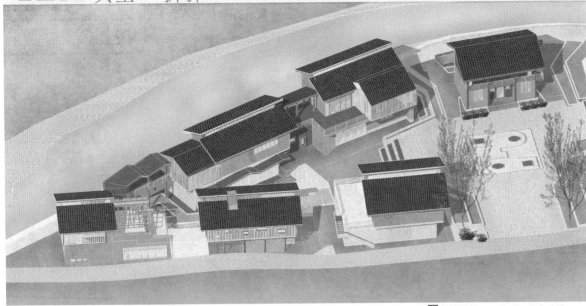

Environment Control System

Solar Water Heating System

Solar Photovoltaic System

PKPM Energy Saving Calculation

Solar Illumination Time

According to the Ladybug software, the sunshine hours of the site on the winter solstice and summer solstice are calculated. As can be seen from the figure, the site has appropriate sunshine hours on the winter solstice and summer solstice

Comparison of Wind Environment Optimization

The form of the centralized building is arranged according to the site, and then calculated and optimized according to Ladybug. After four times of calculation, the final scheme is obtained

Energy Balance Calculation

Landscape Section

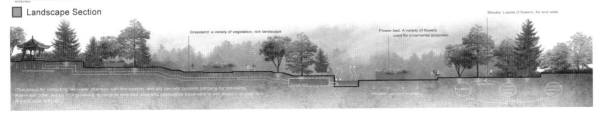

综合奖·入围奖
General Prize Awarded·
Finalist Award

注 册 号：100967
项目名称：循院·浔源
　　　　　Circulating Courtyards, Tracing the Source
作　　者：瞿琳茜、邱潇仪、房璐杰、陈烁杨
参赛单位：重庆大学
指导教师：张海滨

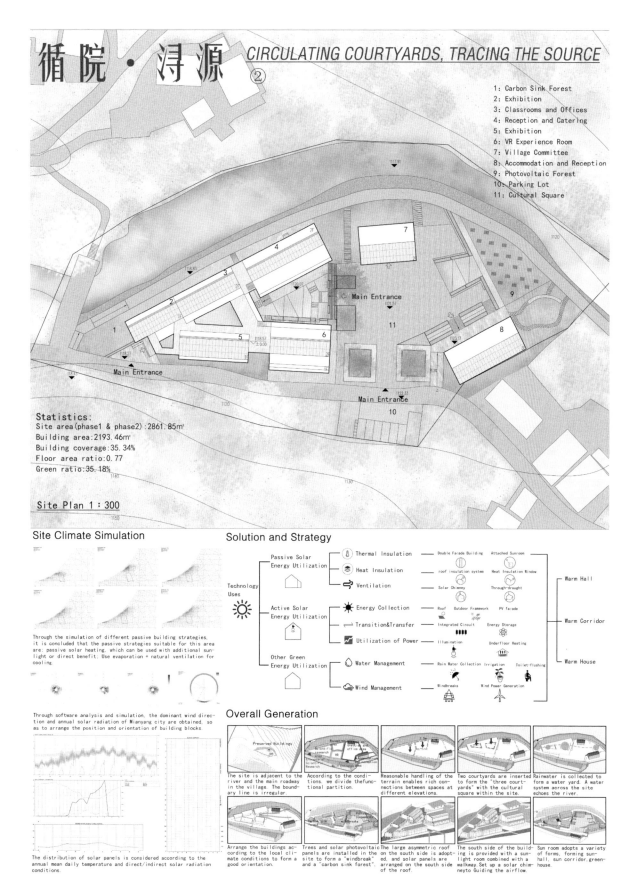

循院·浔源 ③ CIRCULATING COURTYARDS, TRACING THE SOURCE

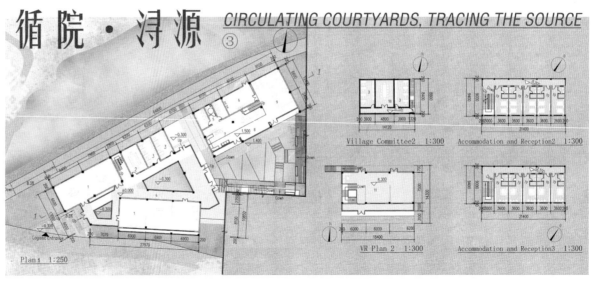

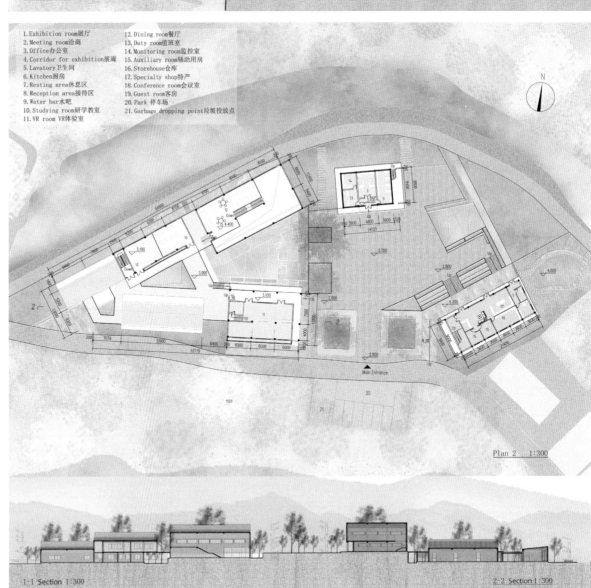

1. Exhibition room 展厅
2. Meeting room 洽商
3. Office 办公室
4. Corridor for exhibition 展廊
5. Lavatory 卫生间
6. Kitchen 厨房
7. Resting area 休息区
8. Reception area 接待区
9. Water bar 水吧
10. Studying room 研学教室
11. VR room VR体验室
12. Dining room 餐厅
13. Duty room 值班室
14. Monitoring room 监控室
15. Auxiliary room 辅助用房
16. Storehouse 仓库
17. Specialty shop 特产
18. Conference room 会议室
19. Guest room 客房
20. Park 停车场
21. Garbage dropping point 垃圾投放点

循院·浔源

CIRCULATING COURTYARDS, TRACING THE SOURCE ④

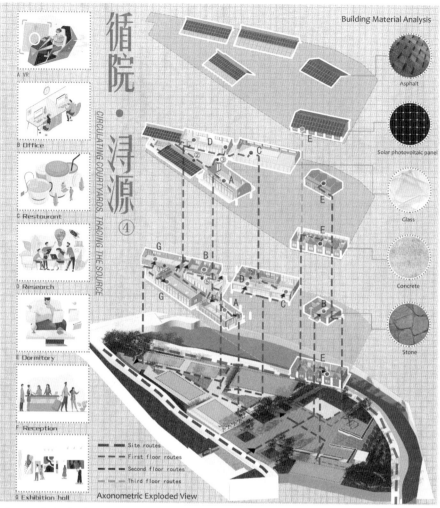

Axonometric Exploded View

- Site routes
- First floor routes
- Second floor routes
- Third floor routes

A VR
B Office
C Restaurant
D Research
E Dormitory
F Reception
G Exhibition hall

Building Material Analysis

- Asphalt
- Solar photovoltaic panel
- Glass
- Concrete
- Stone

Scene Perspective

①

②

③

④

⑤

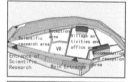

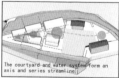

⑥

Energy Calculation

Accommodation 575 square meters, reception 270+115=385 square meters, a total of 960 square meters, according to the electricity consumption index of civil buildings, hotels, restaurants per hour electricity consumption: $40W/m^2$ (1H), accommodation daily electricity consumption: 575 square meters $\times 40W/m^2 \times 24h = 552kW$

The reception part is calculated according to the average daily electricity consumption of 8h, $385m^2 \times 40W/m^2 \times 8h = 123.2kW$

Office and scientific research exhibition part total $1740m^2$, according to the civil building electricity consumption index, the average daily 8 hours of electricity consumption calculation, the daily electricity consumption is $575m^2 \times 30W/m^2 \times 8h = 138kW$

The total amount of VR exhibition in Phase II is $300m^2$. According to the electricity consumption index of civil buildings and the average daily electricity consumption of 8 hours, the daily electricity consumption is $300m^2 \times 50W/m^2 \times 8h = 120kW$

The total daily electricity consumption of the complex is 552kW + 123.2kW + 138kW + 120kW = 933.2kW

The effective sunshine of 1kW module is 6 hours. Considering the 4.2 power generation per day and 30% loss, the daily electricity output of solar panel is:

The area (roof) required for 1kW solar panels to generate electricity is about $6.5m^2$

$1524m^2 \div 6.5m^2 \times 4.2kW/h = 984.7kW/h > 933.2kW/h$ Conclusion:

At present, the electricity generation can basically meet the demand of the building complex.

Courtyard Analysis

Courtyard in Summer

In summer, the sunshade louvered skylight above the courtyard is opened, and the floor-to-ceiling Windows around the atrium are fully opened to form a draft effect. The courtyard pool is filled with water to cool the room.

Courtyard in Winter

In winter, the courtyard water is discharged into the grey water collection box, the doors and Windows are closed, and the corridor becomes the sun room. Meanwhile, the courtyard's cobblestones store heat.

Site Element Analysis

Scientific research area / Reception / Village activities and office area / Entrance of Scientific Research / VR / Main Entrance / Accommodation and reception area

The courtyard and water system form an axis and series streamline.

Green technology in the field

Carbon sink forest / Photovoltaic components / windbreaks

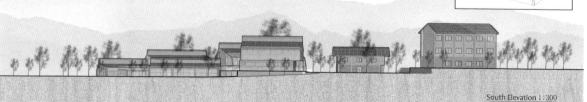

South Elevation 1:300

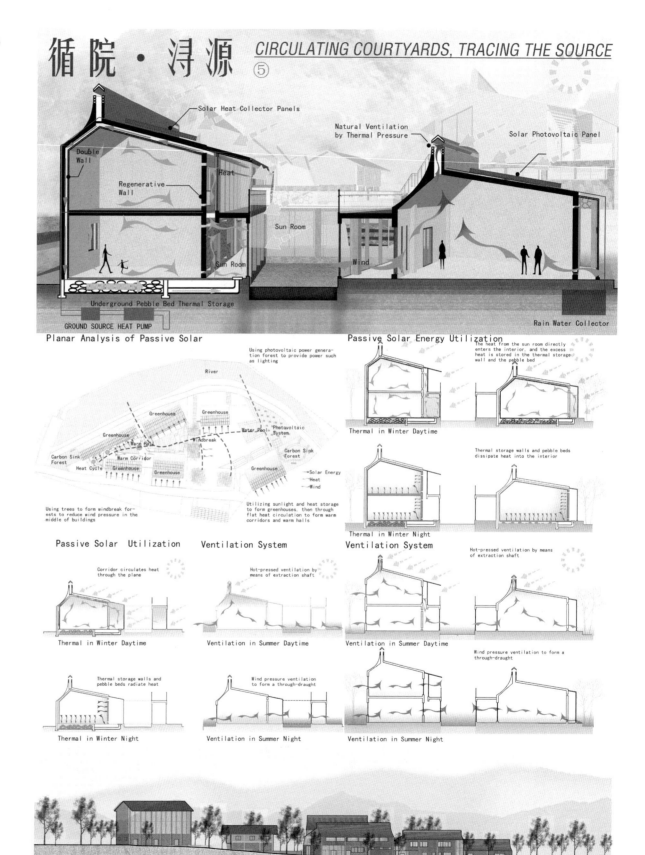

循院·浔源 ⑥ *CIRCULATING COURTYARDS, TRACING THE SOURCE*

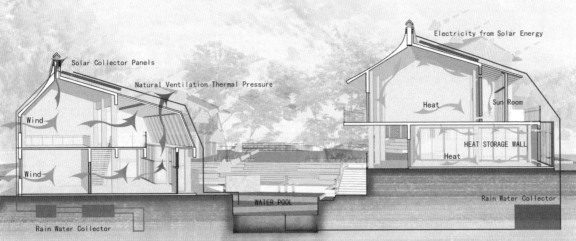

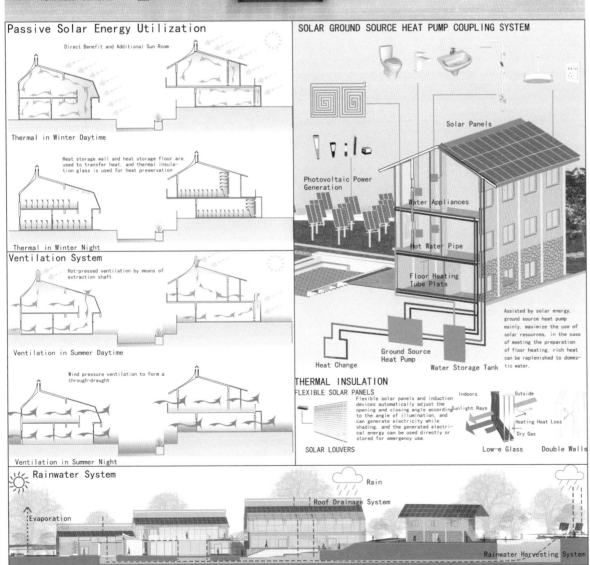

综合奖・入围奖
General Prize Awarded · Finalist Award

注 册 号：100978
项目名称：木林・阳光山水驿
　　　　　Tree Tree Tree
作　　者：杨旖文、王宜杉、曾建木、
　　　　　陈启祯
参赛单位：南京工业大学
指导教师：徐善彬、罗 靖、舒 欣

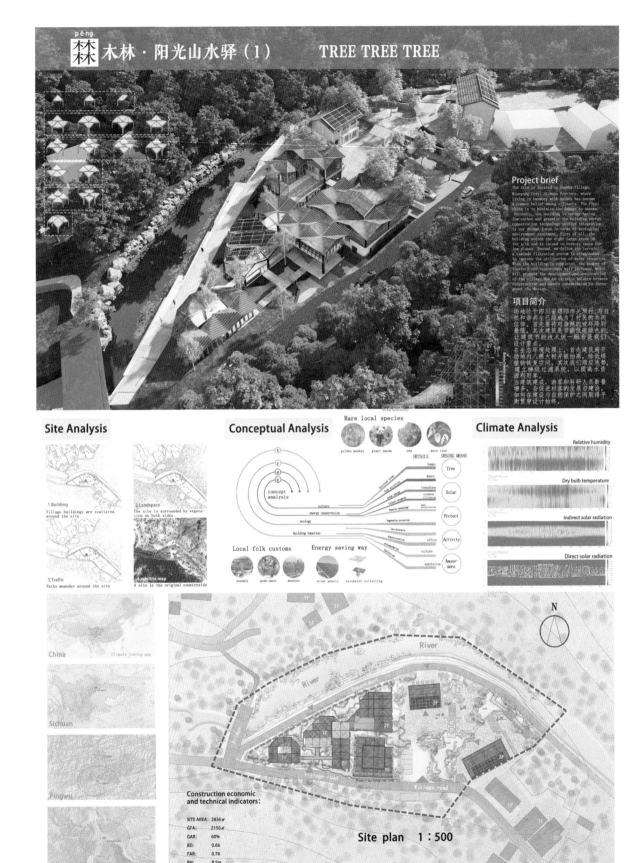

 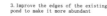

Volume generation

1. Current situation of the original site
2. According to the main building entrance direction, reserved space for parking
3. Improve the edges of the existing pond to make it more abundant
4. Reserve space in front of the village hall for meetings

Site generation

1. original venue
2. Keep eight big trees
3. Based on the 3.6m and 4.5m squares, divide the modulo
4. Adjust the grid arrangement according to the growth radius of the trees
5. Generate blocks according to functional partitions
6. Join the design language to form the final form

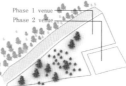  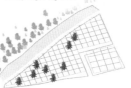   

Construction strategy

Model photo

- Bird's eye view
- Exhibition hall on the second floor
- The south courtyard
- Entrance

Explosive view

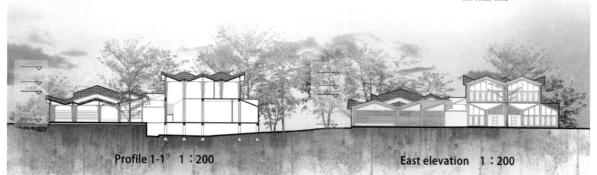

Profile 1-1   1:200   East elevation   1:200

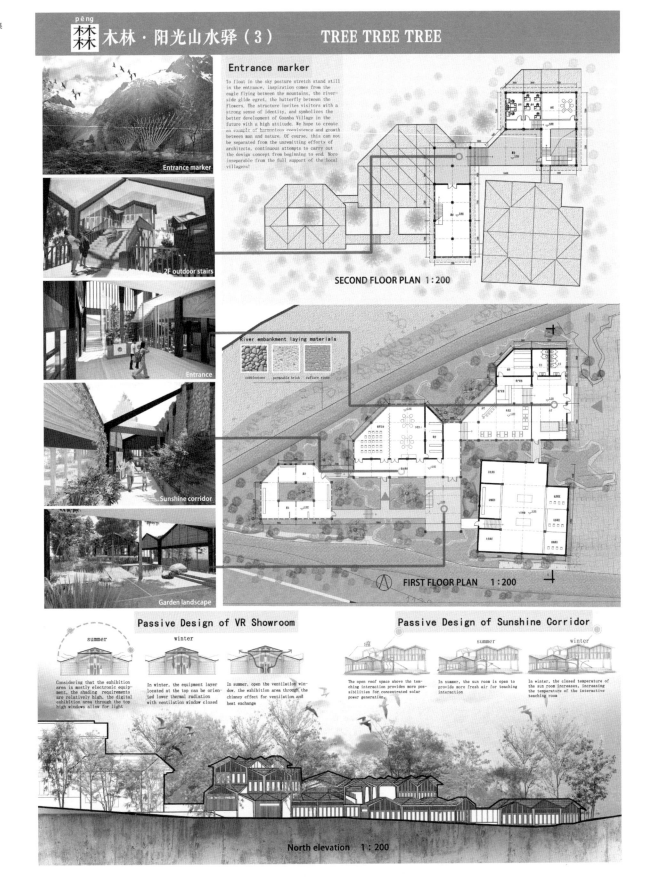

木林·阳光山水驿（4） TREE TREE TREE

Effect picture

Village hall

Exhibition center

Description of the renovation building design

The original buildings in the site are too transparent and the facade materials are difficult to meet the needs of cold in winter and hot in summer. The reconstruction scheme uses masonry as the facade material, which is warm in winter and cool in summer and cheap. The interior experience of the building is optimised through the entryway, and the Trombe wall is utilized to create a pleasant environment.

Building materials

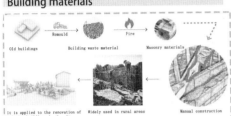

Contrast between old and new buildings

old

The original building is too transparent

new

Retrofitting buildings for shading in summer and insulation in winter

Axial analysis of chimney effect

Natural ventilation in summer

Heating and insulation in winter

Promote ventilation

Elevation of village hall south 1:200

Northwest facade of the exhibition center 1:200

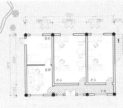
Floor plan of the village hall 1:200

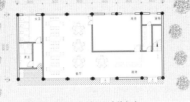
Floor plan of the exhibition center 1:200

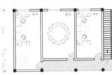
The second floor plan of the village hall 1:200

The second floor plan of the exhibition center 1:200

Ventilation in summer, and winter thermal insulation

Trembus wall windows

The third floor plan of the exhibition center 1:200

Section axonometry of functional relationship

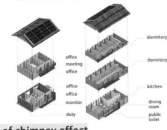

Masonry wall Construction

Display center section fluoroscopic analysis

Profile analysis of chimney effect

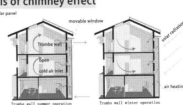

Active solar energy and natural ventilation

Trombe wall summer operation

Trombe wall winter operation

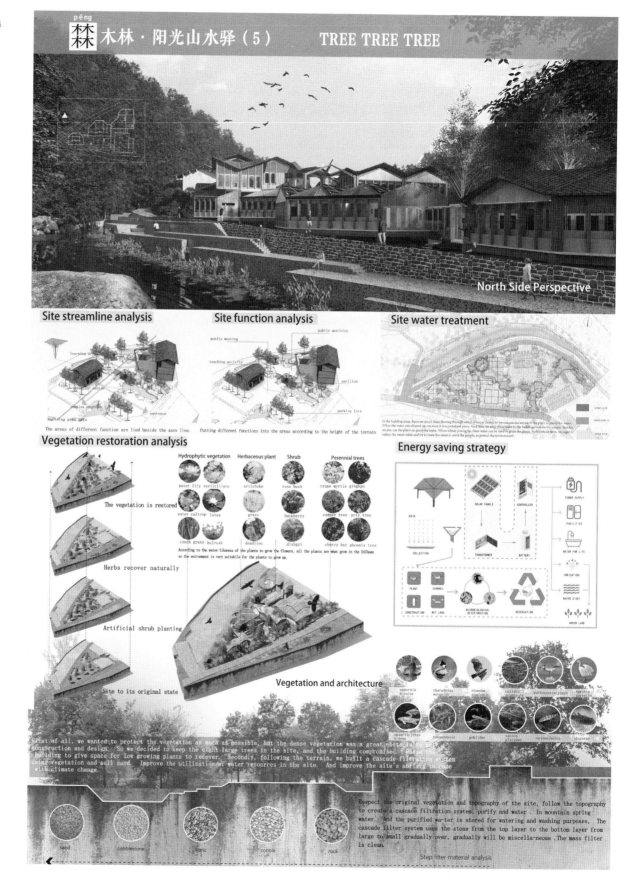

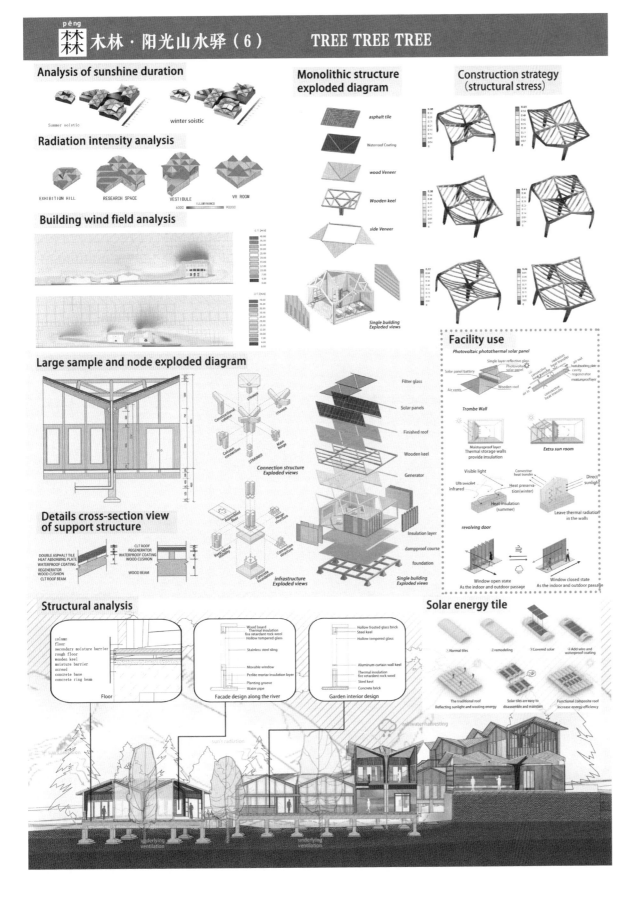

综合奖・入围奖
General Prize Awarded · Finalist Award

注 册 号：100982

项目名称：轻触・自然——装配式建筑理念下自然保护区科考站设计
Light Touch—Design of Scientific Research Stations in Nature Reserves under the Concept of Prefabricated Architecture

作 者：卓金明、李舒祺、许浩川、周超超

参赛单位：西安建筑科技大学

指导教师：李 涛、孔黎明

轻触·自然——装配式建筑理念下自然保护区科考站设计
Light Touch —Design of Scientific Research Stations in Nature Reserves under the Concept of Prefabricated Architecture

1. 设计说明 Design description

本设计从场地的自然、人文出发结合现代的工业化，提出了轻介入、自维持、智能化三个设计概念。在轻介入方面，结合装配式建筑理念，从用户定制到现场供建施工后期运维的全生命周期，简化设计过程，减少对场地的影响，后期智能设备、数据监测，能形成动态的室内调节机制，节约能耗。固体废弃物进行合理的回收利用设计，减少对场地的污染。自维持方面，使用太阳能、生物质能为场地提供能量，结合合理的被动式设计可以减少能量损耗，节约能源。在材料方面，结构选用市面常用的C型轻钢，减少造价同时在四川这样的地震多发区，经钢具有良好的抗震效果。选用碲化镉光伏发电材料，提高阴天的能效率。最后，结合智能化系统，可为远方的自然爱好者提供远程观测，线上科普教学活动，也可为为村民提供特产售卖平台。

This design starts from the nature and humanities of the site combined with modern industrialization, and proposes three design concepts of light intervention, self-maintenance and intelligence. In terms of light intervention, combined with the concept of prefabricated buildings, from user customization to the full life cycle of on-site rapid construction and post-operation and maintenance, the design process is simplified, the impact on the site is reduced, and the later intelligent operation and data monitoring can form a dynamic indoor adjustment mechanism, save energy. Reasonable recycling design of solid waste to reduce site pollution. In terms of self-sustainment, using solar energy and biomass energy to provide energy for the site, combined with a reasonable passive design can reduce energy loss and save energy. In terms of materials, the structure uses C-shaped light steel commonly used in the market to reduce the cost. At the same time, in earthquake-prone areas such as Sichuan, light steel has a good seismic effect. Use telluride photovoltaic power generation materials to improve the utilization rate on cloudy days. Finally, combined with the intelligent system, it can provide remote observation for distant nature lovers, online popular science teaching activities, and a special product sales platform for villagers.

2. 区位分析 Location analysis

关坝自然保护区位于四川省绵阳市平武县、绵阳的北部，四川盆地的西北部，涪江河上游，具有良好的生态环境和多样的珍稀生物，是绵阳市重要的生态腹地和水源涵养地。

Guanba Nature Reserve is located in Pingwu County, Mianyang City, Sichuan Province. It is located in the north of Mianyang, the northwest of the Sichuan Basin, and the upper reaches of the Fujiang River. There is a good ecological environment and rare creatures. It is an important ecological hinterland and water conservation area in Mianyang City.

3. 地域建筑 Regional architecture

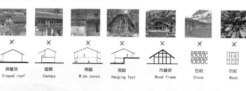

坡屋顶 Sloped roof | 檐廊 Canopy | 宽檐 Wide eaves | 吊脚 Hanging feet | 木檩架 Wood frame | 石前 Stone | 木材 Wood

4. 气候分析 Climate analysis

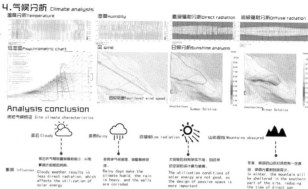

温度分析 Temperature | 湿度 Humidity | 直接辐射分析 Direct radiation | 间接辐射分析 Diffuse radiation

焓湿图 Psychometric chart | 风 Wind | 日照分析 Sunshine analysis

Analysis conclusion

场地气候特征 Site climate characteristics

多云 Cloudy | 多雨 Rainy | 低辐射 Low radiation | 山体遮挡 Mountains obscured

影响 Influences

Cloudy weather results in less direct radiation, which affects the utilization of solar energy. | Rainy days make the climate humid, the rain is heavy, and the walls are corroded. | The utilization conditions of solar energy are not good, so the design of passive space is more important | In winter, the mountain will be sheltered in the southern part of the site, reducing the time of direct sun exposure.

5. 设计概念 Concept of design

自然 Nature | 人文 Humanities | 工业4.0 Industry 4.0

轻介入 Light intervention | 自维持 Self-sustaining | 智能化 Intelligent

轻触·自然 —— 装配式建筑理念下自然保护区科考站设计
Light Touch — Design of Scientific Research Stations in Nature Reserves under the Concept of Prefabricated Architecture

6-2

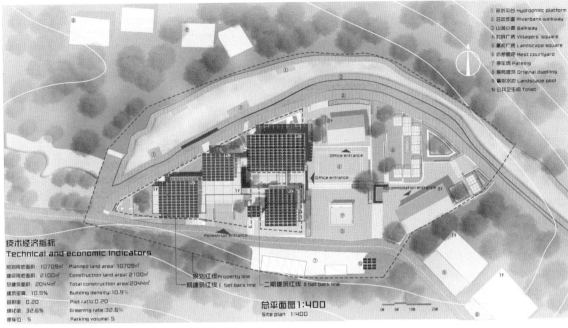

1. 亲水平台 Hydrophilic platform
2. 沿河步道 Riverbank walkway
3. 山涧小道 Walkway
4. 村民广场 Villagers' square
5. 景观广场 Landscape square
6. 休憩庭院 Rest courtyard
7. 停车场 Parking
8. 原有建筑 Original dwelling
9. 景观水池 Landscape pool
10. 公共卫生间 Toilet

技术经济指标 Technical and economic indicators
- 规划用地面积：10709㎡ Planned land area: 10709㎡
- 建设用地面积：2100㎡ Construction land area: 2100㎡
- 总建筑面积：2044㎡ Total construction area: 2044㎡
- 建筑密度：10.9% Building density: 10.9%
- 容积率：0.20 Plot ratio: 0.20
- 绿化率：32.6% Greening rate: 32.6%
- 停车位：5 Parking volume: 5

总平面图 1:400 / Site plan 1:400

场地分析 Site analysis
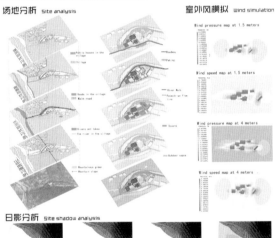

室外风模拟 Wind simulation
方案推演 Scheme deduction
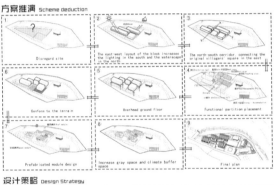

日影分析 Site shadow analysis
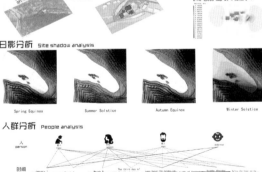
Spring Equinox / Summer Solstice / Autumn Equinox / Winter Solstice

人群分析 People analysis
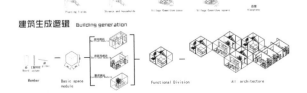

建筑生成逻辑 Building generation

设计策略 Design Strategy
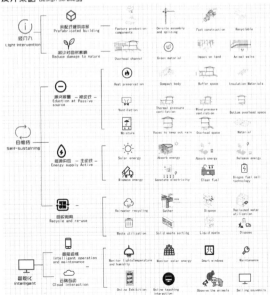

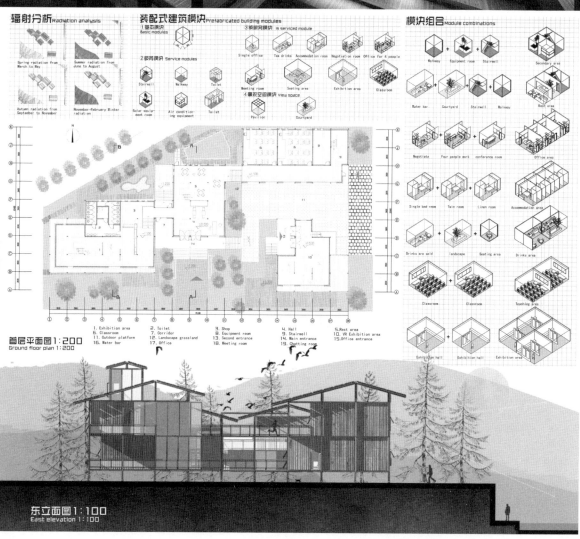

轻触·自然 —— 装配式建筑理念下自然保护区科考站设计

Light Touch — Design of Scientific Research Stations in Nature Reserves under the Concept of Prefabricated Architecture

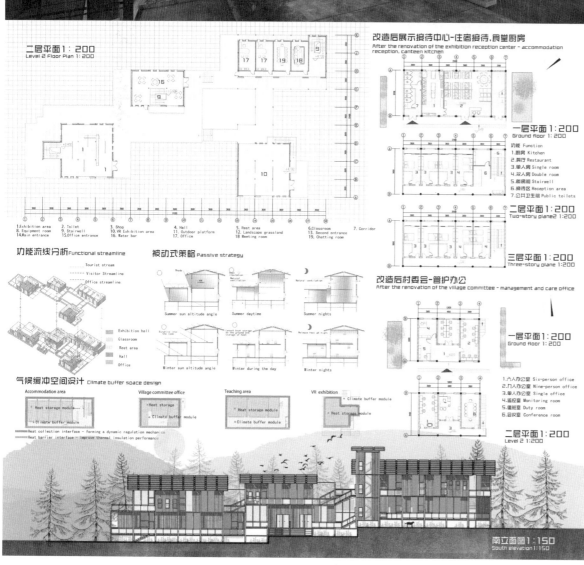

轻触·自然——装配式建筑理念下自然保护区科考站设计
Light Touch—Design of Scientific Research Stations in Nature Reserves under the Concept of Prefabricated Architecture

6-5

效果图 Renderings

装配式建筑定制 Prefabricated building customization

装配式建造过程图 Assembly construction process diagram

智能控制系统 Intelligent control system

能源设备 Energy equipment

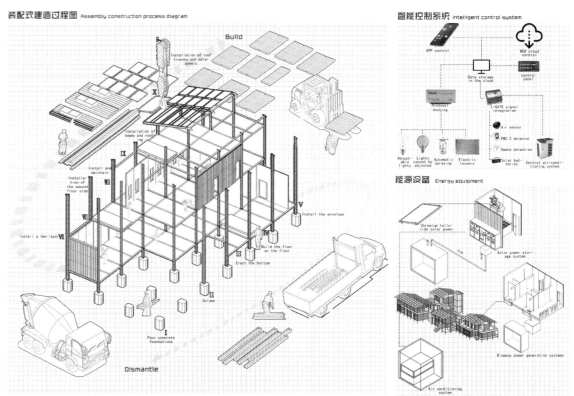

轻触自然——装配式建筑理念下自然保护区科考站设计
Light Touch—Design of Scientific Research Stations in Nature Reserves under the Concept of Prefabricated Architecture

效果图 Renderings

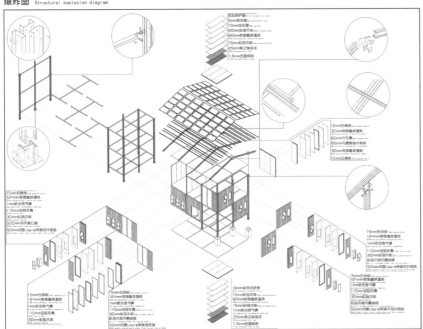

爆炸图 Structural explosion diagram

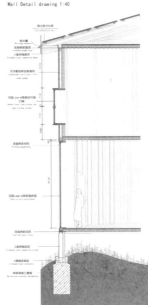

墙身大样图 1:40
Wall Detail drawing 1:40

A-A 剖面图　A-A section

B-B 剖面图　B-B section

综合奖・入围奖
General Prize Awarded・
Finalist Award

注 册 号：100987
项目名称：曲径通幽
　　　　　Winding Path Leading to Seclusions
作　者：张　颖、朱雅萱
参赛单位：河南工业大学
指导教师：张　华、马　静

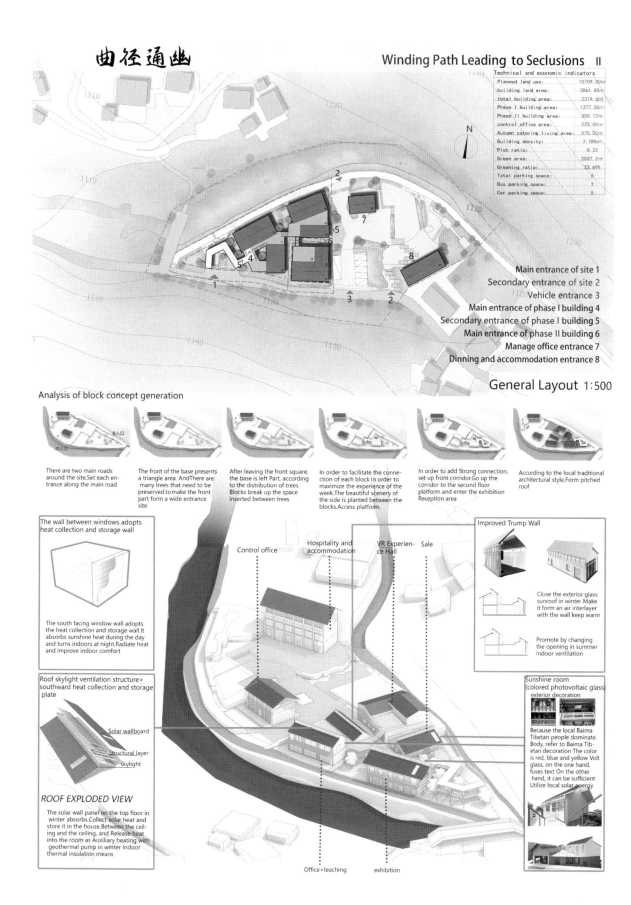

曲径通幽 Winding Path Leading to Seclusions III

Secondary entrance of phase I building 1
Corridor entrance 2
Main entrance of phase II building 3
Secondary entrance hall 4
Exhibition space 5
Teaching interactive room 6
Storeroom 7
Virtual reality experience room 8

First floor plan 1:200

1 Main entrance
2 Reception hall
3 Lounge
4 Water bar
5 Storeroom
6 Sales of specialty
7 Solar house
8 Office
9 Meeting room

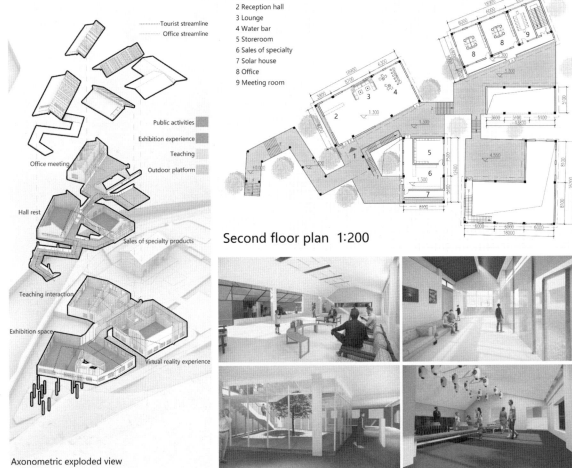

Second floor plan 1:200

Axonometric exploded view

曲径通幽   Winding Path Leading to Seclusions IV

Catering accommodation

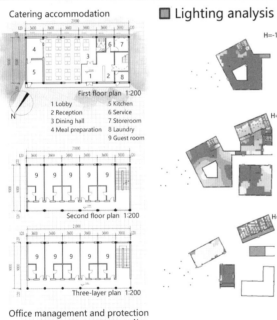

First floor plan 1:200
1 Lobby   5 Kitchen
2 Reception 6 Service
3 Dining hall 7 Storeroom
4 Meal preparation 8 Laundry
9 Guest room

Second floor plan 1:200

Three-layer plan 1:200

Office management and protection

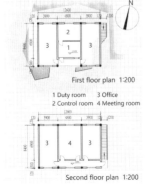

First floor plan 1:200

1 Duty room   3 Office
2 Control room 4 Meeting room

Second floor plan 1:200

Lighting analysis

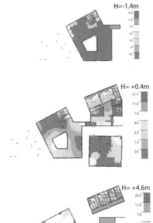

H=-1.4m

H=+0.4m

H=+4.6m

Ventilation analysis

Velocity nephogram   Pressure nephogram

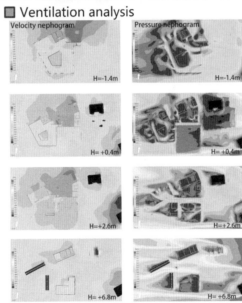

H=-1.4m

H=+0.4m

H=+2.6m

H=+6.8m

East elevation 1:200

曲径通幽 Winding Path Leading to Seclusions V

Working system of heat collecting and storing wall

In winter daytime, the heat collecting and storing wall absorbs heat, so that the heat is stored in the wall.

In winter night, the absorbed heat in the heat collecting and storing wall releases into the room, maintaining the indoor temperature, and keeping the indoor comfortable room temperature.

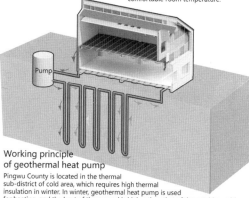

Working principle of geothermal heat pump

Pingwu County is located in the thermal sub-district of cold area, which requires high thermal insulation in winter. In winter, geothermal heat pump is used for heating, and the heat of the ground is higher than that of the outside world. The cold water flows along the ground to increase the temperature and then enters the pipes between floors, thus increasing the indoor temperature.

Design methods of green building at the secondary entrance

Ventilation measures in summer

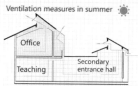

Part of the office is equipped with a solar house. With windows in the solar house and skylight on the roof, it is ventilated by thermal pressure. In the secondary entrance hall, the improved Trombo wall is used to promote indoor ventilation.

Design of photovoltaic system in summer

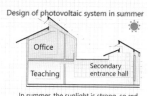

In summer, the sunlight is strong, so red, yellow and blue glass is used on the PV facade of solar houses, which not only plays a decorative role, but also can make use of the sunlight.

Winter daytime

The main heating mode in winter is ground heat pump, and the solar house is set in the office to increase indoor thermal comfort level in winter. The secondary entrance hall is equipped with an improved Trombo wall. While closing the opening of the external glass curtain wall and opening the opening on the Trombo wall, hot air can continuously enter the room and increase the indoor temperature.

Winter night

In addition to geothermal heat pump heat supply in night, the heat collecting and storing wall in the secondary entrance hall releases the heat absorbed during the day into the room.

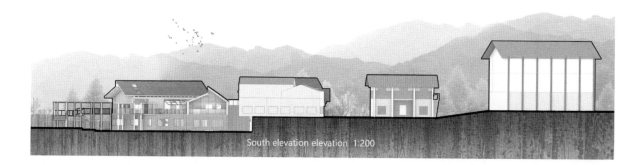

South elevation elevation 1:200

曲径通幽 Winding Path Leading to Seclusions VI

Green building techniques in exhibition area

Structural details of sunlight room

For the greater Make use of sunlight, building Sunshine room is set in the south, Improve indoor Comfort.

Purified water treatment and collection

Rainwater well | Filter unit | Water storage module | Rainwater treatment equipment | Clean water reservoir

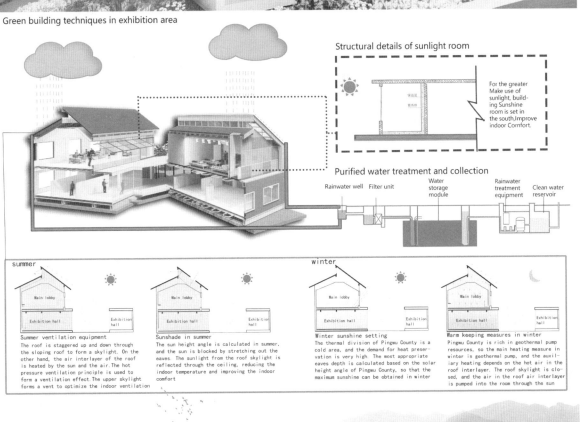

Summer ventilation equipment
The roof is staggered up and down through the sloping roof to form a skylight. On the other hand, the air interlayer of the roof is heated by the sun and the air. The hot pressure ventilation principle is used to form a ventilation effect. The upper skylight forms a vent to optimize the indoor ventilation

Sunshade in summer
The sun height angle is calculated in summer, and the sun is blocked by stretching out the eaves. The sunlight from the roof skylight is reflected through the ceiling, reducing the indoor temperature and improving the indoor comfort

Winter sunshine setting
The thermal division of Pingwu County is a cold area, and the demand for heat preservation is very high. The most appropriate eaves depth is calculated based on the solar height angle of Pingwu County, so that the maximum sunshine can be obtained in winter

Warm keeping measures in winter
Pingwu County is rich in geothermal pump resources, so the main heating measure in winter is geothermal pump, and the auxiliary heating depends on the hot air in the roof interlayer. The roof skylight is closed, and the air in the roof air interlayer is pumped into the room through the sun

Sectional view 1-1 1:200

综合奖 · 入围奖
General Prize Awarded · Finalist Award

注 册 号：100996
项目名称：拾野
　　　　　Return to Nature
作　　者：冯雪珂、柴雨欣、李文潇、
　　　　　秦志宁
参赛单位：兰州交通大学
指导教师：高发文

山水驿·拾野
Return to Nature

The scheme adopts a courtyard-style layout, and the simple and rustic shape forms a rich soul-ink layer through the interspersing of the courtyard and the retraction of space. Each functional room is arranged in four story unitoms. The three main spaces are inserted into three patio-style inner courtyards. The six spaces are visually coherent, forming a repeated superposition of exterior and interior, architecture and nature. Through the organization of deep dimensions, architecture and nature are intertwined.

Surrounding status

Local materials

Building materials for residential buildings in northern Sichuan
Bamboo | Cedar | Stone | Clay

Policy analysis

Climatic conditions

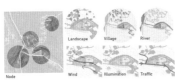

The National Nature Reserve is located in Pingwu County, Sichuan Province. Founded in 1965, it covers an area of 32297 hectares and is one of the first four giant panda nature reserves in China. The establishment of the scientific research station will help to improve people's understanding of the biodiversity of giant pandas and strengthen the overall protection of this typical ecologically fragile area.

According to the analysis of annual average daily temperature and direct/indirect solar radiation conditions, the corresponding lighting area and orientation can be considered to determine the building facade form

The Baima Tibetans mostly live in high mountains and valleys, and the trees around the villages are dense, and nature has endowed the Baima Tibetans with rich building materials. The Baima Tibetans, who depend on nature for their lives, have gradually learned to adapt to local conditions, use local materials, and make full use of the superior conditions of nature to build villages and houses.

Climate strategy research

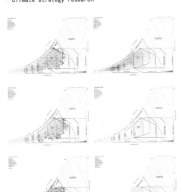

From the calculation results, the average annual temperature of Baima Tibetan Nationality Township is 8 to 12 °C. The average temperature in January is -4 °C, and the extreme minimum temperature is -10 °C; The average temperature in July is 18 °C, and the extreme maximum temperature is 22 °C. The average annual precipitation is 125 mm, the maximum annual rainfall is 155 mm, and the extreme annual minimum rainfall is 75 mm.
By simulating different passive building strategies, it is concluded that the appropriate passive strategies in this field are: natural ventilation and passive solar heating.

Site Plan

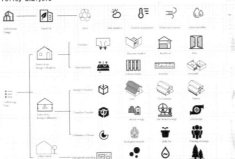

Economic & Technical Indexes:
Site area: 2862 m²
building area: 1980 m² (including the area of Phase II)
Building density: 31%
Plot ratio: 0.69
Building height: 10.2 m
Greening rate: 0.57

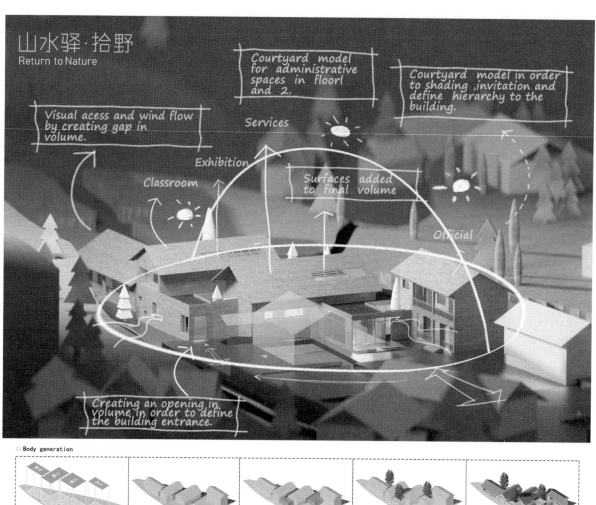
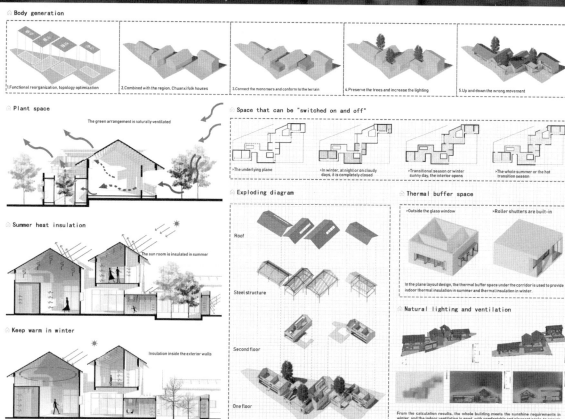

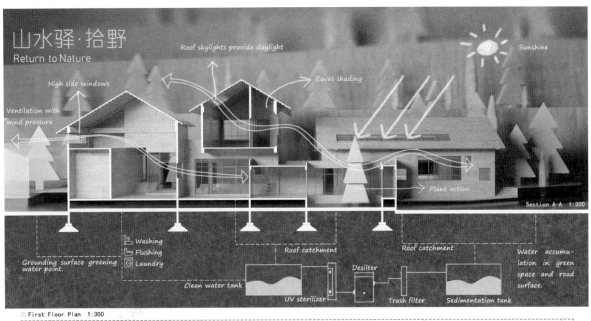

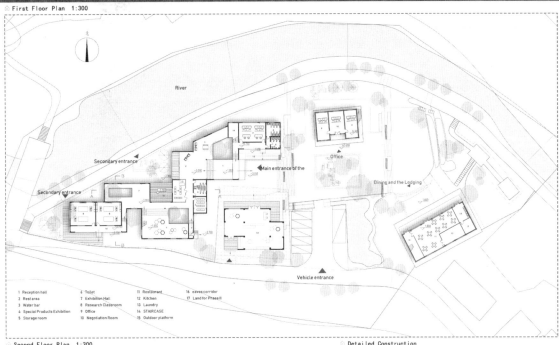

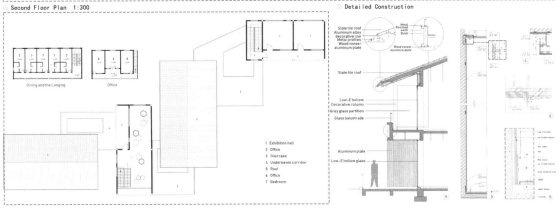

综合奖·入围奖
General Prize Awarded · Finalist Award

注 册 号：101007
项目名称：隐山·观水·纳风
Lifting · Experiencing · Ventilating
作　　者：杨天心、张雨佳、梁骏嘉、程卓然、李圣哲
参赛单位：北京交通大学
指导教师：王　鑫、张　文

阳光·山水驿
关坝自然保护小区科学考察站设计

隐山·观水·纳风 [壹]
LIFTING · EXPERIENCING · VENTILATING

Economic and Technical Indicators
Site Area: 1800 ㎡
Total Construction Area: 1809 ㎡
New Construction Area: 1227 ㎡
Reform Construction Area: 582 ㎡
Floor Area Ratio: 0.68

■ DESIGN SPECIFICATION

本次设计项目选址位于中国四川省绵阳市平武县木皮藏族乡关坝村，当地海拔较高，夏热冬冷，太阳辐射量较低，场地内有东西向较强的山谷风。我们从当地气候和山谷的地形条件出发，旨在创造一个与自然环境共生、适应当地气候、绿色、舒适的科学考察站。

在应对气候方面，为了较好地适应当地潮湿的气候条件，设计通过底层架空、天井和夹层等方式改善建筑通风。为了降低冬季的建筑能耗，我们运用了建筑光伏一体化将光伏构件融入建筑屋顶，并在住宿单元中采用了特朗伯墙这一被动式太阳能技术原型，改善居住环境。

在与自然环境和谐共生方面，我们采用底层架空的策略尽可能减少对地面植物的影响；结构体系采用装配式板墙不仅减少了施工噪声对村民和动物的影响，更让研学教室有一整面墙透明成为可能，实现在林间的学习和研究。此外，针对村落肌理的回应也被纳入设计的考虑范围，新建筑的屋顶形式和建筑体量与周边村落中的民居统一和延续。

The project is located in Guanba Village, Mupi Tibetan Township, Pingwu, Mianyang, Sichuan, China, with high altitude, hot summer, cold winter, low solar radiation, and strong east-west valley winds in the site. Starting from the local climate and the topographic conditions, we aim to create a scientific research station that coexists with the natural environment, adapts to the local climate, is green and comfortable.

In terms of climate, in order to adapt to the humid local climatic conditions better, the design improves building ventilation through overhead ground floor, patios and mezzanines. In order to reduce the energy consumption of the building in winter, we have used the BIPV to integrate photovoltaic components into the roof, and adopted the Trombe wall in the accommodation unit, to improve the living environment.

In terms of harmonious coexistence with the natural environment, the use of overhead underfloor minimize the impact on ground plants; The use of prefabricated walls in the structural system not only reduces the impact of construction noise on villagers and animals, but also makes it possible for the research classroom to have a whole wall transparent, so that people can learn and research beside the forest. In addition, the response to the texture of the village was also taken into consideration in the design. The roof form and volume of the new building were unified and continued with the houses in the surrounding villages.

LIFTING · EXPERIENCING · VENTILATING
隐山·观水·纳风 | 贰

■ SITE PLAN

■ GENERATION PROCESS

Layout the building mass	Lifting the building in order to improve ventilation and reduce impact on plant-covered floor.	Place function blocks according to the surrounding building volume	Connect First floor	Match the roof form with the surrounding buildings

■ NATRUAL & CUTURAL CONTEXT

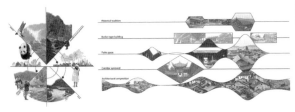

Pingwu County is rich in biodiversity. We extract elements from local buildings in Sichuan then apply ground floor overhead, patio and gallery space into architectural design, which not only protects nature, but also continues the texture and characteristics of Sichuan architecture.

■ CLIMATE SIMULATION

Pingwu has a humid monsoon climate in the cold temperate mountainous region. With high humidity and low temperature, it needs coordinated balance between ventilation and insulation.

■ SITE ANALYSIS

water

habitat

route analysis

■ DESIGN PROCESS

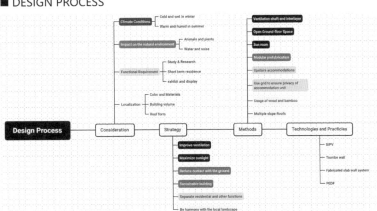

隐山·观水·纳风 | 叁
LIFTING · EXPERIENCING · VENTILATING

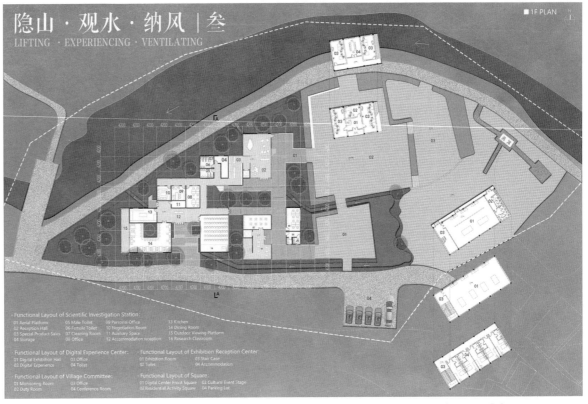

1F PLAN

- Functional Layout of Scientific Investigation Station:
  - 01 Aerial Platform
  - 02 Reception Hall
  - 03 Special Product Sales
  - 04 Storage
  - 05 Male Toilet
  - 06 Female Toilet
  - 07 Cleaning Room
  - 08 Office
  - 09 Personal Office
  - 10 Negotiation Room
  - 11 Auxiliary Space
  - 12 Accommodation reception
  - 13 Kitchen
  - 14 Dining Room
  - 15 Outdoor Viewing Platform
  - 16 Research Classroom

- Functional Layout of Digital Experience Center:
  - 01 Digital Exhibition Hall
  - 02 Digital Experience
  - 03 Office
  - 04 Toilet

- Functional Layout of Exhibition Reception Center:
  - 01 Exhibition Room
  - 02 Toilet
  - 03 Stair Case
  - 04 Accommodation

- Functional Layout of Village Committee:
  - 01 Monitoring Room
  - 02 Duty Room
  - 03 Office
  - 04 Conference Room

- Functional Layout of Square:
  - 01 Digital Center Front Square
  - 02 Residential Activity Square
  - 03 Cultural Event Stage
  - 04 Parking Lot

■ PLAN DRAWINGS

■ 2F PLAN - 1:300

- Functional Layout of Scientific Investigation Station:
  - 01 Rest Area and Water Bar
  - 02 Outdoor Viewing Platform
  - 03 Sun Room
  - 04 Accommodation
  - 05 Roof Platform

- Functional Layout of Digital Experience Center:
  - 01 Digital Exhibition Hall
  - 02 Digital Experience
  - 03 Outdoor Platform

■ OVERHEAD FLOOR PLAN - 1:350

- Functional Layout of Plan of Overhead Floor:
  - 01 First Floor Aerial Platform
  - 02 Drainage Canal
  - 03 Woodland

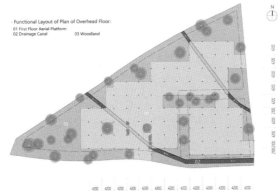

■ STREAMLINE & MATERIALS ANALYSIS

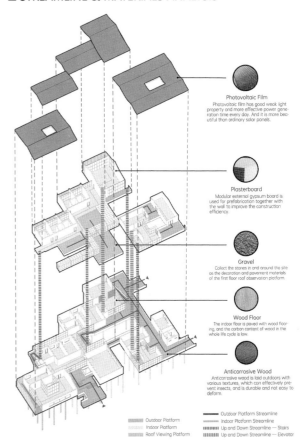

Photovoltaic Film
Photovoltaic film has good weak light property and more effective power generation time every day. And it is more beautiful than ordinary solar panels.

Plasterboard
Modular external gypsum board is used for prefabrication together with the wall to improve the construction efficiency.

Gravel
Collect the stones in and around the site as the decoration and pavement materials of the first floor roof observation platform.

Wood Floor
The indoor floor is paved with wood flooring, and the carbon content of wood in the whole life cycle is low.

Anticorrosive Wood
Anticorrosive wood is laid outdoors with various textures, which can effectively prevent insects, and is durable and not easy to deform.

- Outdoor Platform
- Indoor Platform
- Roof Viewing Platform
- Outdoor Platform Streamline
- Indoor Platform Streamline
- Up and Down Streamline — Stairs
- Up and Down Streamline — Elevator

■ A-A Cutaway perspective

LIFTING · EXPERIENCING · VENTILATING
隐山 · 观水 · 纳风 | 肆

■ PASSIVE DESIGN

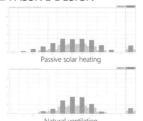
Passive solar heating

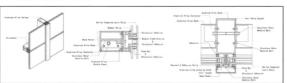
Thermal mass effects

Natural ventilation

Exposed mass + night-purge ventilation

■ NOTE DETAILS

· Glass Curtain Wall

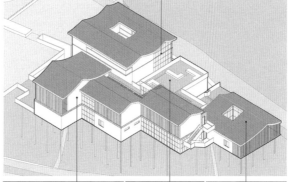

■ SUNSHINE ANALYSIS

Spring   Summer   Autumn   Winter

■ WIND ANALYSIS

Indoor Wind Velocity (1F)

Indoor Wind Velocity (2F)

Outdoor Wind Velocity (1F)

Outdoor Wind Velocity (2F)

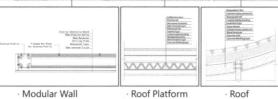
· Modular Wall   · Roof Platform   · Roof

LIFTING · EXPERIENCING · VENTILATING

隐山·观水·纳风 | 伍

■ FUNCTIONAL ANALYSIS

■ CORE TECHNOLOGY

Accommodation	Sunshine Room	Research Classroom	Reception Hall

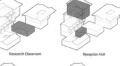

Winter-Day | Winter-Night

 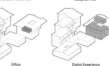

Restaurant and Kitchen	Toilet	Office	Digital Experience

Summer-Day | Summer-Night

■ VIEW ANALYSIS

| Installing the grille | Before installing the grille | Waterscape | Mountain scene |

BIPV + PEDF

High specific heat capacity materials are added into the roof insulation layer for thermal insulation, and the thermal energy generated by solar photovoltaic panels and direct sunlight is stored to keep the indoor temperature warm at night.

TROMBE WALL

By controlling the opening and closing of north-south doors, windows and ventilation valves, and in combination with the sun room, it can play the role of heat preservation in summer and indoor. In winter, heat is trapped between the glass panels and the walls, allowing the Trombe wall to absorb heat efficiently and limit its re-radiation to the environment. The sun's heat passes through the glass, is absorbed by the thermal storage walls, and is then slowly released into the interior of the house.

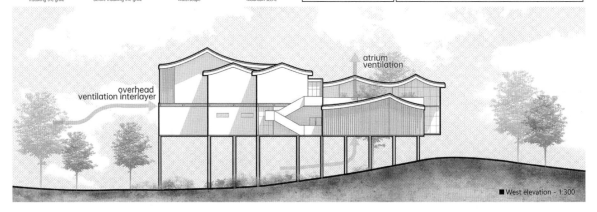

■ West elevation ~ 1:300

LIFTING · EXPERIENCING · VENTILATING
隐山·观水·纳风 | 陆

■ CONSTRUCTION PROCESS

■ MODULAR ANALYSIS

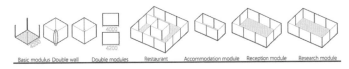

Basic modulus　Double wall　Double modules　Restaurant　Accommodation module　Reception module　Research module

■ ENERGY CALCULATION

· Energy consumption estimation

According to the specification requirements, the constraint value of comprehensive power consumption index with residential buildings is 3100kW*h/(a*H).

The residential area is 140 square meters in total, and the average energy consumption is 5kW*h/(a*H).

The public area is 1087 square meters in total, and the average energy consumption is 10 kW*h/(a*H).

Based on the use time of public area from 8:00 a.m. to 19:00 p.m. in the daytime and the use time of residential area from 19:00 p.m. to 8:00 a.m. the next day, the energy consumption estimation is obtained:

140×5+1087×10=12270 kW·h

· Solar function simulation

Although the power generation efficiency of the solar thin film is not high, it has good weak light. It is estimated that the effective power generation time is 9:00 a.m. to 5:00 p.m. every day, the power generation time is 8h every day, the power generation efficiency is 13%, and the effective light generation area is 830 square meters.

Combined with the average simulated solar radiation power generation, the total simulated power generation is:

830×268×0.18×1/3
=13346.4kW·h>12270kW·h

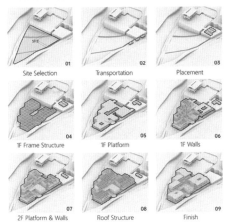

■ South elevation · 1:300

2022 台达杯国际太阳能建筑设计竞赛获奖作品集

综合奖 · 入围奖
General Prize Awarded · Finalist Award

注 册 号：101022
项目名称：落木·云台
　　　　　Timber & Terrace
作　　者：毕睿航、罗志远、任　帅
参赛单位：南京工业大学
指导教师：林杰文

落木·雲台 Timber & Terrace

1. Peception Hall
2. Exhibition Hall
3. Lounge
4. Research & Interaction Room
5. Coffee Bar
6. Specialty Store
7. Storeroom
8. Digital Center
9. Office
10. Management
11. Outdoor Space

FIRST FLOOR PLAN 1:200

1-1 SECTION 1:200

2-2 SECTION 1:200

UNDERGROUND FLOOR PLAN 1:200

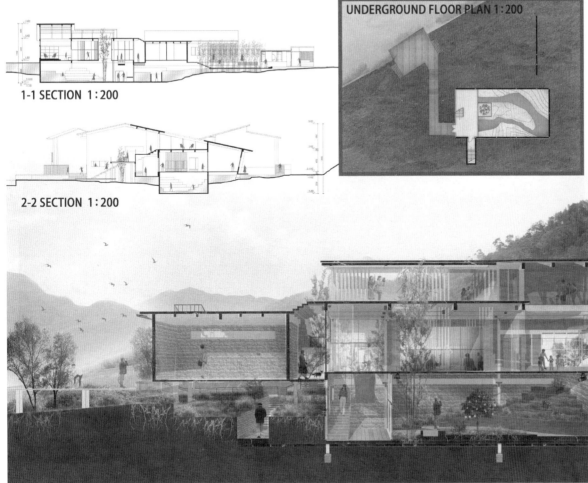

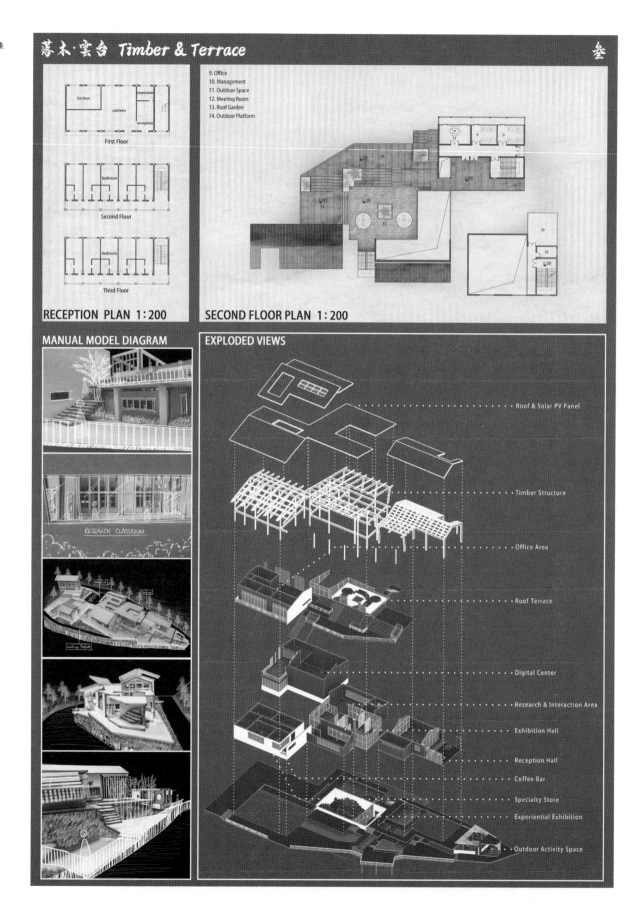

落木·雲台 Timber & Terrace

ELEVATION ANALYSIS
SOLAR RADIATION
SUNSHINE DURATION
SOLAR ORBIT ANALYSIS

10:00 am, Summer Solstice 4:00 pm, Summer Solstice 10:00 am, Winter Solstice 4:00 pm, Winter Solstice

SOLAR PV PANELS ANALYSIS
NODE DETAIL

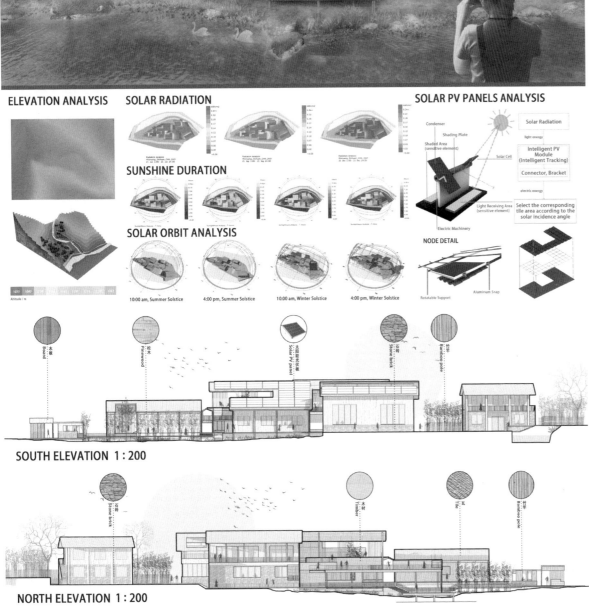

SOUTH ELEVATION 1:200

NORTH ELEVATION 1:200

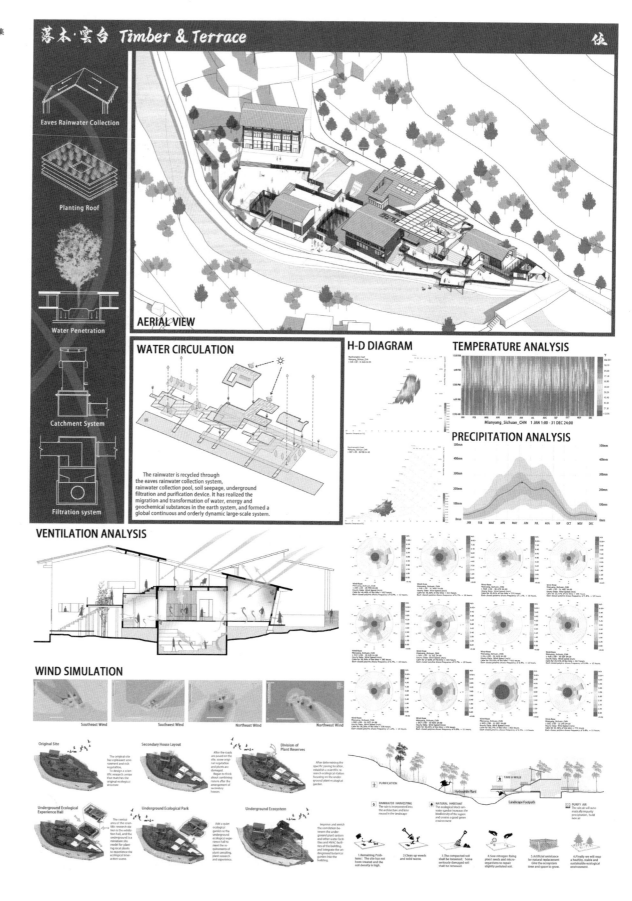

落木·雲台 Timber & Terrace

The exhibition hall on the underground floor tries to create an immersive ecological experience space. In this ecological model, the public can personally understand the local ecosystem and the symbiotic relationship between human and nature. Model shapes the space with the ups and downs of the mountain, and the most typical ecological members are added in this space. By filling infinite details into the limited space, the original ordinary space has been transformed into a "Second Nature".

VEGETATION / ANIMAL TYPES & DISTRIBUTION

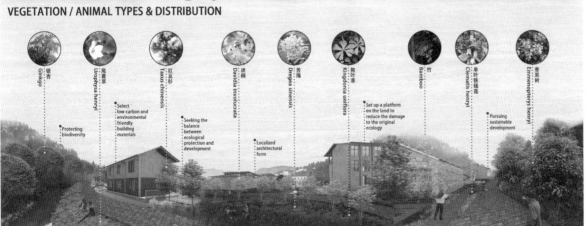

综合奖·入围奖
General Prize Awarded · Finalist Award

注 册 号：101024
项目名称：河谷林驿
　　　　　Valley Villas at Guanba
作　　者：胡一鸣、陈奕宏、黄志毅、
　　　　　刘宇翔、郑渊正、刘日尧
参赛单位：福州大学
指导教师：邱文明

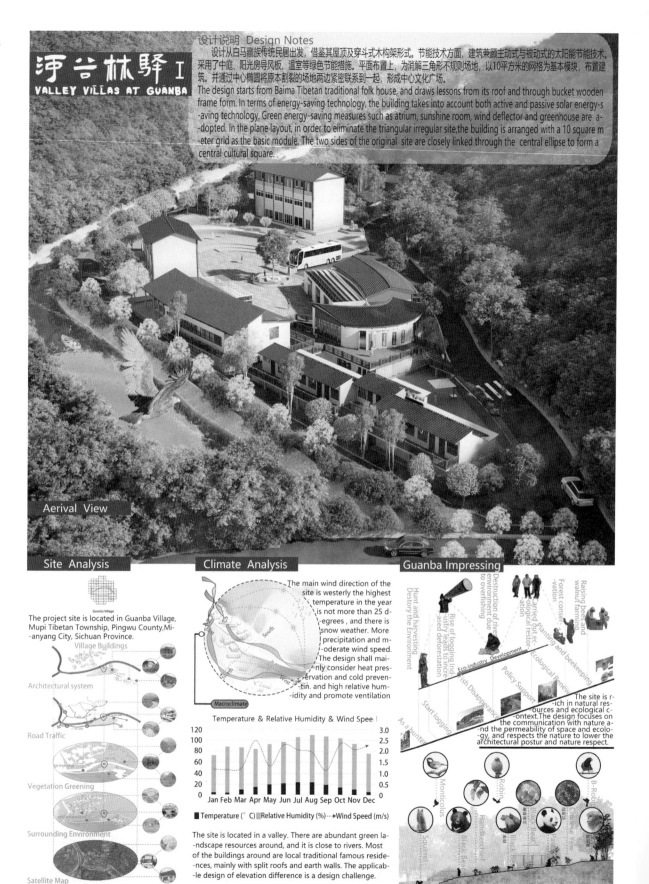

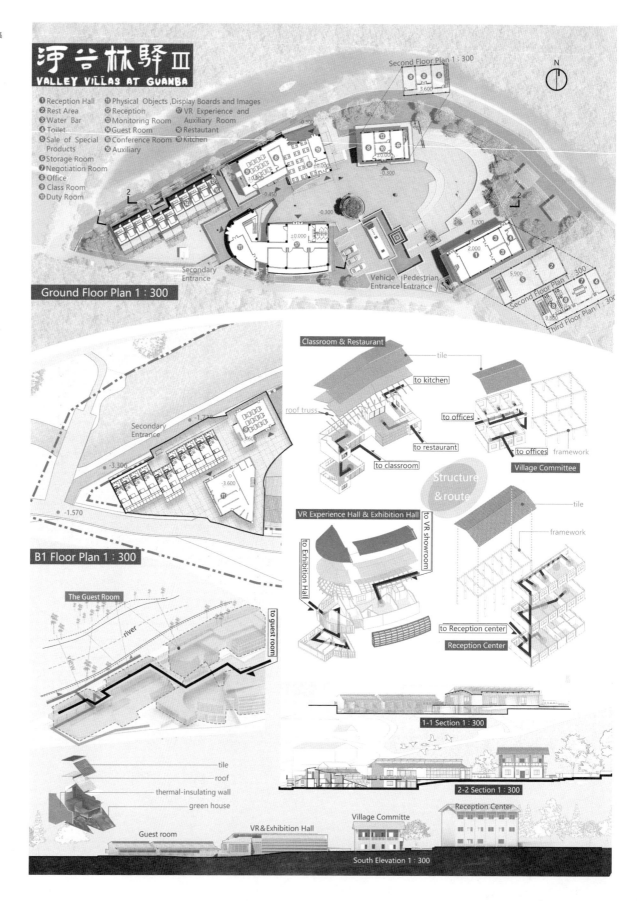

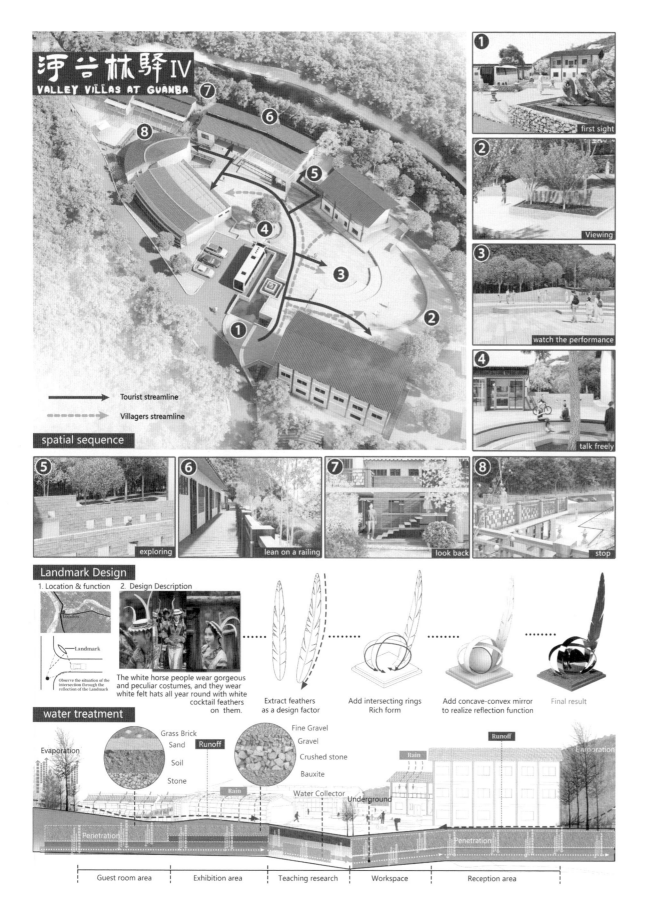

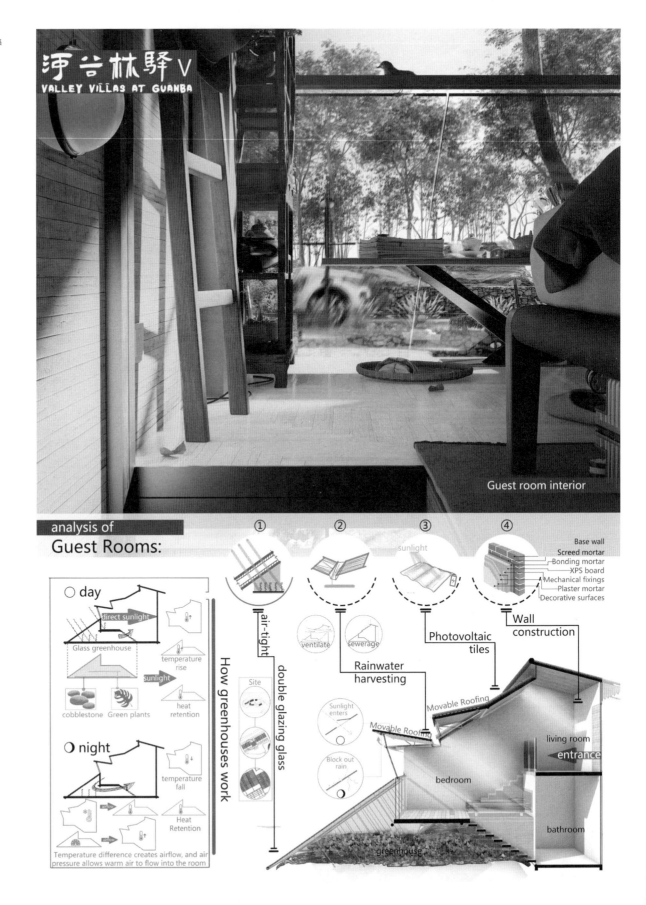

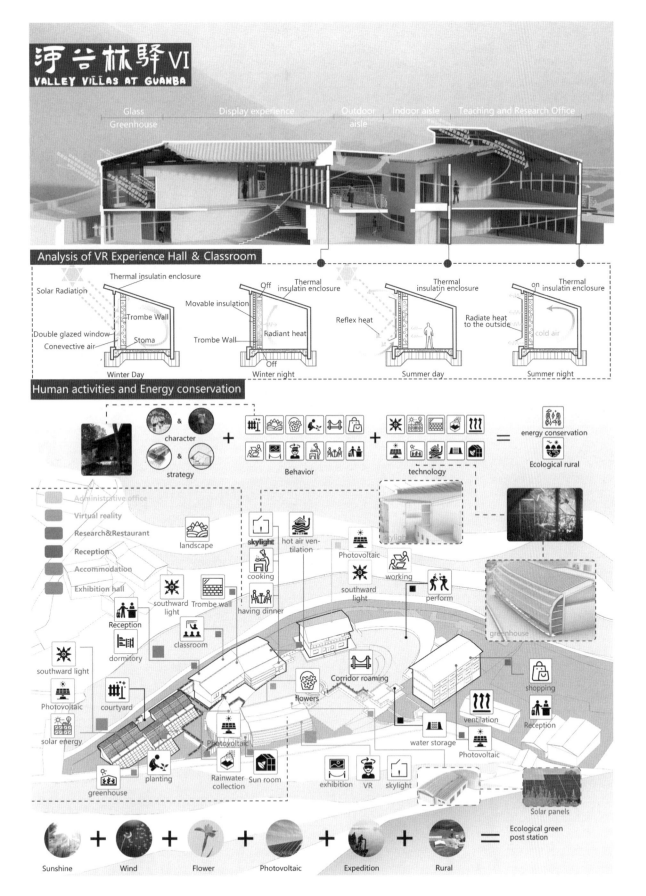

综合奖·入围奖
General Prize Awarded · Finalist Award

注 册 号：101053
项目名称：凉厅新叙
　　　　　Description of New Pavilion
作　　者：杨怡琳、祝浩艺、章子玥、
　　　　　李晨曦
参赛单位：南京工业大学
指导教师：董凌

凉厅新叙 Description of new pavilion 壹
——传承创新 技艺重现
— Inheriting innovative skills and recreating memories

- **设计说明：**

本设计通过不同功能的小体量建筑进行组合，对川渝地区传统的"凉厅子"进行技术改造成为"暖厅子"，结合主被动式太阳能技术，采用轴线布局，创造出丰富的建筑形体，同时营造出具有"街巷感"的公共商业灰空间。我们希望当地工匠能够参与到建设自己家乡的活动中，故采用当地传统穿斗式木结构。我们的初衷——希望创造出类似川渝地区传统场镇的街巷空间模式，唤醒当地人民对家乡最深刻的记忆，构建科考成员与当地居民沟通的桥梁。

- **Description：**

Through the combination of small-scale buildings with different functions, the design transforms the traditional "cool hall" in Sichuan and Chongqing into a "warm hall". It combines both the active and the passive solar energy technology, and adopts the axis layout to create rich architectural shapes, at the same time, it also creates a public commercial gray space with a sense of "Street and lane ". We hope that local craftsmen can participate in the activities of building their own hometown,so we use the local traditional bucket type wood structure. Our original intention is to create a street and lane space pattern which is similar to the traditional towns in Sichuan and Chongqing, awaken the deepest memories of the local people about their hometown, and build a bridge between members of the scientific research team and local residents.

- **District analysis**　- **Prevailing Winds**

Pingwu County, Mianyang City, Sichuan Province

The project is located in Guanba Village, Mupi Tibetan Township, Pingwu County, Mianyang City, Sichuan Province, which close to National Highway G247, and the exit of Jiujiang Mianyang Expressway is 4 kilometers north of the junction of National Highway G247 and the village entrance. Guanba Village lies in the Nature Reserve of Giant Panda Habitat in Minshan District of Giant Panda National Park. It is a village with rich biological resources and high ecological environment quality.

Tourists lack proper space when enjoying the scenery. Photographers lack places for communication.

The original buildings in the site are old and of poor quality

- **Site analysis**

Road
Path
Entrance
Scenery

- **The village**

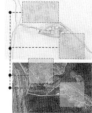

The layout of the villages is relatively scattered, scattered courtyards are loosely distributed, the impact between dwellings is small, and the texture of the original villages is obvious, but there is a lack of public and activity venues, so villagers can get together to celebrate festivals and carry out various activities

- **The problem of local development**

The educational background of local residents is generally low, and the atmosphere of scientific investigation is not strong.

The local economy is relatively backward, which can not meet the job needs of most people. Driven by interests, many indigenous residents choose to leave their hometown.From this, we wondered whether we could provide employment for local villagers through the construction of scientific examination stations.

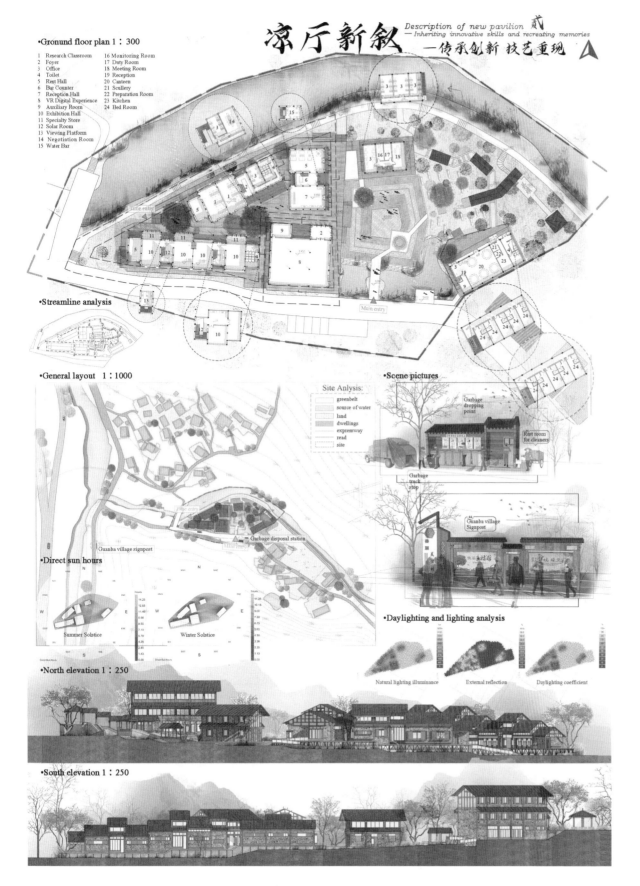

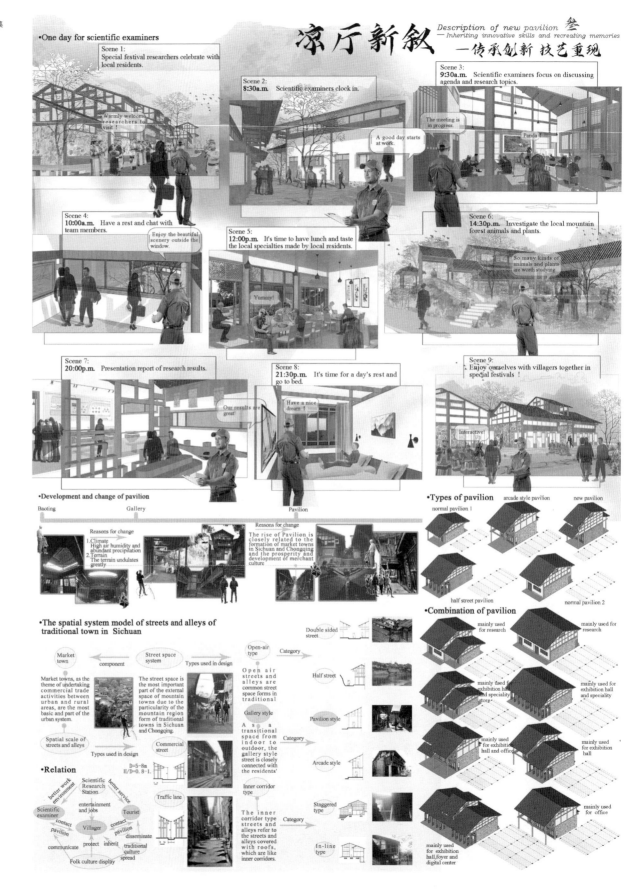

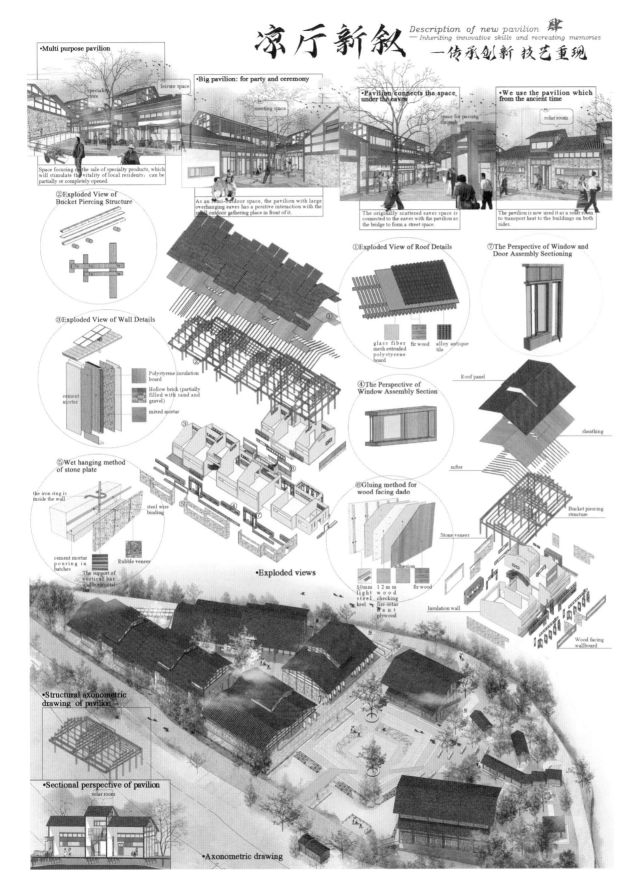

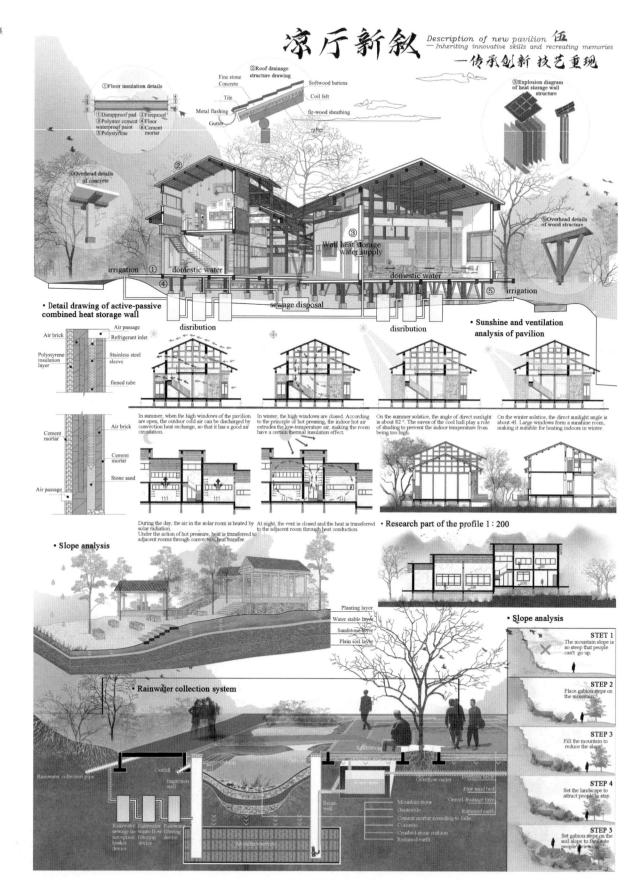

凉厅新叙
Description of new pavilion
—传承创新 技艺重现
Inheriting innovative skills and recreating memories

• **Economic and technical indicators**

New buildings:
Building area of the first floor: 1100 ㎡
Building area of the second floor: 142 ㎡
Existing buildings:
Management and protection building area: 221.88 ㎡
Building area of meals and accommodation: 572.25 ㎡

Total building area: 2036.13 ㎡　　Planned land area: 10709 ㎡
Building density: 48%　　　　　　Building height: 13m
Parking quantity: 6　　　　　　　Plot ratio: 0.19
Green space rate: 64%

• **Analysis of green building technology**

The ventilation system gains heat throughout May-September and is largely negative during the midday period in January, February, March, November and December. Inter-area heat gains are largely negative, with heat gains in the winter quarter concentrated in the morning hours.
The heat gain from the envelope is largely negative, so thermal storage and insulation walls are installed in the south of most buildings. The internal heat gain is more uniform each month.
Ventilation within the site has a low impact on the building, but the direction of the retrofitted building in the southeast corner is not conducive to site ventilation and tends to create larger cyclones. The east-west direction of the site is better, but is less favourable for the movement of people in the activity plaza in the site.

• **Model photos**

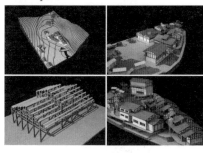

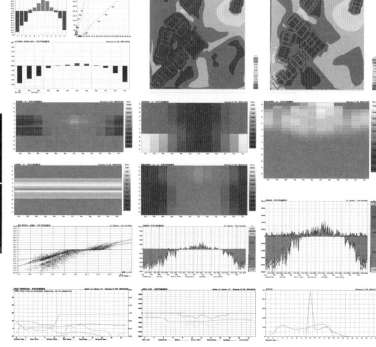

• Elevation of management and protection building 1:250

• Elevation of accommodation building 1:250

• Elevation of Digital Center Building 1:250

• West Elevation 1:250

• Eest Elevation 1:250

综合奖·入围奖
General Prize Awarded · Finalist Award

注 册 号：101078
项目名称：蜀溪春
　　　　　Spring Streams
作　　者：蔡昊哲、李季恒、王湉蝶、
　　　　　欧子怡、朱子跃
参赛单位：昆明理工大学
指导教师：陆　莹

Spring streams ①

设计说明
Design specification

本次设计以场地内的高差和北边广阔的视野为切入点去进行设计，在衔接与保有建筑功能的同时，让建筑本身去连接北边与南边不同高差的道路。

方案考虑到当地气候条件和当地建筑风格，希望建筑在满足使用的情况下融入当地，又不失自己的独特性。

建筑的主入口开在进村路(南侧)部分，因为场地存在高差，为了使建筑能与周围产生有效的连接，建筑层数设置上为地下以一层为主，局部二层，地下一层。

设计采用的被动式太阳能技术：
1. 结构上采用当地民居穿斗式的框架，采用木材建造，提升抗震能力的同时，建造环保低碳。
2. 沿用白马藏族原有民居的"三引一坡"的言语模式，在新建建筑中朝南大面积开窗，最大程度获得太阳能，在其他面上根据功能开小窗。
3. 在材料选择上由于层数较低，对民居中竹架等材料食养，选用白马藏族传统民居材料构成中的夯土进行建造，达到冬暖夏凉的能效。
4. 在通风上通过楼梯间形成风井，楼梯上部形成对流风通过风压对建筑进行通风，通过特兰博墙与玻璃之间加热空间冷凝，同时在静风时路处设有通风间层，热压与风压同时作用下形成室内气流更换。
5. 由于白马藏族民居中出檐较大，地下空间便多为新建筑的阳光周道。
6. 在建筑中南部墙体加入特兰博墙与阳光接，增加建筑对阳光的利用，并在室内运用棉窗帘对夜上进行保温。
主动式太阳能：在屋顶上采用太阳光伏板进行建筑储能。

The design takes the elevation difference within the site and the broad view of the north as the entry point to carry out the design. While connecting with the existing building functions, the building itself connects the roads with different elevation differences between the north and the south.
The plan takes into account the local climate conditions and the local architectural style. It is hoped that the building will blend in with the local area while being suitable for use, without losing its uniqueness. The main entrance of the building is located on Jincun Road (south side). Due to the height difference of the site, in order to effectively connect the building with the surrounding area, the number of building floors is set as one floor above ground, two floors locally, and one floor underground.

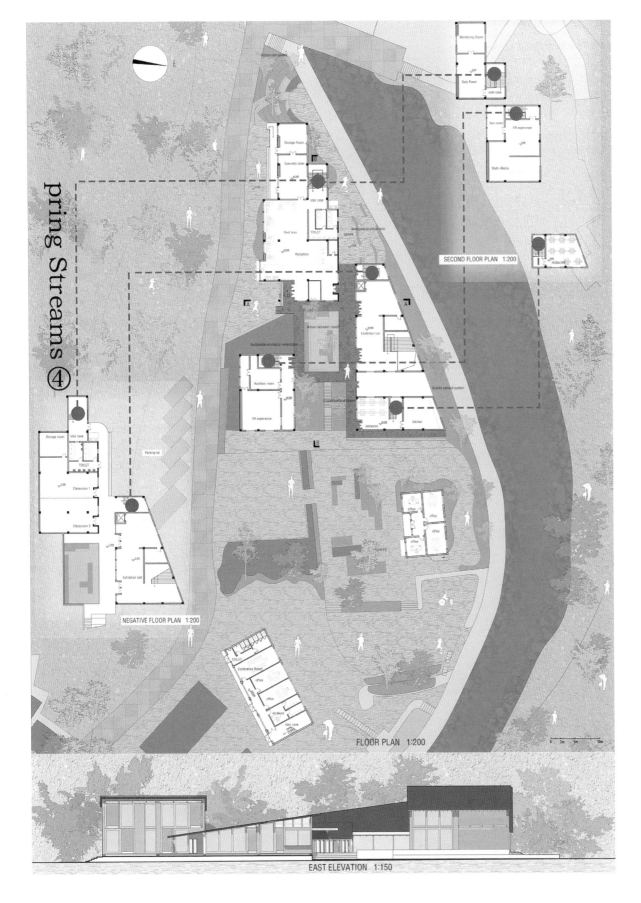

Spring streams ②

Venue location
The project is located in Guanba village, Mupi Tibetan Township, Pingwu County, Mianyang Province, Sichuan province, it is located at latitude 39°33′18″ N and longitude 104°33′58″ E, next to G247. The exit of Jiumian Expressway is 4 km north of the junction of G247 and the village entrance.

Status of land use
The project is planned to be a research station for the Guanbagou Nature Reserve in Pingwu County, Sichuan province, the first phase of the plan combines the existing buildings to build a complex with functions such as scientific research exhibition, Accommodation Courier Station, office management and so on.

Traffic situation
There are two roads to the north and south of the site, the south road is wide and can be driven, the north side is a river trail, along the valley to the east is the road to the Guanbagou Valley Protection Area.

Terrain trends
The project site is located in the valley area of south-east (high) to north-west (low). The stream at the bottom of the valley always has water flowing from east to west. The project site is located in the south side of the stream, and the south side of the valley is steep, the north side of the mountain is

Visual accessibility
The site itself is low in the north and high in the south because of the steep terrain on the south side of the valley and slightly flat terrain on the north side. The north side of the line of sight is relatively open, the line of sight mainly for the Valley Bottom Creek and the northern village.

Population analysis

Scientific researchers — Conduct academic research and field investigation

Tourist — Tourism, sightseeing, visiting exhibition halls and experiencing natural scenery

Local — Leisure, cultural activities, specialty sales

Dry ball temperature map of Mianyang

Dew point temperature map of Mianyang

Wind patterns over Mianyang

Wind charts of Mianyang

Mianyang radiography

Enthalpy and humidity chart of Mianyang
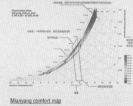

Mianyang comfort map

Mianyang year-round sky radiation model

Mianyang January-March sky radiation model

Mianyang April-June sky radiation model

Mianyang July-September sky radition model

Mianyang October to December sky radiation model

Enthalpy-humidity chart of Mianyang (with passive strategy)
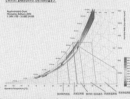

Mianyang solar track from January to December

Mianyang sunshine duration map

The year-round wind and Roses of Mianyang

Radiation Analysis

Shadow map of spring

Shadow map of summer

Shadow map of autumn

Shadow map of winter

Blind spot design

The original trees in the site are preserved

Transplant the number in the field

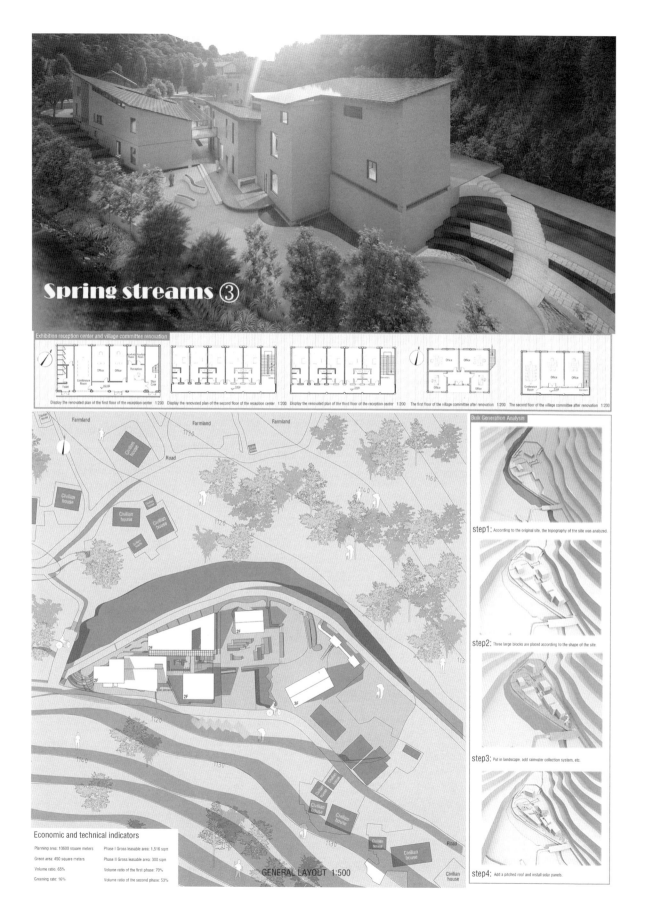

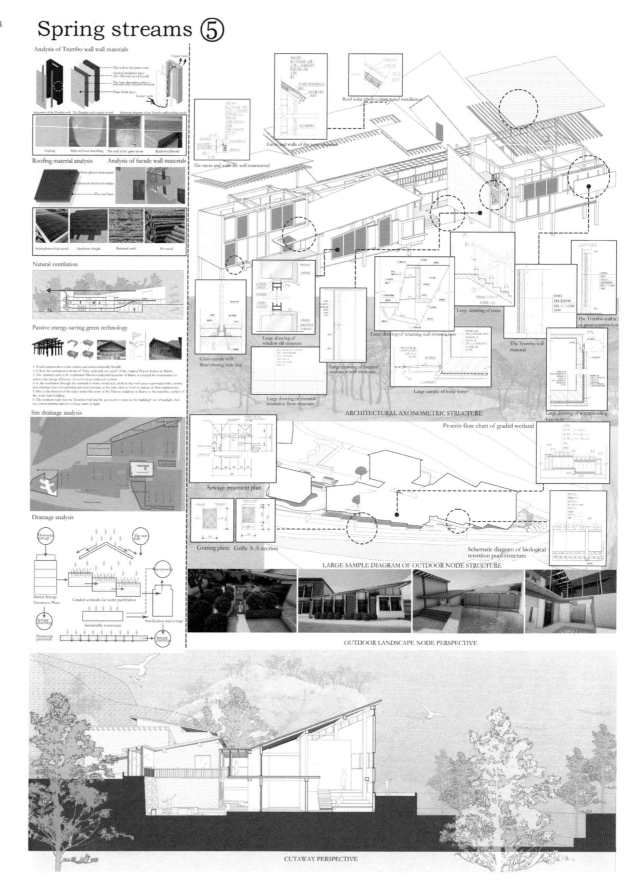

Spring streams ⑥

B-B SECTIONAL VIEW 1:200

A-A SECTIONAL VIEW 1:200

WEST ELEVATION 1:150

综合奖 · 入围奖
General Prize Awarded · Finalist Award

注　册　号：101086
项目名称：生命·守望
　　　　　Life · Care
作　　者：高中岭、张汉昭、刘　鲁、
　　　　　陈绪燕、刘心昊、范　凯
参赛单位：山东建筑大学设计集团有限公司
指导教师：赵学义

生命·守望
LIFE · CARE

DESIGN DESCRIPTION

方案以"生命守护"为设计立意，以守护者隐喻科考站工作者。通过在方案中植入瞭望塔，体现了"守护"的概念，同时展现了良好的标识性。

本方案以可持续发展为设计原则，采用模数化的集装箱模块，通过堆叠、串联、并联等手法营造出丰富的空间层次。将建筑整体架空，既通风防潮，又尊重原有的地形地貌。

集装箱符合可持续发展的"3R原则"，其自身拥有较强的气密性，结合PCM材料增强了建筑的保温、隔热性能，降低了能源需求。由工厂加工、现场安装，可以最大限度减少施工过程中对环境的破坏。另外，后期可以根据需求对建筑进行扩展，实现了建筑与环境共同成长。

Our project scheme is initialized from the idea of 'life care', the 'carer' represents workers at the research station. A watching tower is located within the site, creating a landmark while strengthening our idea of 'life care'.

The project scheme is based on the design principle of sustainable development. Containers are utilized as modular units to construct the research station. By stacking, connecting and combining the containers, abundant and sophisticated architectural spaces are created. Moreover, all the building units are lofted from ground, enhancing the performance of ventilation and dampness resistance, while respecting the original geographic condition.

The utilization of containers complies with '3R Principle - Reducing, Reusing, Recycling' of sustainable development, the material itself is of decent airtight, and its insulation performance can be further strengthened with PCM material, therefore costs for energy could be reduced. Fabricated in manufacturer, assembled on site, the damage to natural environment is significantly reduced during the construction phase. Moreover, containers can be flexibly extended in following days, thus the cooperative growth between building and nature is now accomplished.

CLIMATE ANALYSIS

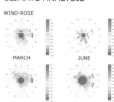

DESIGN PRINCIPLE

DESIGN STRATEGY
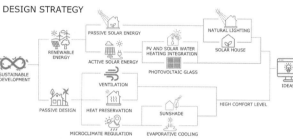

生命·守望
LIFE·CARE

SITE PLAN 1:500

ECONOMIC & TECHNICAL INDICATOR
SITE AREA: 1799.92 m²
NEW BUILDINGS AREA: 1369.89 m²
BUILDINGS AREA(SUM): 2164.01 m²

CONCEPT GENERATION

1. THE SITE IS DIVIDED TO 2 TERRACE LEVELS

2. LOFTED BUILDINGS TO AVOID DAMPNESS

3. THE CONTAINERS ARE STACKED IN A MODULAR APPROACH

4. THE CONTAINERS FORM A PARTIAL 2-STOREY BUILDING

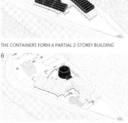
5. POND AND LANDSCAPE INSTALLATIONS ARE INTRODUCED

6. THE VR HALL PLAYS A CENTRAL ROLE OF THE SITE

PROGRAMME & ROUTINE

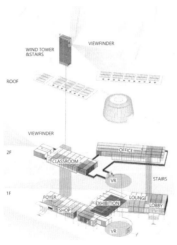

SCENE

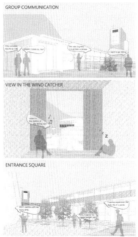

GROUP COMMUNICATION

VIEW IN THE WIND CATCHER

ENTRANCE SQUARE

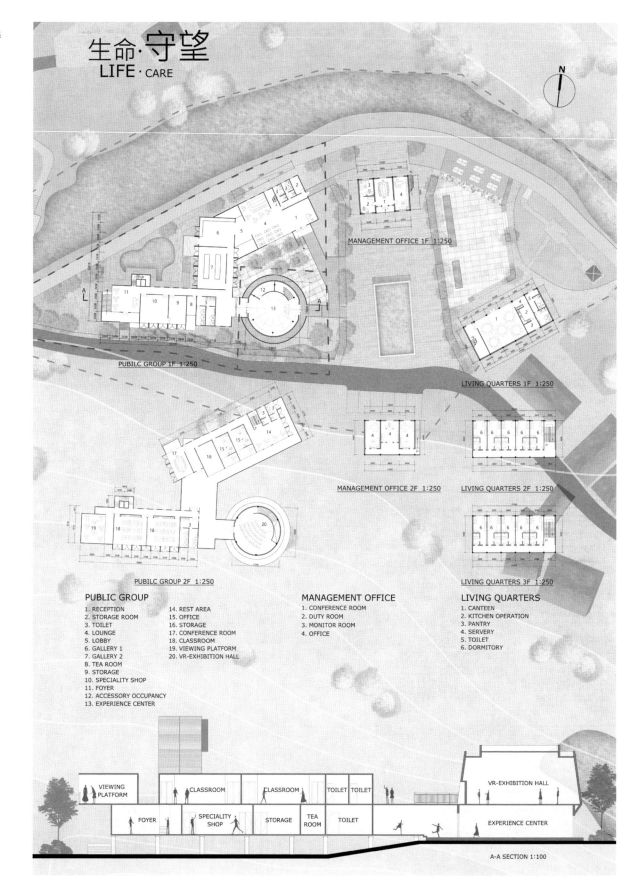

生命·守望
LIFE·CARE

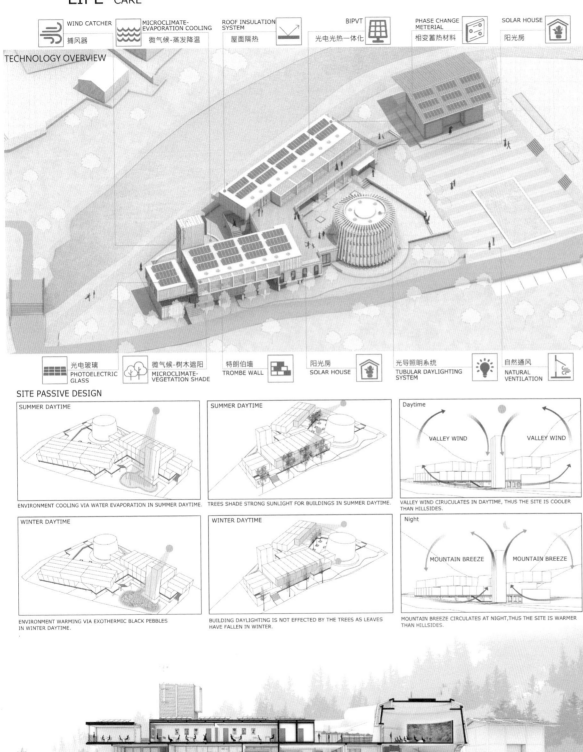

生命·守望
LIFE·CARE

TEACHING AREA

OFFICE & VISITOR

EXHIBITION HALL

SPECIAL ENERGY DESIGN

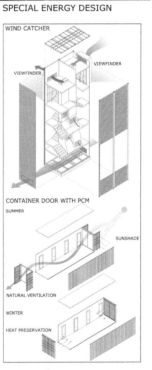

SOUTH ELEVATION 1:200

生命·守望
LIFE·CARE

SOLAR HOUSE COMPONENTS

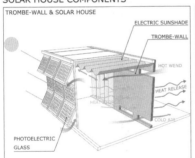
TROMBE-WALL & SOLAR HOUSE

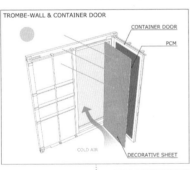
TROMBE-WALL & CONTAINER DOOR

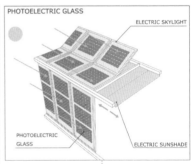
PHOTOELECTRIC GLASS

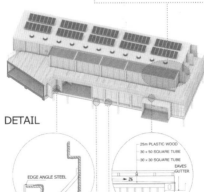

DETAIL

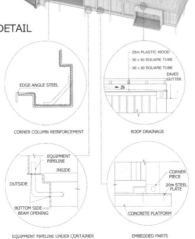

WATER TREATMENT

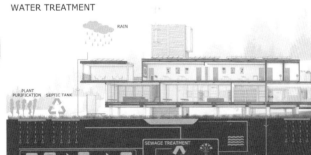

TUBULAR DAYLIGHTING

FUTURE EXPANSION
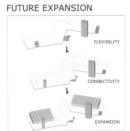

THE SIGNPOST ANALYSIS

综合奖・入围奖
General Prize Awarded・
Finalist Award

注 册 号：101106
项目名称：山・风・舞
　　　　　Mountain Breeze Dancing
作　　者：郭思洁、李若楠
参赛单位：河南工业大学
指导教师：张　华、马　静

山·风·舞 Mountain Breeze Dancing
四川省自然科学考察站设计 02

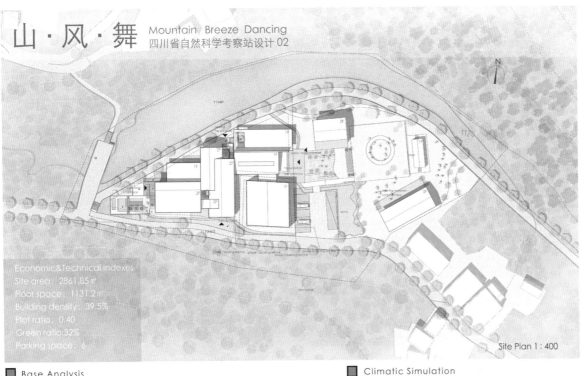

Economic&Technical Indexes
Site area: 2861.85㎡
Floor space: 1131.2㎡
Building density: 39.5%
Plot ratio: 0.40
Green ratio: 32%
Parking space: 6

Site Plan 1:400

■ Base Analysis

The base is located in Guanba Village, Mupi Tibetan Township, Pingwu County, Mianyang City, Sichuan Province. It is a nature reserve of giant panda habitat, with a forest coverage rate of 96.3%. There are more than 70 species of rare animals and plants under national protection, such as giant panda and golden monkey, in the area.

Project with status in east south west north towards the valley area, the bottom stream of water all year round. Project construction land is located in the south stream, a steep valley, south to the mountain, the north side slightly flat the cool rainy summer, winter cold and snowy. January average temperature of 4 degrees Celsius, annual average temperature of 13.9 degrees Celsius, annual average rainfall of 800 millimeters.

Guanba Village

Massif | Road | House | River
Sunlight | Wind | Noise | Vegetation

■ Local Characteristics

Production Mode
Local food crops to corn and taro, the main cash crops are walnut and yam, animal husbandry to feed pigs, cattle, sheep based.

Architectural Features
The building are roughly rectangular on the plane and the main building materials are earth, stone, wood, bamboo and rattan.

Religious Belief
The white horse Tibetan worship the natural god, mainly mountain god, land god, tree god, water god and so on, among which the mountain god is the most admired.

Local Conditions And Customs
White horse people like singing and dancing, have their own strong style of national customs and folk culture form, will be held in the square beekeepers competition every year.

■ Climatic Simulation

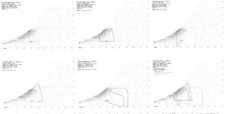

Through the simulation of different passive building strategies find the passive building strategies suitable for the area are passive solar heating, night purification ventilation, natural ventilation and evaporative cooling.

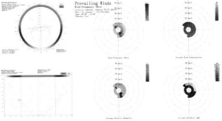

Through simulation analysis, it is found that the best orientation of the building is south, with the wind from northeast to north, the average temperature does not change much annually, the humidity is high, and the precipitation is relatively abundant.

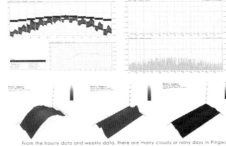

From the hourly data and weekly data, there are many cloudy or rainy days in Pingwu County, Mianyang City, which is very unfavorable to the utilization of solar energy. According to the annual average daily temperature and direct and indirect solar radiation conditions, the distribution of solar panels is considered.

■ Technological Analysis

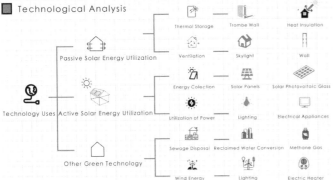

Technology Uses
- Passive Solar Energy Utilization
  - Thermal Storage
  - Trombe Wall
  - Heat Insulation
  - Ventilation
  - Skylight
  - Wall
- Active Solar Energy Utilization
  - Energy Collection
  - Solar Panels
  - Solar Photovoltaic Glass
  - Utilization of Power
  - Lighting
  - Electrical Appliances
- Other Green Technology
  - Sewage Disposal
  - Reclaimed Water Conversion
  - Methane Gas
  - Wind Energy
  - Lighting
  - Electric Heater

山·风·舞 Mountain Breeze Dancing
四川省自然科学考察站设计 03

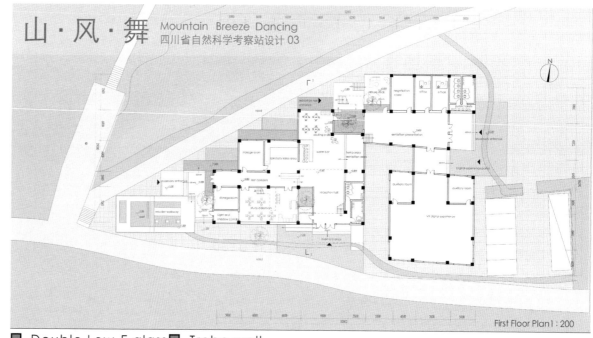

First Floor Plan 1:200

■ Double Low-E glass ■ Trobe wall

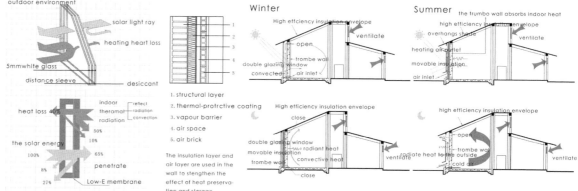

1. structural layer
2. thermal-protrctive coating
3. vapour barrier
4. air space
5. air brick

The insulation layer and air layer are used in the wall to stengthen the effect of heat preservation and storage

■ Trees shade ■ Pas solar ■ solar water heating system

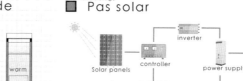

Photovoltaic power generation, using the photogenerated volt effect of semiconductor interface, converts sunlight intuition into electric energy, which is truly clean energy without noise and pollution.

Solar water heater system insulation performance is good, not only energy-saving and efficient, will not cause pollution to the environment, but also has the function of sewage purification, water source clean, no pollution.

■ Soil energy ■ Biogas system analysis

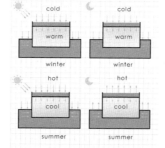
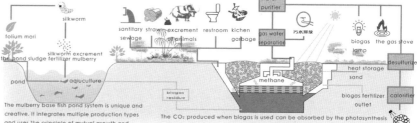

The mulberry base fish pond system is unique and creative. It integrates multiple production types and uses the principle of mutual growth and mutual cultivation of organisms to achieve "zero" pollution to the ecological environment

The CO_2 produced when biogas is used can be absorbed by the photosynthesis of the same amount of growing plants, which can achieve "zero" CO_2 emission. A biogas digester of 10 to 12 cubic meters can save two tons of standard coal per year for a household of three to five people

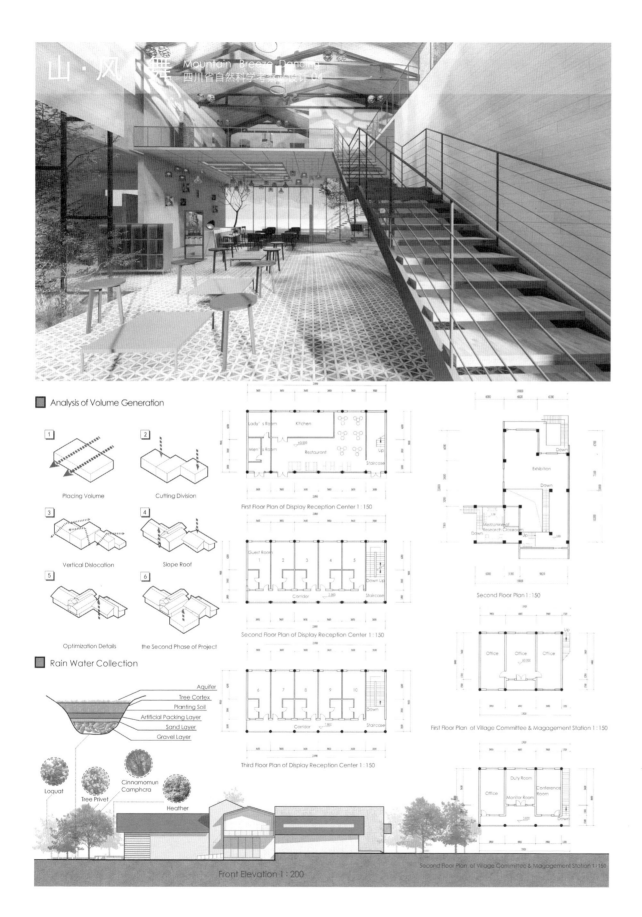

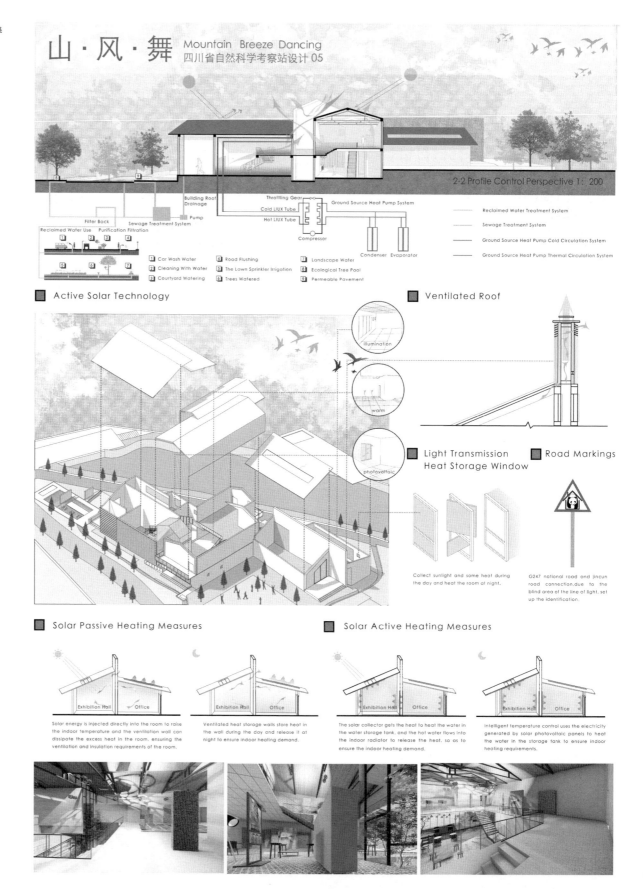

山·风·舞 Mountain Breeze Dancing
四川省自然科学考察站设计 06

- Indoor Scene
- Outdoor Scene

- Shadow Analysis
- Sunshine & Ventilation Analysis
- Venting Practice

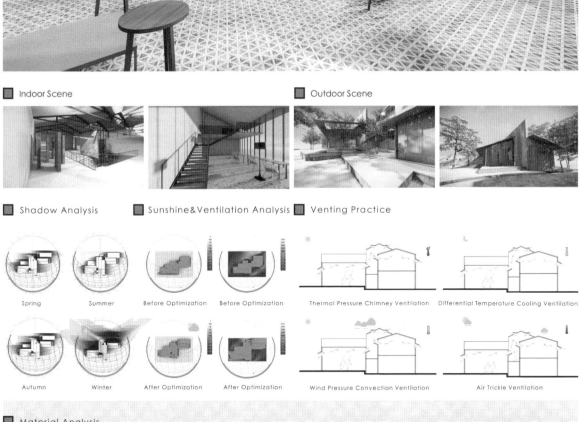

Spring | Summer | Before Optimization | Before Optimization | Thermal Pressure Chimney Ventilation | Differential Temperature Cooling Ventilation

Autumn | Winter | After Optimization | After Optimization | Wind Pressure Convection Ventilation | Air Trickle Ventilation

- Material Analysis

glass | metal | wood | brick | concrete | plastic

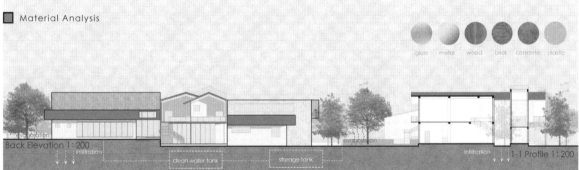

Back Elevation 1:200   1-1 Profile 1:200

综合奖・入围奖
General Prize Awarded · Finalist Award

注 册 号：101107
项目名称：处幽篁兮
　　　　　The Bamboo Forest
作　　者：张一鸣、王晓燕、宋　婕、
　　　　　杨敏敏、邹宇飞、杨云辉
参赛单位：西北工业大学
指导教师：刘　煜

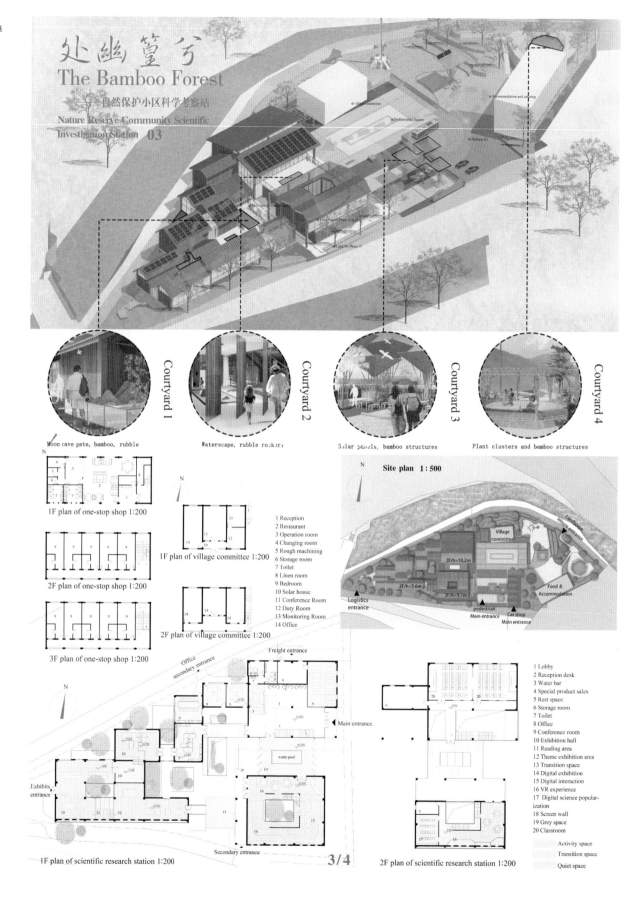

处幽篁兮
The Bamboo Forest
—— 自然保护小区科学考察站
Nature Reserve Community Scientific Investigation Station_04

Using BIPVT technology and passive heat storage can reduce **126.62 t CO_2** per year

综合奖·入围奖
General Prize Awarded · Finalist Award

注　册　号：101123
项目名称：万物生·归一
　　　　　　Living with Nature
作　　　者：李若娴、束子玥、周宇航、
　　　　　　王　璇
参赛单位：南京工业大学
指导教师：刘　强、薛　洁

萬物生·歸一
Living with Nature
关坝沟流域自然保护小区科学考察站1

■ Site condition analysis

Distance from the periphery　　Analysis of site climate conditions　　The point of landscape analysis

■ Econo-technical norms
First-stage project
用地面积：2293.96㎡　　建筑占地面积：885.3㎡　　总建筑面积：1793.3㎡
容积率：0.78　　　　　建筑密度：37%　　　　　建筑高度：9.700m
Phase II project
用地面积：567.89㎡　　建筑占地面积：269.1㎡　　总建筑面积：269.1㎡
容积率：0.47　　　　　建筑密度：47%　　　　　建筑高度：9.700m

■ Design instruction

该设计的概念意象为关坝村当地的神树，希望建筑像树木一样融入自然，建筑像树一样在不同高度为动物做出设计，来形成良好的生态系统。达到以"屋相融、人观望"的姿态守护自然，呼应"保护生态多样性"的设计主题。

在被动式技术方面，我们结合川西传统民居，把"天井、抱厅、挑檐和架空底层"作为空间原型，不仅形成了丰富的空间层次，同时也达到了除湿排风的效果。关于建筑的平面，我们采取了川西民居天井抱团的组合方式，同时顺应地形加入了落差的处理手法；关于内部空间，我们加入了顶部可活动的阳光房对内部温度进行调节；关于建筑立面，我们使用了可调节的立面模块，希望在立面上达到鸟类与植物共生的效果。

在这一片似大树一般延绵起伏的大屋顶下，形成了人类观览、科研、办公的空间，希望人们在大山深处与动物、植物产生更多互动交流。

The conceptual image of the design is the local sacred tree in Guanba Village. It is hoped that the architecture will blend into nature like a tree, and the architecture will design animals at different heights like a tree to form a good ecosystem. To protect nature with the attitude of "harmony between houses and people", and to echo the design theme of "protecting ecological diversity".

In terms of passive technology, we take "patio, holding hall, overhanging eaves and overhead ground floor" as the space prototype, which not only forms rich space levels, but also achieves the effect of dehumidification and ventilation. With regard to the plane of the building, we adopted the combination of courtyard houses in western Sichuan, and at the same time, we added poor treatment methods in accordance with the terrain. Regarding the interior space, we added a movable sun room at the top to adjust the interior temperature; Regarding the building facade, we use the adjustable facade module, hoping to achieve the symbiotic effect of birds and plants on the facade.

Under the roof of this big house, which stretches like a big tree, there is a space for human viewing, scientific research and office work. It is hoped that people will have more interaction with animals and plants in the depths of the mountains.

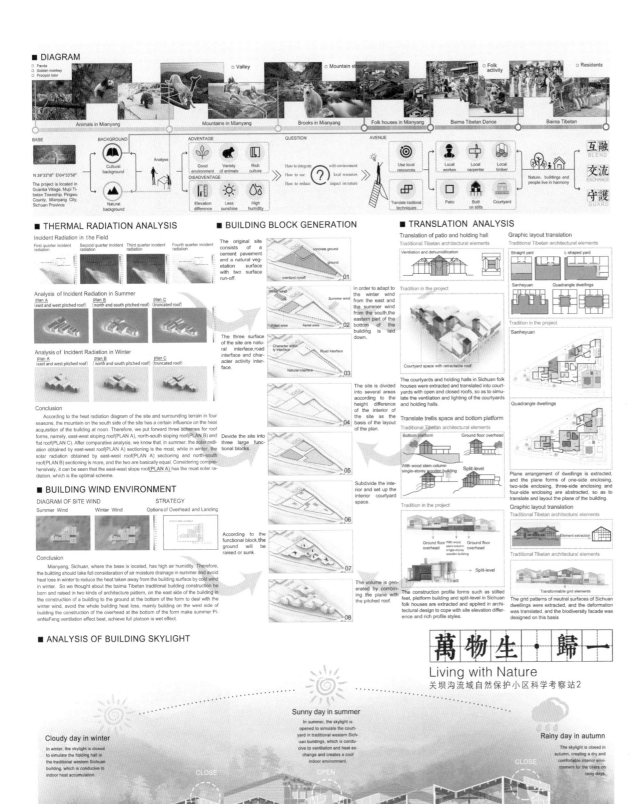

萬物生·歸一
Living with Nature
关坝沟流域自然保护小区科学考察站3

Site Plan 1:500

■ Small perspective between the first and second buildings

First Floor Plan 1:200
1-Dining Room 3-Toilet 5-Laundry Room
2-Kitchen 4-Linen Room 6-Commodity Shelf

Section 1-1 1:200

Second Floor Plan 1:200
1-Guest Room
2-Bathroom

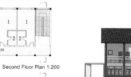

West Facade 1:200

■ Small perspective of entrance square

Third Floor Plan 1:200
1-Guest Room
2-Bathroom

First Floor Plan 1:200
1-Monitor Room 3-Conference Room
2-Duty Room 4-Office

Second Floor Plan 1:200

Section 2-2 1:200

East Facade 1:200

■ Small perspective of the renovated exhibition

■ Use of Space Under Stair

Staircases of metal

The newly added metal staircase is placed under the existing staircase to expand the space under the stairs and use the space to make shelves. The space under the stairs is divided into linen room, laundry room and shelving.

Linen room
Shelving
Laundry room
Washing machine

■ Drainage of The Same Floor

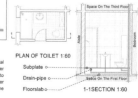

PLAN OF TOILET 1:60

Subplate
Drain-pipe
Floorslab

1-1 SECTION 1:60

In order to avoid the dining room under the second floor guest rooms, we use the bedding method in the same floor drainage technology. Pipes avoid leaks and are easier to clean.

萬物生·歸一
Living with Nature
关坝沟流域自然保护小区科学考察站4

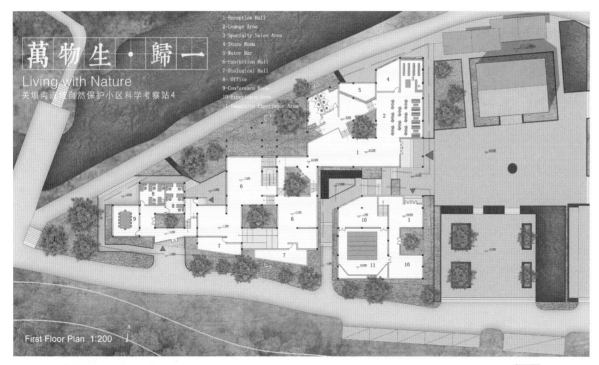

1-Reception Hall
2-Lounge Area
3-Specialty Sales Area
4-Store Room
5-Water Bar
6-Exhibition Hall
7-Biological Hall
8-Office
9-Conference Room
10-Experience Area
11-Interactive Experience Area

First Floor Plan 1:200

■ Small perspective of patio

■ Small perspective of staggered floor

■ Small perspective of interactive teaching space

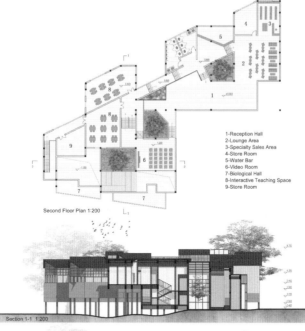

1-Reception Hall
2-Lounge Area
3-Specialty Sales Area
4-Store Room
5-Water Bar
6-Video Room
7-Biological Hall
8-Interactive Teaching Space
9-Store Room

Second Floor Plan 1:200

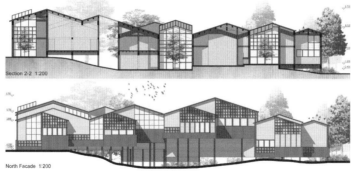

Section 1-1 1:200

Section 2-2 1:200

North Facade 1:200

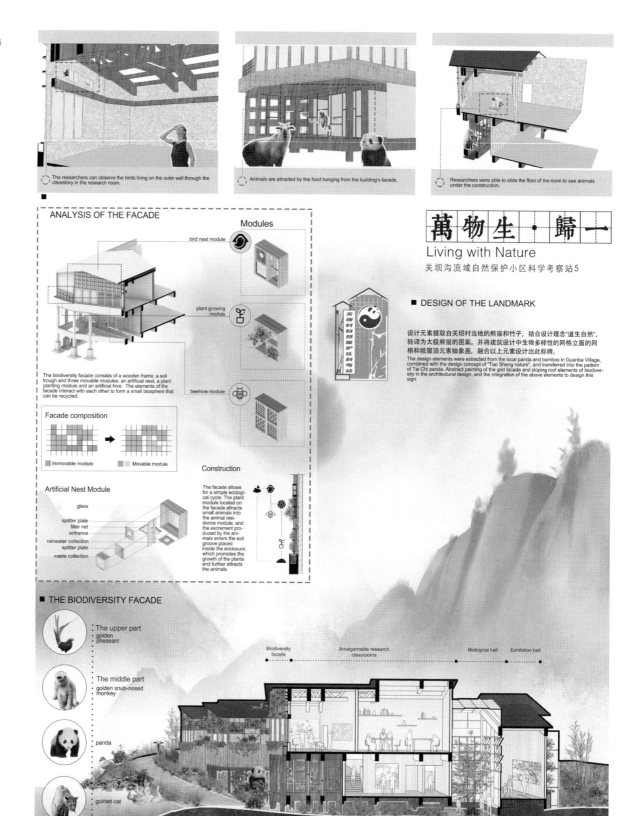

萬物生·歸一
Living with Nature
关坝沟流域自然保护小区科学考察站6

■ TARGET ANALYSIS

Carbon Emission Factor of Building Materials

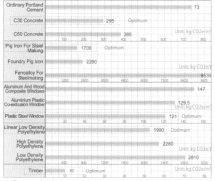

The choice of building materials mainly takes the carbon emission of building materials as the main consideration, and takes into account the stability, cost and comprehensive performance of building materials, so as to determine the more green and environmentally friendly materials. The above table mainly compares the carbon emissions of concrete, steel, window and steam insulation materials commonly used in construction. Since timber has less carbon footprint, we will use a lot of it in our buildings.

Selection of Partial Building Materials

Roofing materials we use polymer linoleum tiles, hoping to reduce the cost, reduce the construction time, enhance the waterproof performance of the building. We use straw insulation layer in the middle of the outer wall, which can be directly produced locally, greatly reducing the transportation cost and the overall cost of the building. The interior and exterior panels of floors and walls are made of cypress wood from Mianyang, Sichuan, which reduces the cost to a certain extent.

Evaluation Standard for Green Building

The Result of Assessing			
		Pre-Evaluation	Self-Evaluasion
Basic Score Of Control Item	①	400	400
Evaluation Indicators Score items	Safety And Durability ②	100	75
	Healthy And Comfortable ③	100	67
	Convenience ④	100	70
	Material Saving ⑤	200	161
	Environmentally Livable ⑥	100	73
	Improvement And Innovation ⑦	100	81
Total		1100	927
Final		110	93
Grade			★★★

According to the green building evaluation standard, we evaluated the seven scoring items (Basic Score of Control Item, Safety and Durability, Healthy and Comfortable, Convenience, Material Saving, Environmentally Livable, Improvement and Innovation) in detail and concluded that our building should be a three-star green building.

■ TECTONIC ANALYSIS

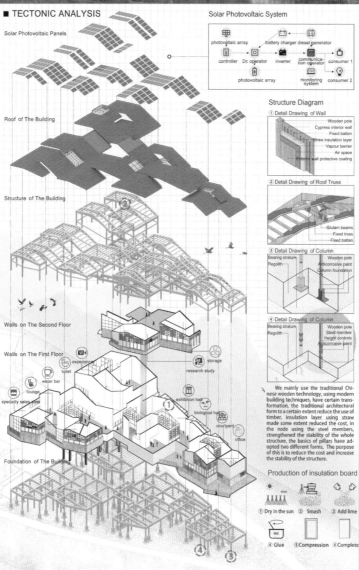

We mainly use the traditional Chinese wooden technology, using modern building techniques, have certain transformation, the traditional architectural form to a certain extent reduce the use of timber, insulation layer using straw made some extent reduced the cost, in the node using the steel members, strengthened the stability of the whole structure, the basics of pillars have adopted two different forms, The purpose of this is to reduce the cost and increase the stability of the structure.

Production of insulation board
① Dry in the sun ② Smash ③ Add lime ④ Glue ⑤ Compression ⑥ Complete

Preface of Painting Landscape 1/6
画山水序 壹

综合奖・入围奖
General Prize Awarded · Finalist Award

注　册　号：101128
项目名称：画山水序
　　　　　Preface of Painting Landscape
作　　者：卢秋羽、潘明慧、徐曾娜、
　　　　　俞嘉敏
参赛单位：南京工业大学
指导教师：郭　兰、彭克伟

Site analysis
The project site is located in Guanba Village, Mupi TibetanNationality Township, Pingwu County, Mianyang City, Sichuan Province, at latitude 39°33′18″N, longitude 104°33′58″E, close to G247 National Highway, and 4 km north of the junction of G247 National Highway and Village Entrance.

Status analysis
01. The site is located in a mountainous area with limited traffic conditions. Only a road in Mupiba Street turns into the village road to enter the site.

02. The current buildings around the site are all residential buildings, with rugged roads and scattered freedom.

Biodiversity analysis

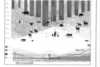
Local context analysis
Analysis of folk culture

Extraction of site features

Tree　Water　House　Mountain
Inspire

Econo-technical norms
Land use area:10709 m²
Overall floorage:2100 m²
Building floor area:1243m²
Building density:0.59
Volume fraction:0.29
Height of existing building: 13.5m
Height of new-built building :9.6m

The scenery of Guanba Village has great charm. People living here are enthusiasm and earnest. What will be happened between man and nature in the future?

Design description
通过对中国山水画的绘制进行分析与解读，学习山水画中蕴含的生态智慧与叙事手法，不断从山水画中汲取灵感源泉，寻找空间设计的生成逻辑，我们从中国山水画中受到了启发，采用自由式构图法来进行空间序列的设计，有机组合屋、廊、台、径、林、塘、溪等元素，打造出了可观、可游、可居的漫游路径；同时在散点透视的框架下，采用游动视点的绘画策略并结合资源循环、柔性生态等生态策略，设计了丰富的室内外活动空间，以此来展示多重空间叙事重合的可能性，营造诗意山水、步移景异的意境。

By analyzing and interpreting the drawing of Chinese landscape painting, learning the ecological wisdom and narrative techniques contained in landscape painting, constantly drawing inspiration from landscape painting, and looking for the generation logic of space design, we have been inspired by Chinese landscape painting. We use the free structure method to design the space sequence, organically combining the elements of houses, corridors, platforms, paths, forests, ponds, streams, and so on, to create a considerable, livable Navigable roaming path; At the same time, under the framework of scattered perspective, using the painting strategy of moving viewpoint and combining the ecological strategies such as resource circulation and flexible ecology, we designed a rich indoor and outdoor activity space to show the possibility of multiple space narrative overlap, and create a poetic landscape and an artistic conception of moving scenery.

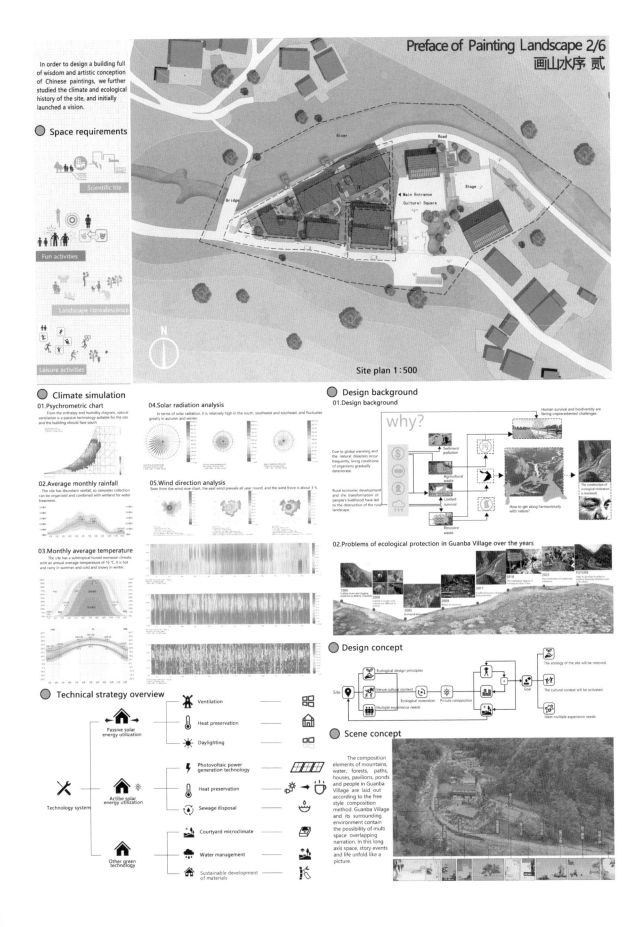

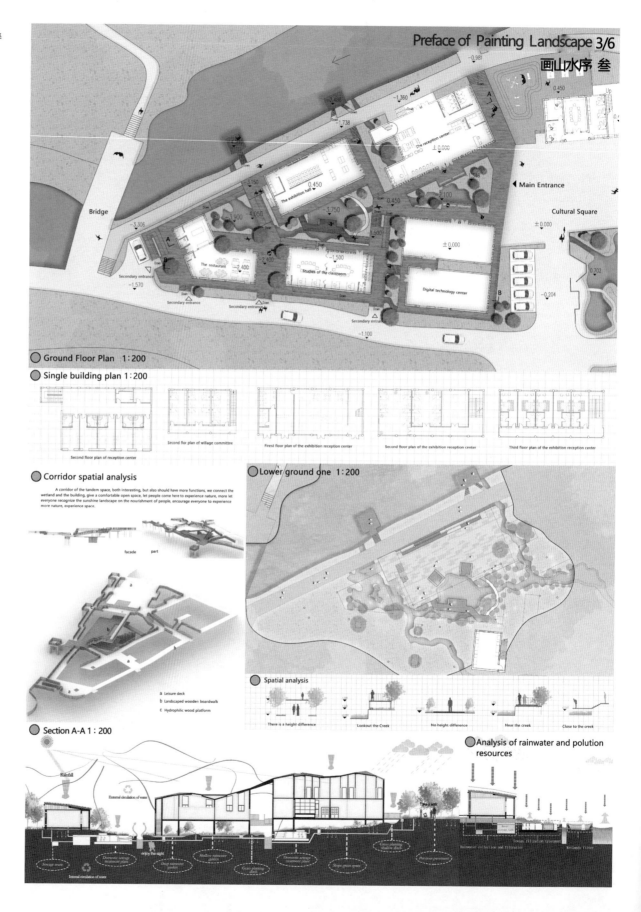

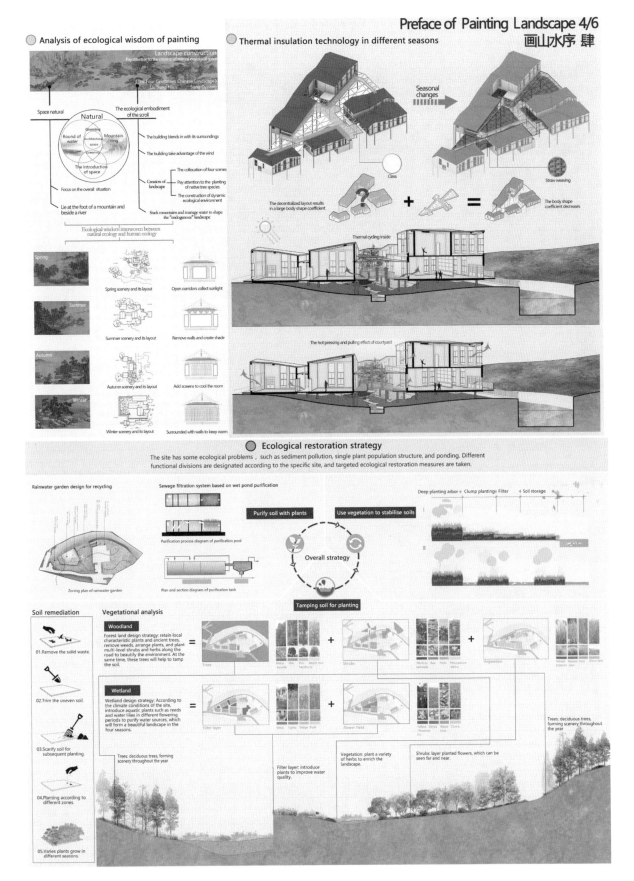

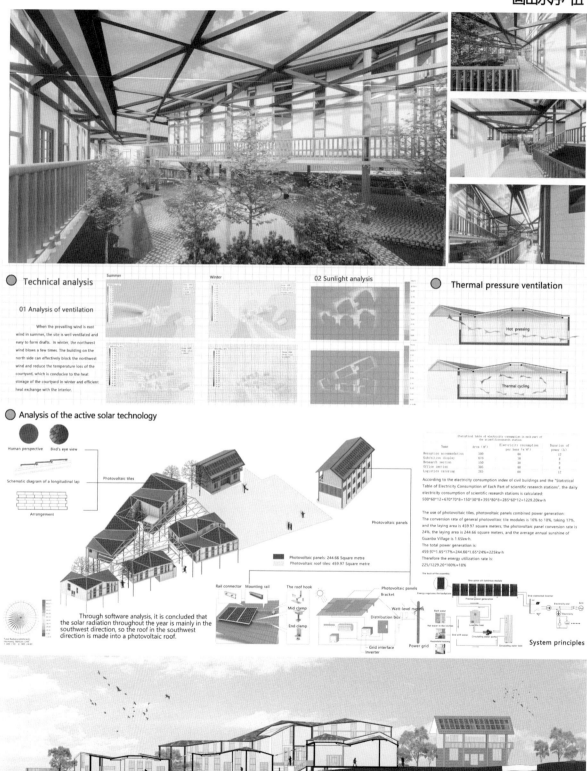

Preface of Painting Landscape 6/6
画山水序 陆

○ Evaluation of green building

○ Structural decomposition of the envelope structural
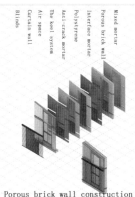

Porous brick wall construction

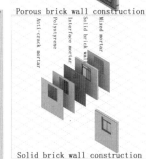

Solid brick wall construction

○ Exploding diagram of a monomer
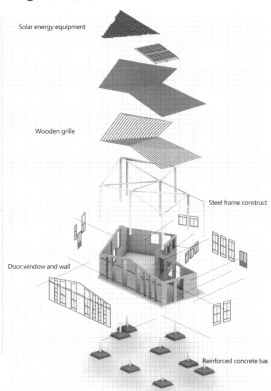

- Solar energy equipment
- Wooden grille
- Steel frame construct
- Door, window and wall
- Reinforced concrete bas

○ Landscape signage design

The sign is located at the junction of G247 National Road and Jincun Road and is 20 meters high.

The logo is based on the towering diaolou of Qiang nationality in northern Sichuan. Through the transformation of the form, solar photovoltaic power generation version and modern technology such as lighting are inserted, which makes the logo have local characteristics, striking and unique.

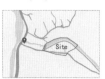

The location of the identity

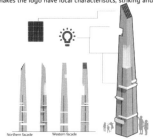

Northern facade   Western facade
Southern facade   Eastern facade

○ Analysis of traditional houses

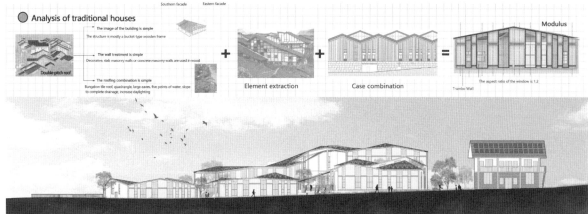

- The image of the building is simple
- The structure is mostly a bucket-type wooden frame
- Double-pitch roof
- The wall treatment is simple
  Decorative, slab masonry walls or concrete masonry walls are used in wood
- The roofing combination is simple
  Bungalow tile roof, quadrangle, large eaves, five points of water, slope to complete drainage, increase daylighting

Element extraction + Case combination = Modulus

The aspect ratio of the window is 1:2
Trumbo Wall

综合奖・入围奖
General Prize Awarded・Finalist Award

注 册 号：101158
项目名称：风游光塔
　　　　　Roaming in the Sun Tower
作　　者：田泽轩、吴　月、唐梓榕、
　　　　　赵一凡、孙翊宸
参赛单位：西南民族大学
指导教师：熊健吾、王晓亮、毛　刚

风游光塔 -1
Roaming in the Sun Tower

Economic and technical indicators
planned land area: 10709m²
building land area: 2293.76m²
total building area: 2100m²
building area: 1895.23m²
building density: 13.7%
base area: 2293.96m²
floor area ratio: 0.65
greening rate: 37.8%

Design Description
The theme of this scheme is "Roaming in the Sun Tower", which is used to deliver heat by wind and gather light into a tower. Taking passive solar energy technology as the starting point of design, combined with the local Tibetan traditional residential language, local buildings are raised to form a heating chamber, which is connected by a double-layer air duct. A warm corridor is arranged on the south side, and a heat storage tank is implanted, giving consideration to the day heat storage and night heat release, moisture-proof ventilation and room insulation, so as to fully realize building energy conservation. At the same time, in terms of active technology, solar photovoltaic system and building roof are used for integrated design. In addition, rainwater garden and reclaimed water system are introduced to play the role of water circulation together with the regulation and storage tank.

本方案主题为"风游光塔"，乘风送热，汇光成塔。以被动式太阳能技术为设计的出发点，结合当地藏族传统的民居语言，利用局部建筑升高形成加热腔体，以双层风道贯通，南侧布置暖廊，并植入蓄热水池，兼顾日间得热蓄能与夜晚放热供能、防潮通风与房间保温，充分实现建筑节能。同时，在主动技术上，利用太阳能光伏系统与建筑屋顶进行一体化设计。此外，引入雨水花园及中水系统，与调蓄池共同发挥水循环的作用。

Location
The project is located in Guanba Village, Mupi Tibetan Township, Pingwu County, Mianyang City, Sichuan Province. Guanba Village is located in the Nature Reserve of Giant Panda Habitat in Minshan District of Giant Panda National Park, and is the Sichuan Forest Nature Education Base and Guanbagou Valley Nature Reserve. The village is an important ecological hinterland and water conservation area in Mianyang City.

Diagram of Design Process

Region Feature

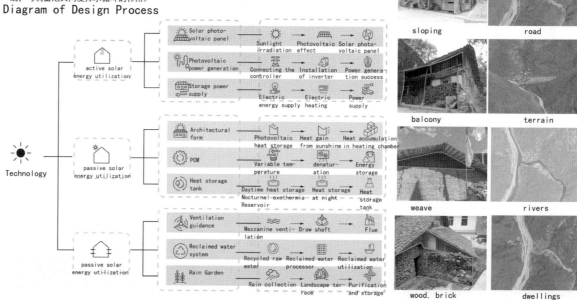

sloping　　road
balcony　　terrain
weave　　rivers
wood, brick　　dwellings

风游光塔-2
Roaming in the Sun Tower

Climate

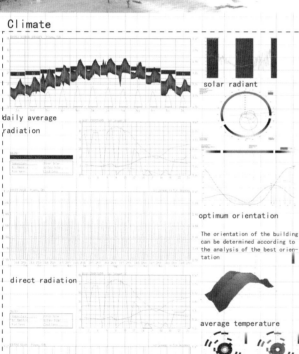

daily average radiation

direct radiation

diffuse radiation

According to direct radiation, indirect radiation and daily average radiation, combined with active and passive solar energy utilization, analyze the configuration of solar photovoltaic panels.

solar radiant

optimum orientation

The orientation of the building can be determined according to the analysis of the best orientation

average temperature

wind environment analysis

Design ventilation interlayer and flue to optimize building wind environment with natural ventilation.

Site Plan 1:800

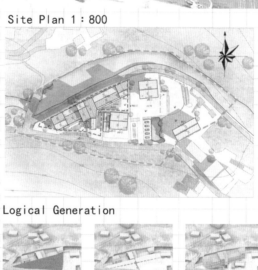

Logical Generation

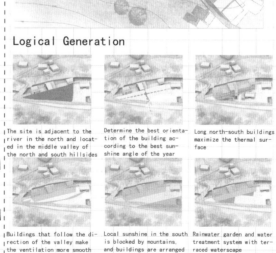

The site is adjacent to the river in the north and located in the middle valley of the north and south hillsides

Determine the best orientation of the building according to the best sunshine angle of the year

Long north-south buildings maximize the thermal surface

Buildings that follow the direction of the valley make the ventilation more smooth

Local sunshine in the south is blocked by mountains, and buildings are arranged near the north

Rainwater garden and water treatment system with terraced waterscape

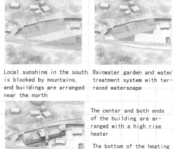

The center and both ends of the building are arranged with a high rise heater

The bottom of the heating chamber is introduced to the north water body for heat storage

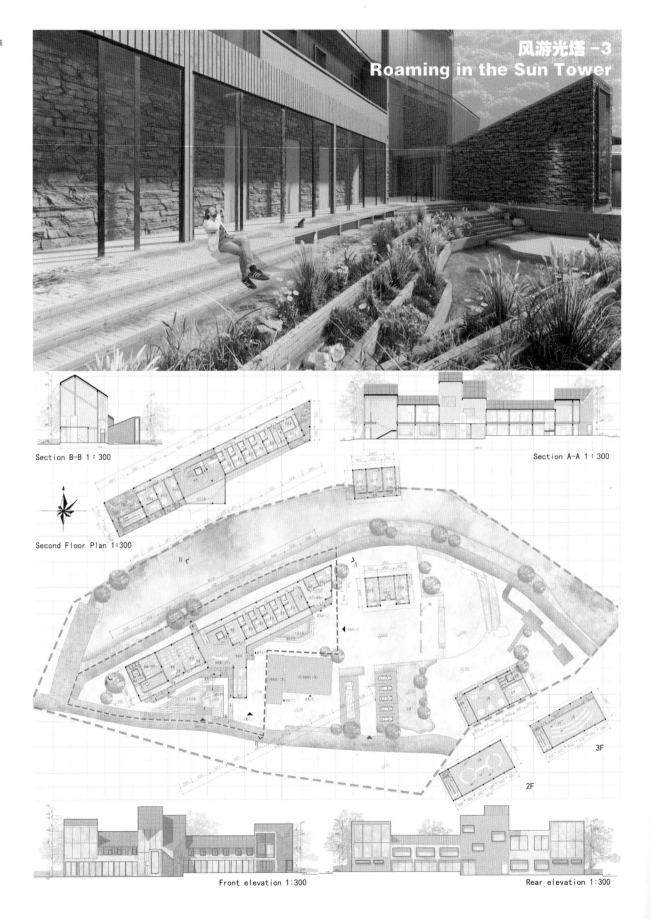

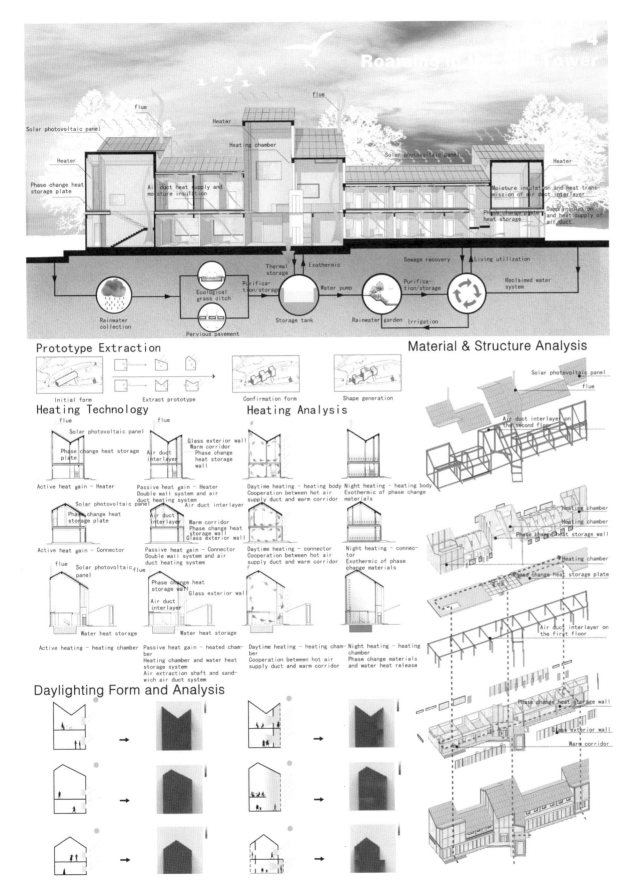

风游光塔 -5
Roaming in the Sun Tower

Thermal storage during the daytime

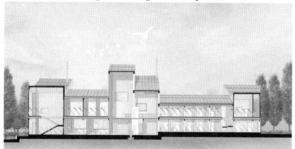

Heat release and energy use at night

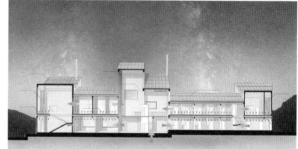

Detail drawing

Activity space

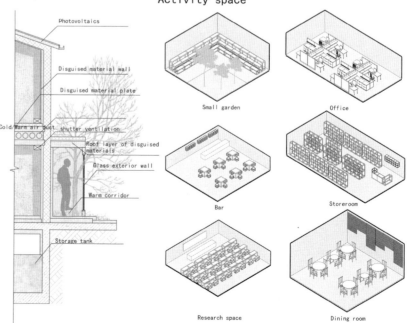

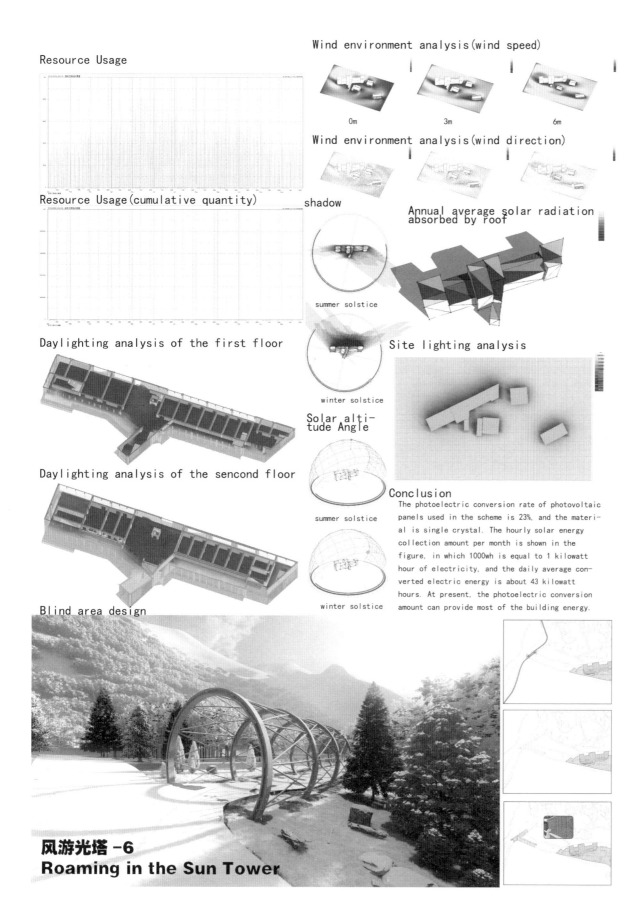

Roaming in the Sun Tower

风游光塔 -6

Conclusion

The photoelectric conversion rate of photovoltaic panels used in the scheme is 23%, and the material is single crystal. The hourly solar energy collection amount per month is shown in the figure, in which 1000wh is equal to 1 kilowatt hour of electricity, and the daily average converted electric energy is about 43 kilowatt hours. At present, the photoelectric conversion amount can provide most of the building energy.

综合奖·入围奖
General Prize Awarded · Finalist Award

注　册　号：101178
项目名称：檐下丘语
　　　　　Hill Language under the Roof
作　　者：王兰、李佳、于洽卿、
　　　　　孙剑、张文馨、马炳蔚
参赛单位：山东建筑大学
指导教师：薛一冰、杨倩苗

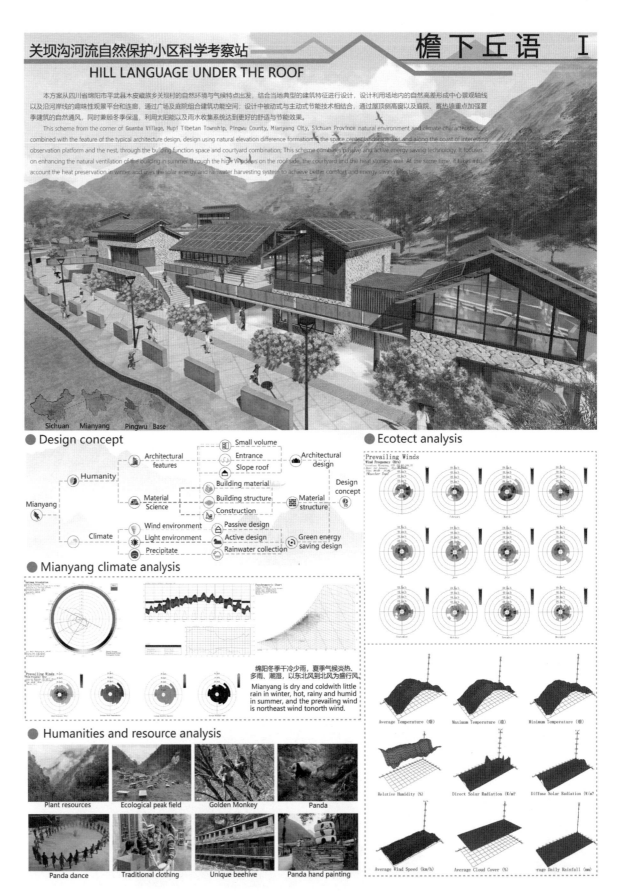

关坝沟河流自然保护小区科学考察站

檐下丘语 II

HILL LANGUAGE UNDER THE ROOF

2022 台达杯国际太阳能建筑设计竞赛获奖作品集

● Site plan 1:500

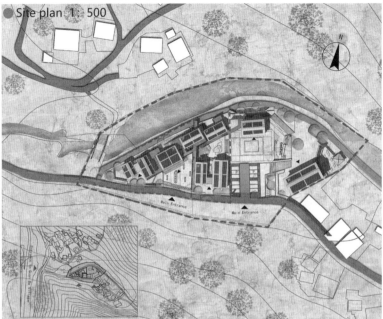

● Base analysis

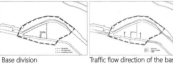

Base division

Traffic flow direction of the base

Dominant wind direction of the base

Best orientation

● Site design concept

The site covers an area of 10709 m² and is irregular in shape.

Functional blocks are divided according to the current situation and conditions of the site.

Basic building blocks are formed according to existing buildings and wind environment conditions of the site.

The building blocks are divided according to the topographic height difference.

Make reasonable green layout.

The sloping roof is designed according to the surrounding environment.

The buildings are connected through the observation corridor.

Create a landscape axis.

● Design elements

Sloping roof | Photovoltaic system | Local timber | Local stone

Planned land area	10709㎡
Site area of Phase I	2294㎡
Site area of Phase II	568㎡
overall floorage	2100㎡
Total Floor space	1268㎡
Building area	2203㎡
Floor area ratio	0.2
Density of building	11.8%
Greening rate	38.4%
Coach parking	1
Car parking	5

● Site environment analysis

Wind speed at -1.8m

Wind speed at 0.2m

Wind speed at 1.5m

Wind speed at 3.2m

Natural lighting

Solar radiation

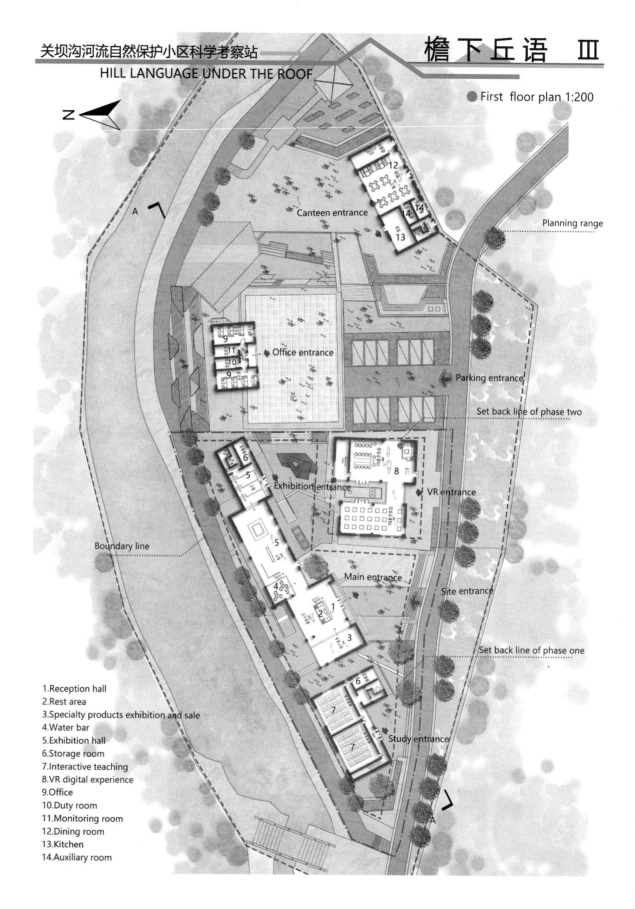

檐下丘语 IV

关坝沟河流自然保护小区科学考察站
HILL LANGUAGE UNDER THE ROOF

Cultural square

Residential square

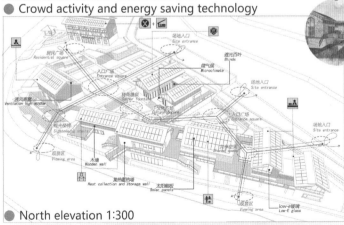
sightseeing platform

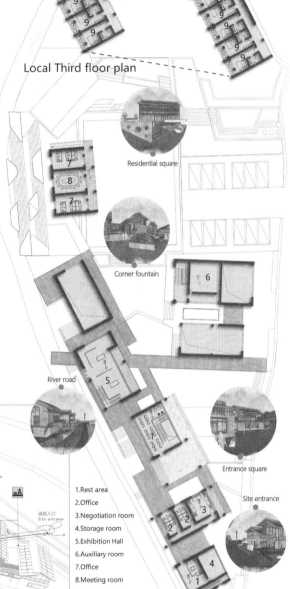

Second floor plan 1:200

Local Third floor plan

Residential square

Corner fountain

River road

Entrance square

Site entrance

1. Rest area
2. Office
3. Negotiation room
4. Storage room
5. Exhibition Hall
6. Auxiliary room
7. Office
8. Meeting room
9. Guest room

● Crowd activity and energy saving technology

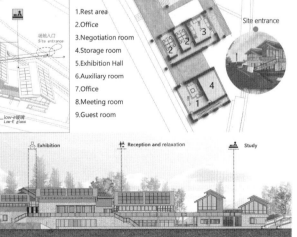

● North elevation 1:300

关坝沟河流自然保护小区科学考察站 檐下丘语 V

HILL LANGUAGE UNDER THE ROOF

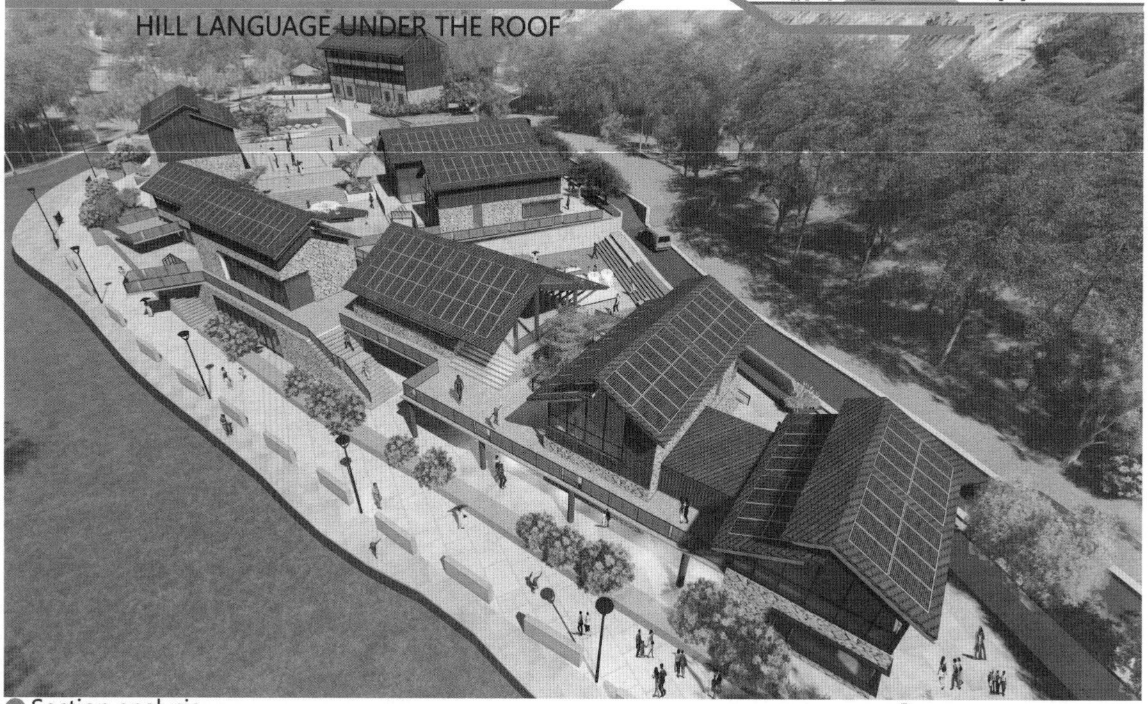

Section analysis

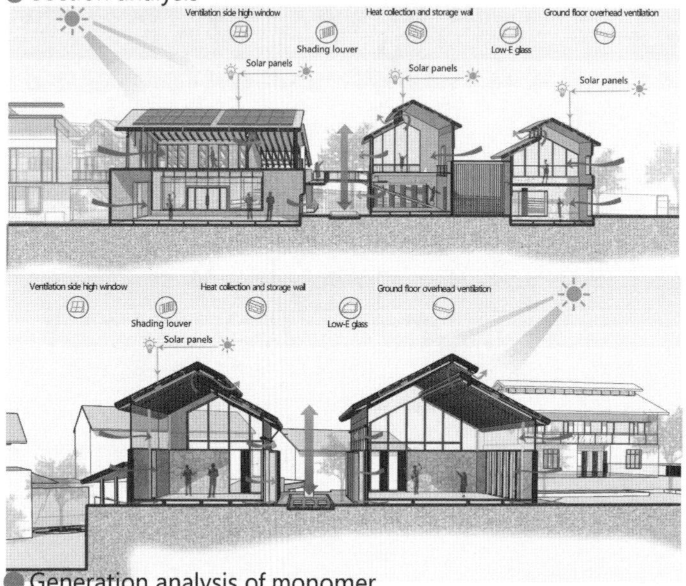

Ventilation side high window
Heat collection and storage wall
Ground floor overhead ventilation
Shading louver
Low-E glass
Solar panels
Solar panels
Solar panels
Ventilation side high window
Heat collection and storage wall
Ground floor overhead ventilation
Shading louver
Low-E glass
Solar panels

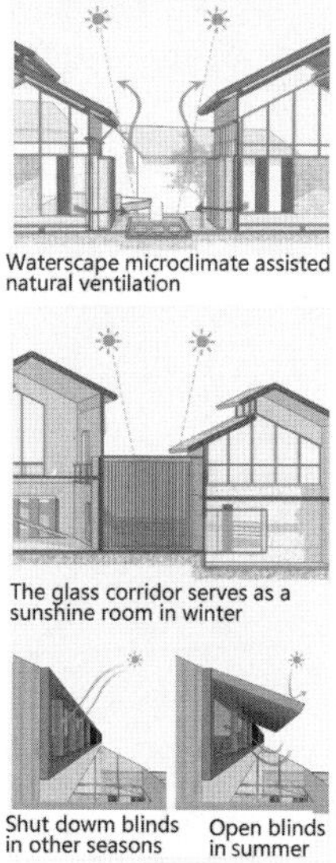

Waterscape microclimate assisted natural ventilation
The glass corridor serves as a sunshine room in winter
Shut dowm blinds in other seasons
Open blinds in summer

Tecnonic node

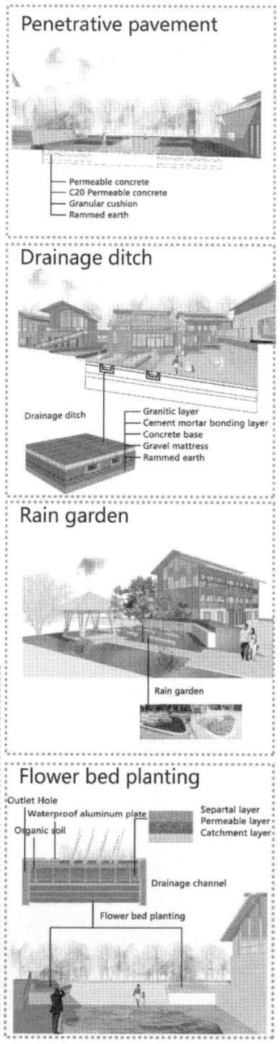

Penetrative pavement
Permeable concrete
C20 Permeable concrete
Granular cushion
Rammed earth
Drainage ditch
Drainage ditch
Granitic layer
Cement mortar bonding layer
Concrete base
Gravel mattress
Rammed earth
Rain garden
Rain garden
Flower bed planting
Outlet Hole
Waterproof aluminum plate
Organic soil
Separtal layer
Permeable layer
Catchment layer
Drainage channel
Flower bed planting

Generation analysis of monomer

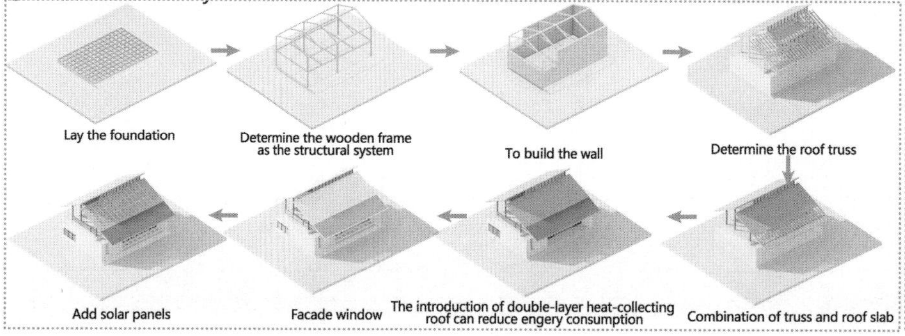

Lay the foundation
Determine the wooden frame as the structural system
To build the wall
Determine the roof truss
Add solar panels
Facade window
The introduction of double-layer heat-collecting roof can reduce engery consumption
Combination of truss and roof slab

檐下丘语 VI
HILL LANGUAGE UNDER THE ROOF

关坝沟河流自然保护小区科学考察站

2022 台达杯国际太阳能建筑设计竞赛获奖作品集

Section strategy

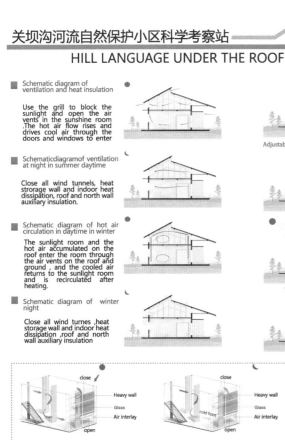

- **Schematic diagram of ventilation and heat insulation**
  Use the grill to block the sunlight and open the air vents in the sunshine room. The hot air flow rises and drives cool air through the doors and windows to enter.

- **Schematic diagram of ventilation at night in summer daytime**
  Close all wind tunnels, heat storage wall and indoor heat dissipation, roof and north wall auxiliary insulation.

- **Schematic diagram of hot air circulation in daytime in winter**
  The sunlight room and the hot air accumulated on the roof enter the room through the air vents on the roof and ground, and the cooled air returns to the sunlight room and is recirculated after heating.

- **Schematic diagram of winter night**
  Close all wind turnes, heat storage wall and indoor heat dissipation, roof and north wall auxiliary insulation.

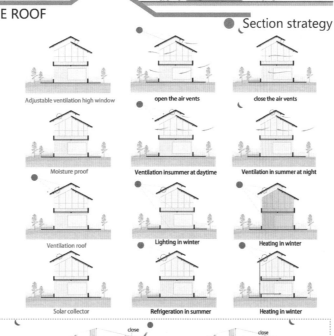

Technique

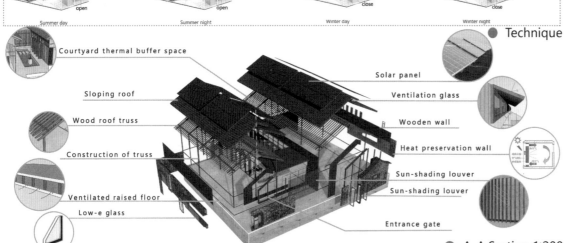

A-A Section 1:200

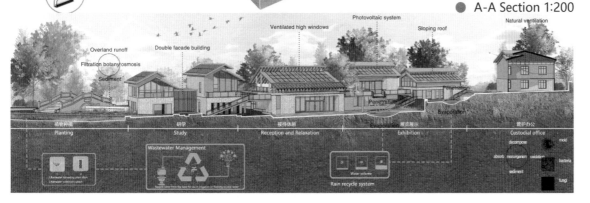

综合奖・入围奖
General Prize Awarded・Finalist Award

注 册 号：101181
项目名称：山山而川・生生不息
　　　　　The Circle of Life
作　　者：戴滢婕、关怡平、肖　霄
参赛单位：南京工业大学
指导教师：刘　强、薛　洁

山山而川・生生不息・壹
The Circle of Life

Economic Indicators
Land Area: 10709 m²
Building Area: 2130 m²
Building Density: 10%
Floor Area Ratio: 0.20

Design Specification

本次设计将生态和文化作为切入点，旨在保护生物多样性、延续白马藏族文化。

建筑造型在借鉴当地传统民居的基础上，加以创新，探索坡屋顶组合的多种可能，同时穿插平屋顶，丰富建筑造型。为保护场地原有树木，建筑置入了多个有趣的庭院空间。

方案采用主动式与被动式相结合的节能方式，由于当地的气候湿润，太阳辐射较弱，因此重点考虑建筑的保温蓄热与通风，西向屋顶安装太阳能板，捕捉有限的自然光照，从而达到建筑生态、绿色、节能的目的。

The design takes ecology and culture as the entry point, aiming to preserve biodiversity and perpetuate the Tibetan culture of the White Horse.

The building's form is based on the traditional dwellings of the area, but with an innovative twist. The combination of sloping roofs is explored, while flat roofs are interspersed to enrich the building's form. Interesting courtyard spaces have been incorporated to protect the original trees on the site.

The scheme adopts a combination of active and passive techniques. Due to the humid climate and weak solar radiation in the area, the focus is on thermal insulation and ventilation, installing solar panels on the west-facing roof to capture the limited natural light, thus achieving an ecological, green and energy-saving building.

Climatic Simulation

Wind rose diagram

Dry Bulb Temperture

Dew Point Temperture

Realitive Humidity

Analysis Conclusion

· Wind speeds are generally low in Mianyang. Northeast to north winds are the prevailing winds.
· The air here is humid and rainfall is abundant.
· The site is located in a valley, surrounded by rugged terrain with mountains.

Enthalpy diagram

Terrain Analysis
Site
① Elevation
② Slope
③ Concavity
④ Roughness

Site Analysis

The site is located in a valley. It has a good ecological environment and is inhabited by Baima Tibetan residents.

Natural environment | Peripheral buildings | Land development planning

Diagram of Design Process

Design = Protecet the nature + Local culture + Solar technology

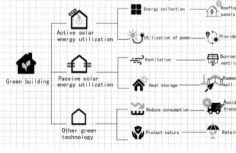

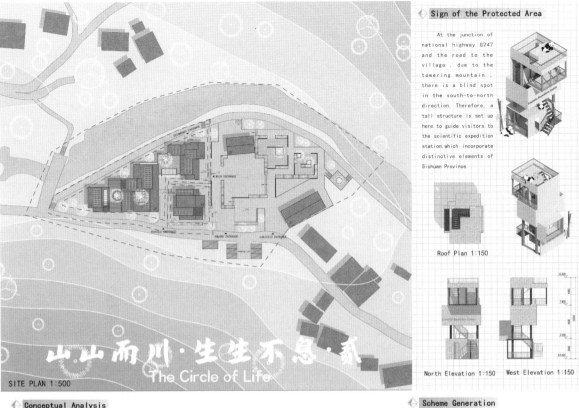

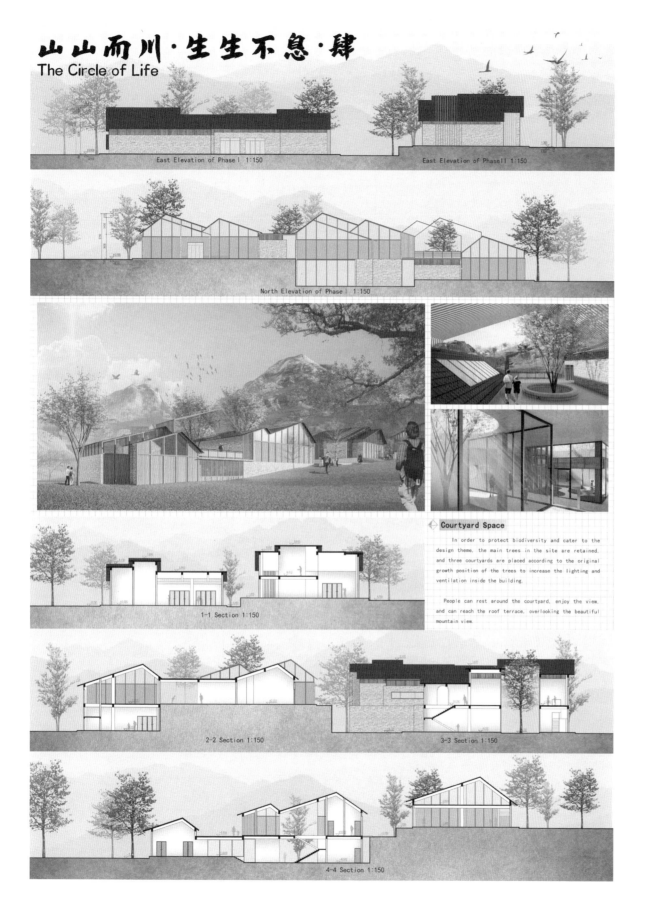

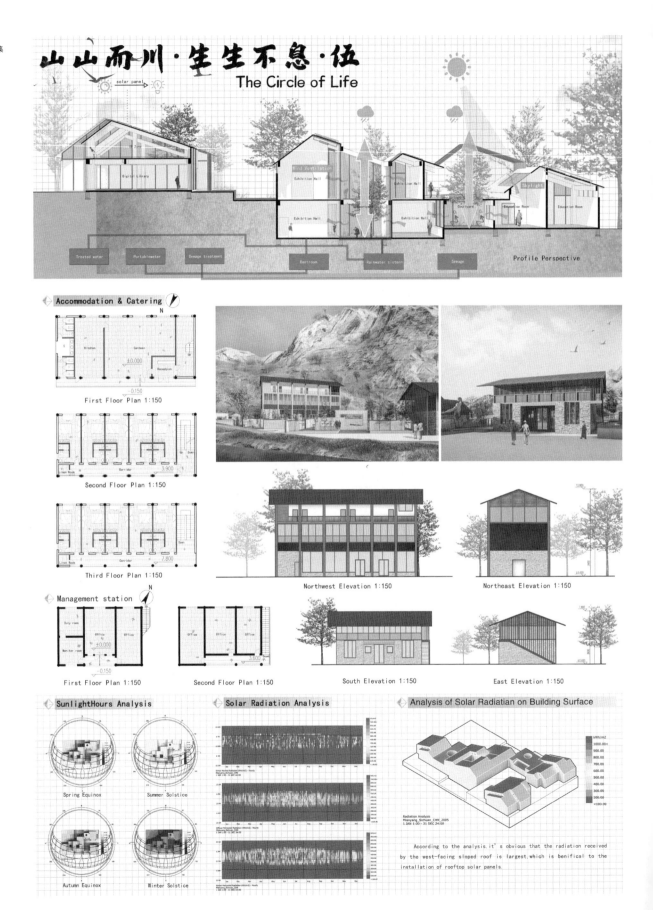

山山而川·生生不息·陆
The Circle of Life

Wind Simulation

Relative Elevation: -2.0m   0.0m   2.0m

The valley wind blows from west to east towards the base, the wind on the south side of the base is blocked by the high mountains and the wind speed is low. At the same time, the wind speed increases with the increase in altitude.

Rammed Earth Process

Dig out the base groove | Add the soil | Set a mold | Add the soil | Ram the earth

Ramming complete | Change the position | Ram the earth | Ramming complete | Complete

Rammed Earth Analysis

rall soil material

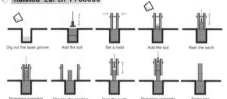

earth : crushed stone : sand = 5 : 3 : 2

Rammed earth wall / Mixture compacted / Rammed earth foundation / Earth / Baseboard / Floor / Separated layer / Insulating layer

Rammed earth wall / Compacted soil (sprinkle water) / Compacted soil

Passive and Active Technique Analysis

Natural lighting | Pressure ventilation

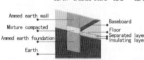

Natural thermal insulation | Natural ventilation

Attached sunspace and courtyard | The solar panel

Evaluation Standard for Greening Building

Evaluation results of green buildings		
Project	Standard evaluation	Evaluation score
Control basic score	400	400
Safety and durability	100	80
Healthy and comfortable	100	75
Life convenience	100	60
Resource savings	200	160
Livable environment	100	81
Improvement and innovation	100	85
Total	110	94
Rank		Three-star

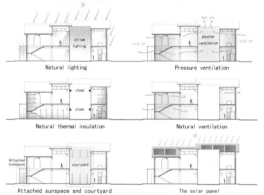

Green building——Public building score calculation					
Evaluation indicator	Energy-saving outdoor environment	Energy conservation energy utilization	Water saving resources utilization	Material saving resources utilization	Indoor environmental quality
Serial number	1	2	3	4	5
Control Result	comply √	comply √	comply √	comply √	comply √
Project Ratio Wi	0.16	0.28	0.18	0.19	0.19
Applicable score	100	100	100	100	100
Actual score	80	75	90	85	85
Score Qi	80	75	90	85	85
Weighted score	12.8	21	16.2	16.15	16.15
Extra point			9		
Total			91		

Exploded Views

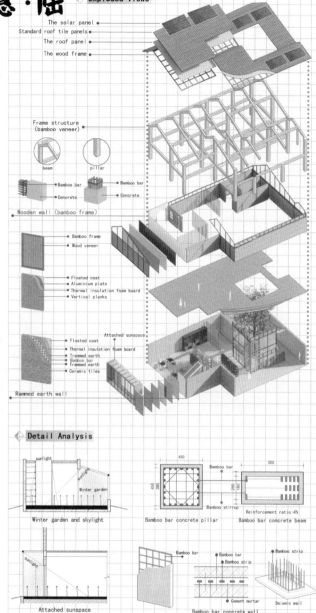

The solar panel / Standard roof tile panels / The roof panel / The wood frame

Frame structure (bamboo veneer) / beam / pillar

Bamboo bar / Concrete / Bamboo bar / Concrete

Wooden wall (bamboo frame)

Bamboo frame / Wood veneer

Floated coat / Aluminium plate / Thermal insulation foam board / Vertical planks

Floated coat / Thermal insulation foam board / Bamboo frame / Rammed earth / Ceramic tiles / Attached sunspace

Rammed earth wall

Detail Analysis

Winter garden and skylight | Bamboo bar concrete pillar | Bamboo bar concrete beam

Attached sunspace | Bamboo bar concrete wall | Seismic wall

The External Elevation and Architectural Form Details Analysis

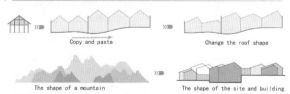

Copy and paste → Change the roof shape

The shape of a mountain → The shape of the site and building

The Roof Form Analysis

 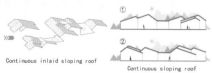

Continuous inlaid sloping roof | Continuous sloping roof

2022 台达杯国际太阳能建筑设计竞赛获奖作品集

综合奖 · 入围奖
General Prize Awarded · Finalist Award

注　册　号：101182
项目名称：山与火之歌
　　　　　A Song of Mountain and Fire
作　　者：霍　然、张小曼、尹　航、
　　　　　邓晓颖
参赛单位：山东建筑大学
指导教师：侯世荣、杨倩苗

山与火之歌 I
A Song of Mountain and Fire

DESCRIPTION OF DESIGN

本次方案设计是从白马藏族人传统生活中频繁出现的"火塘"为线索，深入挖掘了白马藏族人的传统习俗与文化内涵，在仔细研究了基地的地形地貌以及当地人民对于民居的传统习俗后，选定围合式的建筑布局来表现"火塘"这个传统空间和交流属性。基地中心下挖，且顺应地势逐层下沉，整体形成了回环流畅的布局流线。由于一些限制，建筑单体分布较为分散，为了整合形体，我们采用了起伏的坡屋顶，一是呼应传统民居中的坡屋顶元素，二是起伏的形态呼应白马藏族人对山川的崇拜；另外，环形的屋顶又进一步加强了围合的形态。采用夯土、杉木板等传统材料，使得整个方案良好地嵌入自然环境而不会使人有突兀之感。

The scheme design is frequent the Baima Tibetan traditional life "fireplace" as the clue, dig the Ba--ima Tibetan tradition and culture connotation, after carefully study the topography of the base and the local people for residential traditions, selected use round shaped building layout to e--xpress the "measures" of the traditional space to surround close and exchange properties. Dig under the center of the base, and comply with the terrain layer by layer, the overall formation of a loop smooth layout streamline. Due to some lim--itations, the individual buildings are scattered. In order to integrate the shapes, we adopt the ro--lling slope roof, which echoes the elements of the slope roof in the traditional residential houses, and the rolling form echoes the worship of the White Horse Tibetans to the mountains and rivers. In addition, the circular roof further strengthens the enclosed form. Using traditional materials su--ch as rammed earth and fir boards, the scheme is well embedded in the natural environment wi--thout being obtrusive.

CLIMATE ANALYSIS

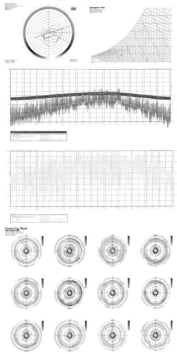

山与火之歌 II
A Song of Mountain and Fire

LOCATION ANALYSIS

The base is located in Mupi Tibetan Township, Pingwu County, Mianyang City, Sichuan Province, at the junction of multiple geographical boundaries in China, and in the transition zone between the Qinghai Tibet Plateau and Sichuan Basin.

CONCEPTUAL PRODUCTION

Mountain God Worship — 山

The primitive way of life has formed Baima people's worship of nature, worship of mountains and rivers, and also bred Baima people's primitive national culture.

Fire Circle Dance — 火

The festival starts on the eighth day of the twelfth lunar month. Every night, everyone in the village gathers firewood, heats up a fire, sings and dances in the fire circle.

Panda Dance — 灵

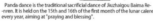

Panda dance is the traditional sacrificial dance of Jiuzhaigou Baima Renren. It is held on the 15th and 16th of the first month of the lunar calendar every year, aiming at "praying and blessing".

ECOLOGICAL RESOURCES

Panda — Ginkgo

Golden Monkey — Cycas / Handkerchief tree

Takin — Cercidiphyllum japonicum / Eucommia ulmoides

TRADITIONAL MATERIALS

Cedar board

Bamboo woven earth wall

Bamboo wall

Rubble foundation

BASE CHARACTERISTICS

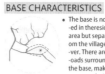

- The base is not located in the residential area but separated from the village by a river. There are two roads surrounding the base, making it connected with the outside world.

- The main house is generally oriented towards the upper part of the river, with its back facing the lower part of the river and its seat facing east and west. That is, the left wing of the Han house is taken as the main house.

- The project land is located in the valley from east to south to west to north. There is water flowing from east to west in the stream at the bottom of the valley all the year round. The village site belongs to the piedmont valley type.

Satellite map of the base

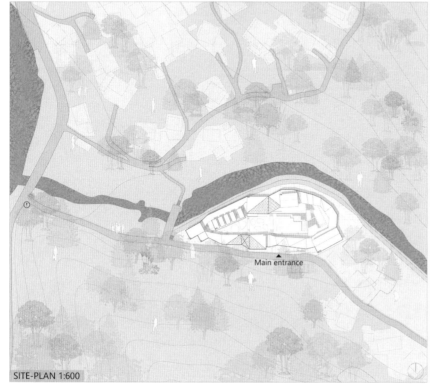

SITE-PLAN 1:600

Main entrance

山与火之歌 III
A Song of Mountain and Fire

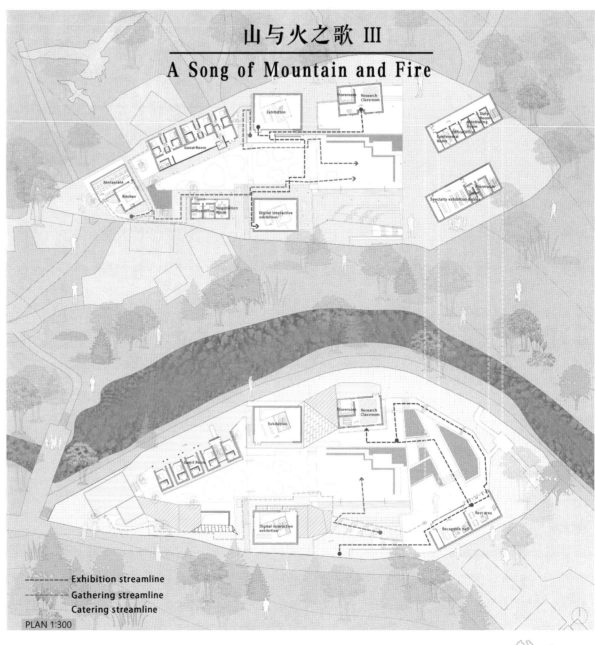

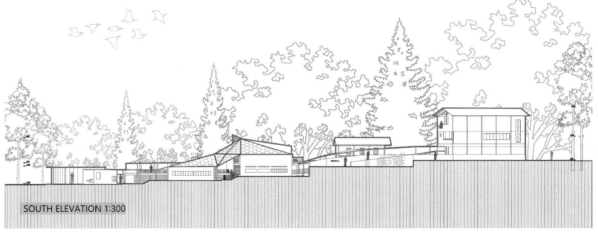

山与火之歌 IV
A Song of Mountain and Fire

FUNCTIONAL ANALYSIS

Entrance reception area · Exhibition research area · Accommodation

DESIGN DEVELOPMENT

The original base conditions are relatively primitive, with large elevation difference.

Arrange the terrain to form a sinking state layer by layer.

Pull up the building volume around to enclose the middle site.

Determine the volume and form of single building as required.

Determine the annular roof according to the building trend.

Add details to form the final plan.

PROFILE ANALYSYS

Depending on the elevation difference, the undulating roof form is formed, which is combined with the indoor space to form a rich spatial experience. Staggered relationship, promoting.

SECTION 1:300

山与火之歌 V
A Song of Mountain and Fire

TRUMBERT WALL

The Trumbert wall is a passive system, which is heated by the sun. There is a layer of air several tens of centimeters thick between the glass on the sunny side and the heat collecting wall. In different seasons, the opening switch on the glass and the wall can achieve different performance gains.

Summer

Transition season

Winter

A standard Trumbo wall places the glass about 2 to 5 cm away from a 10-41 cm thick masonry wall. Wall are often made of brick, stone or concrete. The sun's heat passes through the glass, is absorbed by the heat storage walls, and is slowly released into the interior of the house.

Winter day
Summer day
Winter night
Summer night

- Winter night
- Hot air
- Sun rays
- Heat gains/loss

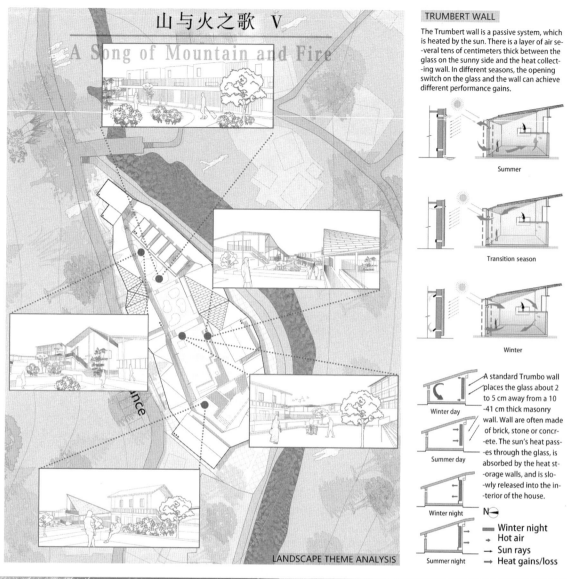

LANDSCAPE THEME ANALYSIS

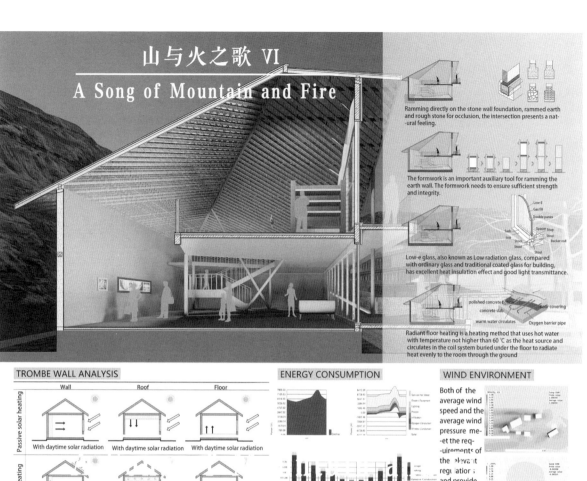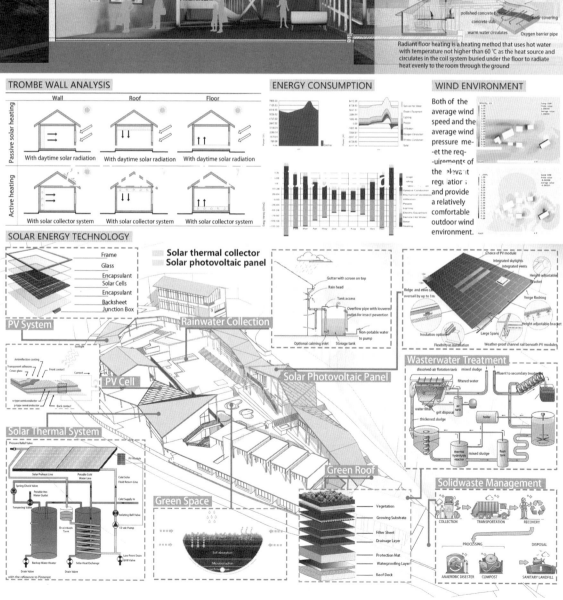

综合奖·入围奖
General Prize Awarded · Finalist Award

注 册 号：101216
项目名称：山谷间的呼吸
　　　　　Breathing in the Valley
作　者：费　凡、罗明宇、叶　清、
　　　　程子瀚
参赛单位：合肥工业大学
指导教师：李　早、王　旭

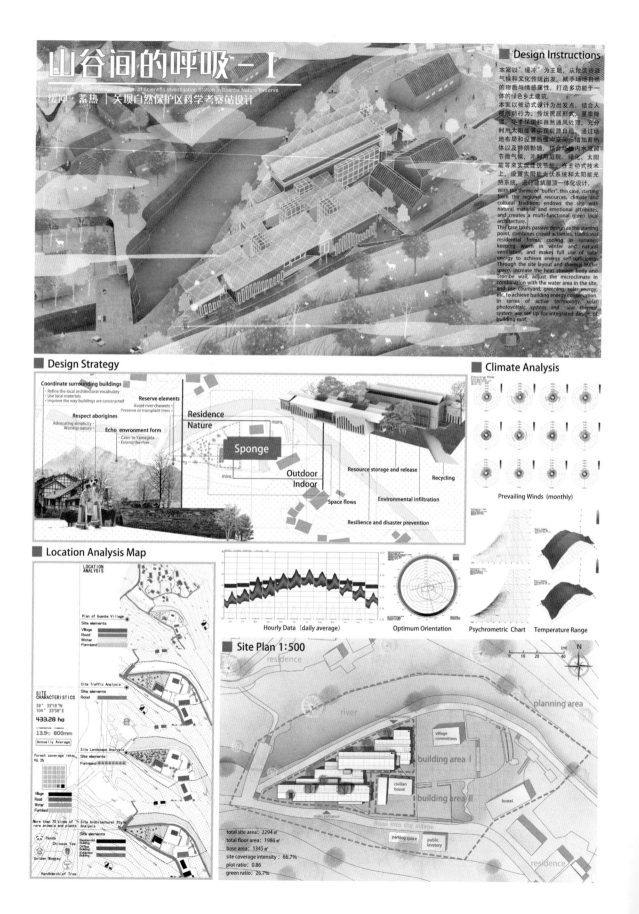

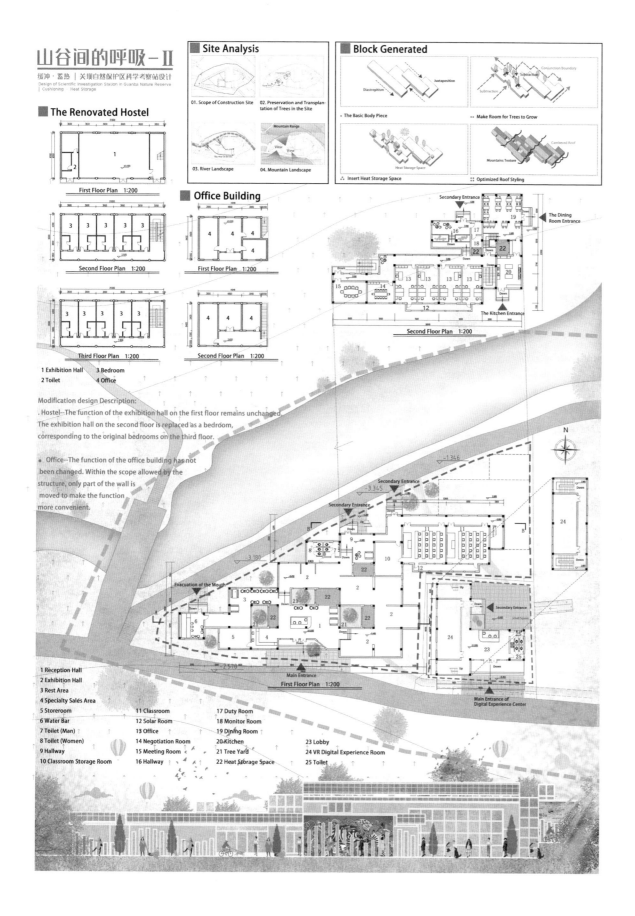

山谷间的呼吸 – III

缓冲·蓄热 | 关坝自然保护区科学考察站设计

Cushioning | Heat Storage | Design of Scientific Investigation Station in Guanba Nature Reserve

8:00-spring-sunny
12:00-summer-cloudy
16:00-autumn-rainy
20:00-winter-snowy

■ Crowd Behavior Analysis

GROUP STRUCTURE
The area is a typical ethnic minority settlement, which will attract more tourists and researchers in the future.
- researcher 15%
- tourist 15%
- aborigine 75%

Age composition
The area has a large elderly population, which will attract more young people in the future.
- >60 years old 22%
- <6 years old 6%
- 6-12 years old 12%
- 12-21 years old 15%
- 30-60 years old 45%

CROWD ACTIVITIES
Aborigine
Tourist
Researcher
0:00 3:00 6:00 9:00 12:00 15:00 18:00 21:00 24:00

PEOPLE DEMAND
Present Situation: Provide the functions of scientific research and demonstration, reception and accommodation.
Demand

Present Situation: Improve the infrastructure in the site and provide passive energy supply.
Demand

Present Situation: Use solar photovoltaic, photothermal and active solar energy technologies.
Demand — ECO

Present Situation: Inherit the local traditional cultural characteristics and echo the panda protection base.
Demand

■ Wall Construction Details

winter daytime heating

winter night heating

summer daytime heating

summer night heating

Tromb Wall With Solar Room

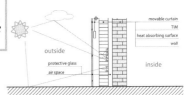

Transparent Insulated Wall

■ Roof Improve Strategy

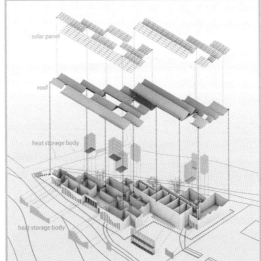

- solar panel
- roof
- heat storage body
- heat storage body

normal roof
curved surface
Increase south-facing surface

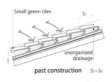
Small green tiles
unorganized drainage
past construction S1 < S2

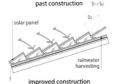
solar panel
rainwater harvesting
improved construction

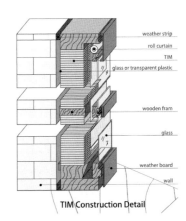

- weather strip
- roll curtain
- TIM
- glass or transparent plastic
- wooden fram
- glass
- weather board
- wall

TIM Construction Detail

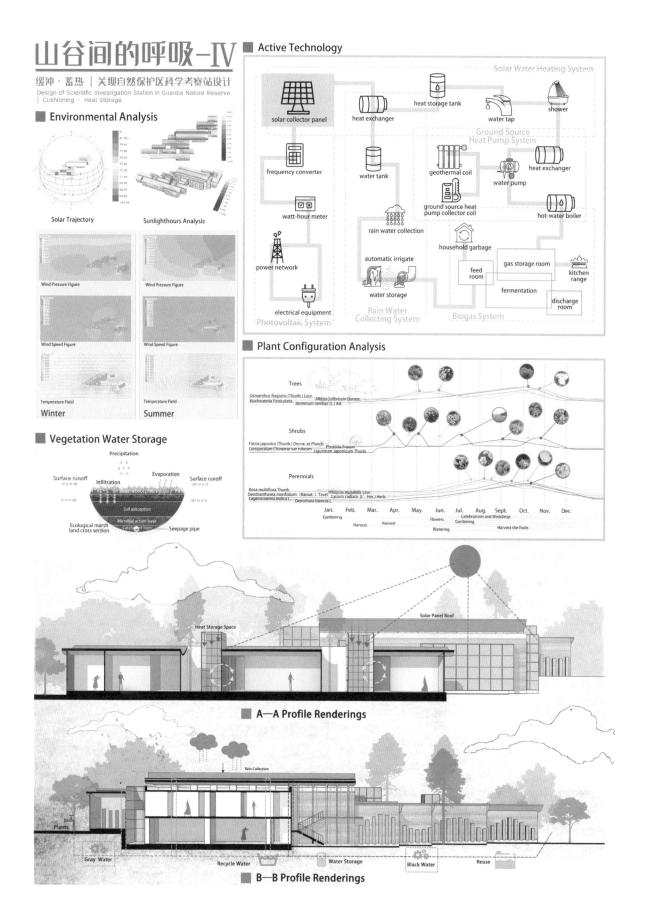

综合奖·入围奖
General Prize Awarded · Finalist Award

注 册 号：101237
项目名称：归巢来兮
　　　　　Back to Nature
作　　者：陈璐倩、王　可、韦寒雨、
　　　　　杨　宇
参赛单位：南京工业大学
指导教师：罗　靖、刘晓光

归巢来兮 Back to Nature

四川省平武县关坝沟流域自然保护小区科学考察站
Scientific research station of Guanbagou Watershed Nature Reserve in Pingwu County, Sichuan Province

The project is located in the Guanba Village Nature Reserve, which is an important ecological hinterland and water conservation area of Mianyang City. Based on the principle of low-profile intervention in the ecological environment, we found that many biological features have Tyson polygon patterns, such as the pattern of dragonfly wings, dried land, leaf veins, cells, etc., so it is also called "cell algorithm". We use this as a base point to form a basic pattern of the site. The four "nuclei"-themed atriums are enclosed, each of which retains the original ecological foundation of the site, namely the flower realm, the tea realm, the rain realm, the four seasons realm, and the forest realm. Using the algorithm of Tyson Polygon, a reasonable building boundary is determined, and the penetrating space of the site is retained as an air channel, maintaining the temperature and humidity of the site, and ensuring the comfort of the microclimate environment. The architectural design uses traditional energy-saving technologies such as piercing hall wind, patio courtyard, pole railing elevation and bamboo sunblind, combined with solar photovoltaic systems, to create a biological research base that integrates the natural environment and humanistic characteristics of the place and fully reflects the healthy living environment.

总建筑面积：1473 ㎡　容积率：0.48　建筑高度 9.43m
Total construction area: 1473 ㎡　Plot ratio: 0.48　Building height: 9.43m

■ Basic venue features

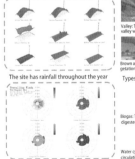

■ Site meteorological data analysis

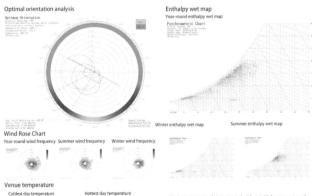

The best orientation of Mianyang city buildings is 20 degrees west-south. Mianyang has a humid, mild and rainy climate, with an average temperature of 17.9℃, 279 days of excellent air quality, and an average annual precipitation of 800mm. The interannual variation in the average annual temperature is small, the general wind speed is small, and the city is dominated by northeast to north winds.

Block generation

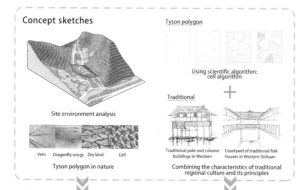

Concept sketches

Tyson polygon

Using scientific algorithm: cell algorithm

Traditional

Site environment analysis

Vein | Dragonfly wings | Dry land | Cell
Tyson polygon in nature

Traditional pole and column buildings in Western | Courtyard of traditional folk houses in Western Sichuan

Combining the characteristics of traditional regional culture and its principles

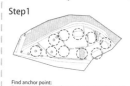

Step1
Find anchor point:
Retain trees, canal landscape, original buildings and structures

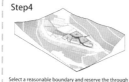

Step4
Select a reasonable boundary and reserve the through as the air duct to achieve the ventilation and moisture removal of the whole site

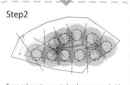

Step2
Tyson polygon is generated at the same speed with the positioning point as the center of gravity to form functional zones and roads

Step5
Keep the courtyard space centered on the point and create a landscape space to adjust the microclimate to achieve uniform ventilation in summer and heat radiation in winter

Step3
Tyson polygon principle:
All points inside a single polygon reach the center of gravity of the polygon, i.e. the positioning point, which is the closest.

Step6
a. The whole building is lifted without damaging the vegetation layer, providing traffic space for animals
b. Form a ventilation channel to isolate the moisture of the site

Regional features translation

Bamboo weaving
Sichuan is rich in civet bamboo, which is mainly used for weaving technology, not for building structures. Sichuan is the hometown of bamboo weaving, such as Qingshen bamboo weaving and Daoming bamboo weaving are listed as intangible cultural heritage. In combination with the local characteristics and the principle of the Columbus wall, we use the 800 * 800 bamboo woven wall as the external wall, and let the local people participate in the construction process of this building. Secondly, bamboo itself has the advantages of warm winter and cool summer.

Frame height
The reason for the multi pole column buildings in Western Sichuan is that the local climate is humid and hot. The elevation of the building can make the wind flow through, which can not only insulate the moisture, but also take away the moisture and heat.

Draught
In order to eliminate the moisture in summer, the traditional dwellings in Sichuan usually have open rooms, extending in all directions, and have a reasonable air flow organization in the halls.
The external walkway serves as the traffic space between buildings. We also integrate this feature into the building.

Patio
The biggest feature of Sichuan folk houses is the courtyard. This is mainly because Sichuan is hot and rainy, with heavy fog in autumn and winter and long cloudy days.
The patio meets the needs of sun protection and ventilation in summer and more sunshine and light in winter.
We combine the principles of Tyson polygon and courtyard to make the building evenly ventilated in summer and evenly heated as a sunshine room in winter, and plant deciduous plants to play a microenvironment regulation.

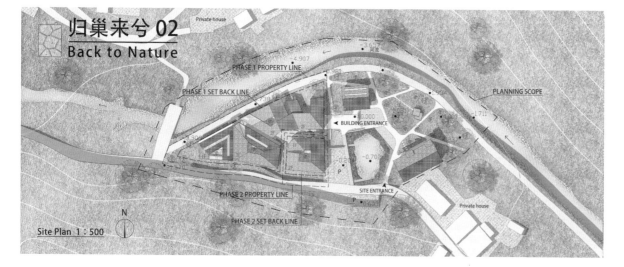

归巢来兮 02
Back to Nature

Site Plan 1 : 500

归巢来兮 03
Back to Nature

Y1-Y5:
The five courtyards combine the ecological site conditions and the environmental atmosphere of different themes to give people different experiences.

Yard 1: Flower realm
It is located in the accommodation area. We have planted many flowers blooming in different seasons here. We hope that the residents can observe different kinds of flowers.

Yard 2: Tea realm
It is located in a wide communication public space in the accommodation area. We have set up a tea tasting area here. We hope you can enjoy the nature while drinking tea.

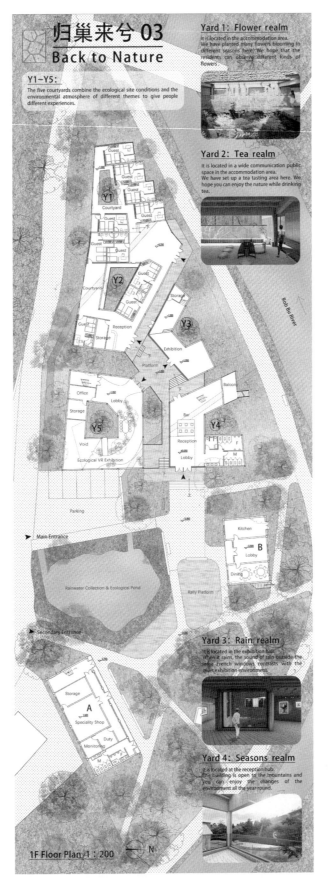

1F Floor Plan 1:200

2nd Floor plan 1:200

Yard 5: Forest realm
In this "second nature", the original architectural spatial framework is diluted, and the model and exhibition content become the focus.

The ups and downs of the ecological model shape the space, which fluctuates up and down with the mountains.

The most typical ecological members are implanted in the model, from soil to fungi, plants, animals, trees, insects and mosses. The finite space is folded into infinite details.

Reconstructed buildings
We have put the functions of the administrative area in the reconstruction of building a to provide a concentrated and quiet environment for the staff.

We put the kitchen and canteen in the entrance square of the building, hoping that when there are festivals in the village or meetings in the base, we can hold cold meals and other activities here.

Building B 2nd Floor Plan 1:200

Building A 2nd Floor Plan 1:200   Building A 3rd Floor Plan 1:200

Yard 3: Rain realm
It is located in the exhibition hall. When it rains, the sound of rain outside the large French windows contrasts with the quiet exhibition environment.

Yard 4: Seasons realm
It is located at the reception hub. The building is open to the mountains and you can enjoy the changes of the environment all the year round.

归巢来兮 04
Back to Nature

South elevation 1:150

■ Seasonal events

Spring — Spring water begins to flow and spring woods furiously grow. Observe the habitat of birds.

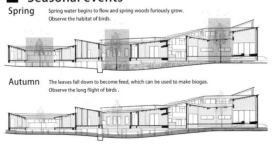

Summer — Open windows for ventilation and insulation. Shade from trees. Reserved air duct for dehumidification.

Autumn — The leaves fall down to become feed, which can be used to make biogas. Observe the long flight of birds.

Winter — Use the sun room for heating. Use double walls and roof storage for heat storage.

■ Analysis of thermal radiation and wind environment

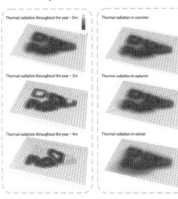
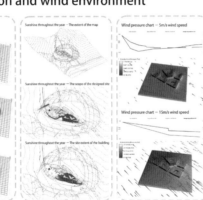

■ Trombe wall

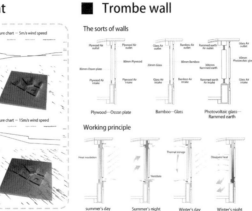

■ Partial sectional perspective

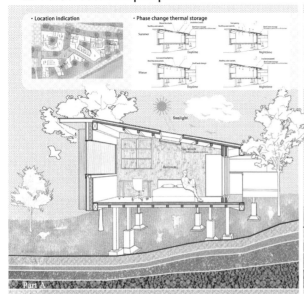

Part A

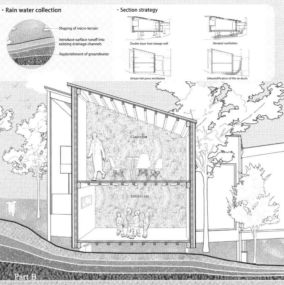

Part B

归巢来兮 05
Back to Nature

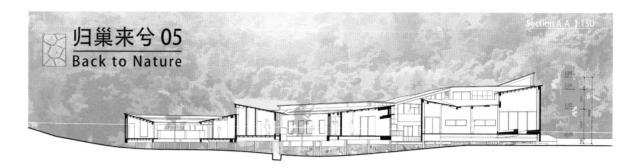

Section A-A 1:150

■ Bamboo exterior wall

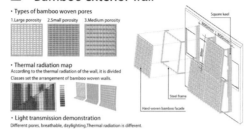

■ Construct nodes

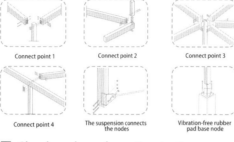

■ Shock and condensation tactics

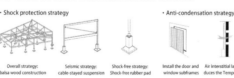

■ Structure and materials

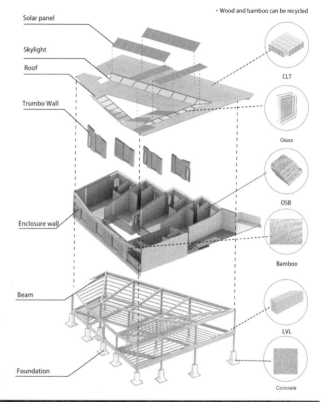

归巢来兮 06
Back to Nature

Active strategies

1. Intelligent air monitoring system
The intelligent fresh air system monitors the indoor environment in real time, reducing energy consumption while maximizing the comfort of the user experience.

2. Rainwater harvesting technology
Roof rainwater collection, site with water treatment to meet the water recycling replenishment lake water and groundwater.

3. Solar photovoltaic technology
The use of solar power generation, power storage to meet daily power supply needs, save energy consumption.

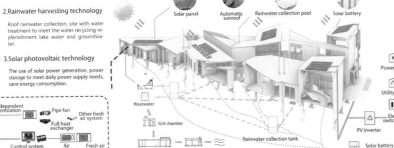

Reform strategies

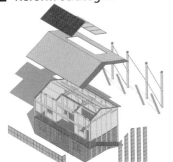

Three-bay building
1. The roof is partially opened to increase the lighting area.
2. Solar panels are added to the roof.
3. The enclosure structure is made using the local bamboo weaving process.
4. Retain the bucket structure, adding suspension cables to wooden columns.

Five-bay building
1. Under the original steel-concrete structure, a steel structure lightweight sun room was added.

Construction: Nest

The structure is located at the junction of the expressway and the village road. It is intended as a warning sign for passing vehicles. At the same time, it implements the concept of ecological construction and the concept of homing, creating a place for birds to inhabit.

Institutional ecological system

The design concept of hierarchical treatment is adopted in site design. And four ecological levels are established according to the influence of the building. Environment to meet the ecological needs to the greatest extent.

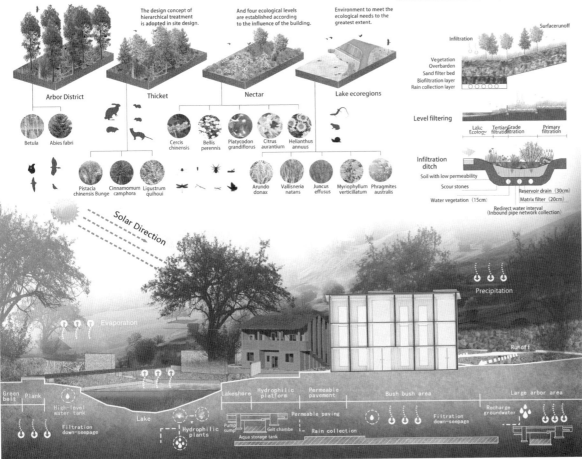

技术专项奖
Technical Special Award

注　册　号：100725
项目名称：共生之径——光孕众生
　　　　　The Path of Symbiosis—
　　　　　Light Gives Birth to Life
作　　者：郑何山、何秀敏、李子昂、
　　　　　张景辉
参赛单位：西安建筑科技大学
指导教师：何文芳、何　泉、李　帆

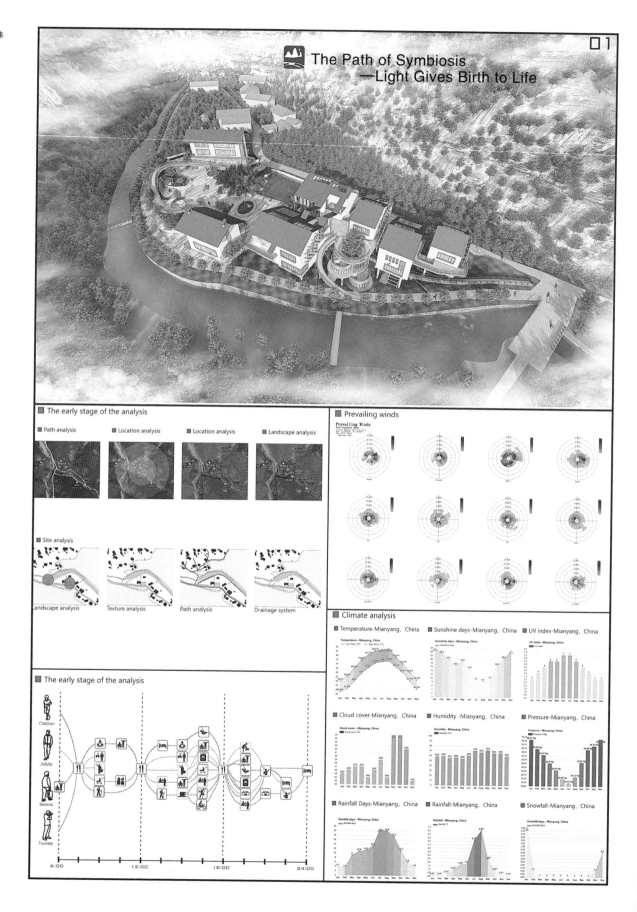

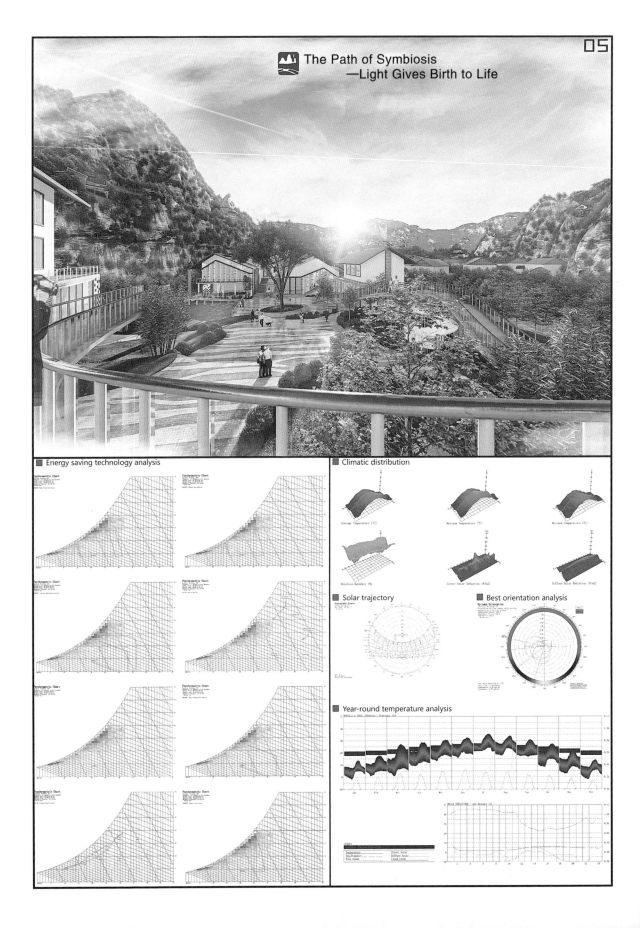

技术专项奖
Technical Special Award

注 册 号：101014
项目名称：光泽·栖居
　　　　　Sunshine·Habitat
作　　者：王成君、董丽英、刘青彪、
　　　　　朱巧凤、彭连飞、毛兴媛
参赛单位：潍坊科技学院
指导教师：周　超、武海泉

光泽·栖居 Sunshine·Habitat 1

Design Desription

本设计题名为"光泽·栖居"，"光泽"意为阳光普照大地，润泽了万物，同时阳光也给予了世界源源不断的能源。"栖居"意为岷山片区作为大熊猫栖息地自然保护区，给大熊猫提供了优越的居住环境。

本设计充分考虑了地域特点，结合当地自然条件、气候特征与建筑能源需求，融合使用多样化太阳能建筑技术，将太阳能光热技术、光伏发电技术、光伏光热一体化（PVT）技术等主动式节能技术，与太阳能集热空气墙、中庭通风技术、遮阳设计等被动式太阳能建筑技术结合，融合形成了本设计特色的太阳能建筑体系。

This design is entitled "Sunshine·Habitat." "Sunshine" means that sunshine shines on the earth, while giving the world constant energy. "Habitat" means the Minshan area as a habitat nature reserve for pandas, providing a superior living environment for pandas. This design fully consider the regional characteristics, combined with the local natural conditions, climate characteristics and building energy demand. The use of diversified active solar building technology including solar thermal technology, photovoltaic power generation technology, photovoltaic thermal integration (PVT) technology, as well as the passive solar building technology including solar trumbe air wall, atrium ventilation technology, shading design, thus to form the integration of the design characteristics of the solar building.

Region Feature

Courtyard｜Panorama｜Roof style｜Window opening mode｜Rivers｜Local house｜River corridor｜Periphery｜Cultural customs

Solar Energy Design Diagram

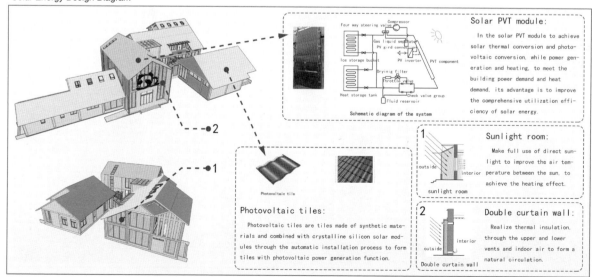

Solar PVT module:
In the solar PVT module to achieve solar thermal conversion and photovoltaic conversion, while power generation and heating, to meet the building power demand and heat demand, its advantage is to improve the comprehensive utilization efficiency of solar energy.

1 Sunlight room:
Make full use of direct sunlight to improve the air temperature between the sun, to achieve the heating effect.

Photovoltaic tiles:
Photovoltaic tiles are tiles made of synthetic materials and combined with crystalline silicon solar modules through the automatic installation process to form tiles with photovoltaic power generation function.

2 Double curtain wall:
Realize thermal insulation, through the upper and lower vents and indoor air to form a natural circulation.

光泽·栖居 Sunshine · Habitat 2

Logical Generation

The base area is 10709 m², close to the residential area.

According to the shape of the site, a single building combination was adopted.

In order to enrich the architectural form and meet the functional needs.

Set aside part of the atrium to allow the building to breathe.

The dominant orientation of the building faces south.

The base of the building is designed with sidewalks and driveways.

Reasonable setup of entrances and exits, blend into the environment.

Retain the original green plants and increase the green reasonably.

Solar photovoltaic panels are placed on the roof and surface of the building.

The shape of the building minimizes wind damage to the building itself.

Incorporate Local Features

Technical-Economic Indices

Site area: 10709m²
Building area: 2180m²
Building density: 13%
Floor-area ratio: 20.36%
Greening rate: 67%
Parking: 6
Square area: 895m²

Site-Plan 1:500

Incorporate panda element

Incorporate the local Tibetan architectural style

光泽·栖居 Sunshine · Habitat 3

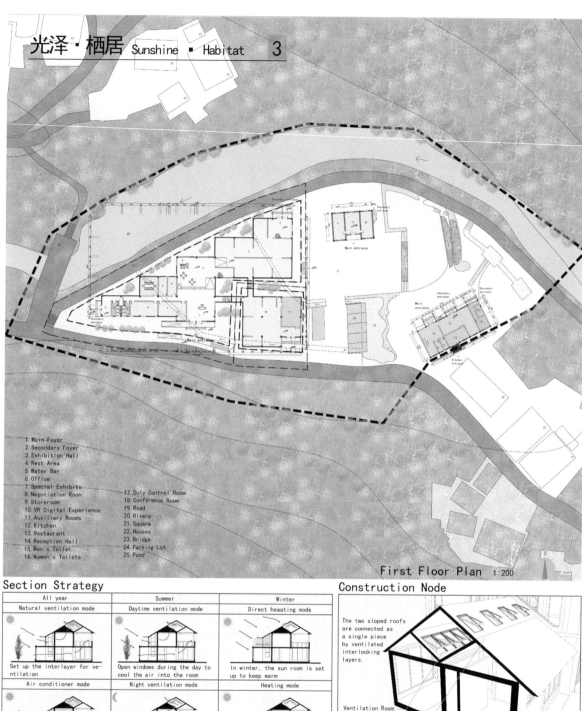

1. Main Foyer
2. Secondary Foyer
3. Exhibition Hall
4. Rest Area
5. Water Bar
6. Office
7. Special Exhibits
8. Negotiation Room
9. Storeroom
10. VR Digital Experience
11. Auxiliary Rooms
12. Kitchen
13. Restaurant
14. Reception Hall
15. Men's Toilet
16. Women's Toilets
17. Duty Control Room
18. Conference Room
19. Road
20. Rivers
21. Square
22. Houses
23. Bridge
24. Parking Lot
25. Pond

First Floor Plan 1:200

Section Strategy

All year	Summer	Winter
Natural ventilation mode	Daytime ventilation mode	Direct heaating mode
Set up the interlayer for ventilation	Open windows during the day to cool the air into the room	In winter, the sun room is set up to keep warm
Air conditioner mode	Night ventilation mode	Heating mode
Set up the interlayer for thermal insulation	Open the window at night and hot air flows outdoors	Air conditioning is used for heating in winter

Construction Node

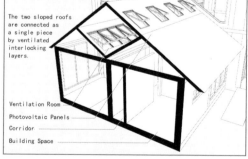

The two sloped roofs are connected as a single piece by ventilated interlocking layers.

- Ventilation Room
- Photovoltaic Panels
- Corridor
- Building Space

South Elevation 1:200

光泽·栖居 Sunshine · Habitat 4

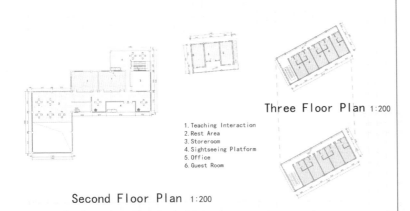

1. Teaching Interaction
2. Rest Area
3. Storeroom
4. Sightseeing Platform
5. Office
6. Guest Room

Three Floor Plan 1:200

Second Floor Plan 1:200

Structure Decomposition Diagram

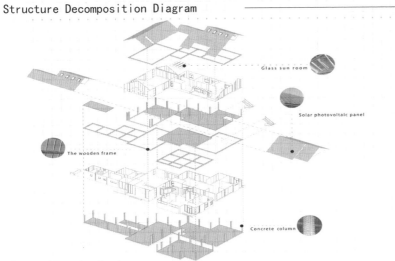

- Glass sun room
- Solar photovoltaic panel
- The wooden frame
- Concrete column

Streamline Analysis

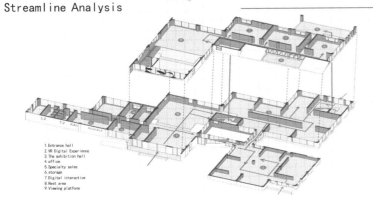

1. Entrance hall
2. VR Digital Experience
3. The exhibition hall
4. office
5. Specialty sales
6. storage
7. Digital interactive
8. Rest area
9. Viewing platform

Behavioral Activity Analysis

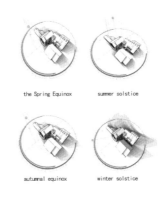

- the Spring Equinox
- summer solstice
- autumnal equinox
- winter solstice

Architectural Shadow Analys

North Elevation 1:200

光泽·栖居 Sunshine · Habitat 5

Heating in Winter

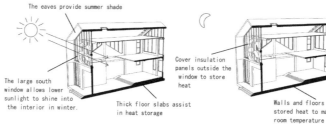

- The eaves provide summer shade
- The large south window allows lower sunlight to shine into the interior in winter.
- Thick floor slabs assist in heat storage

Daytime

- Cover insulation panels outside the window to store heat
- Walls and floors radiate stored heat to maintain room temperature

Nighttime

Entrance to Identify

At the road intersection of the site, due to the high terrain, the sight of motorists is blocked. Therefore, the speed limit and convex road mirror are spray-painted at the approaching intersection.

Passive Solar Thermal Storage

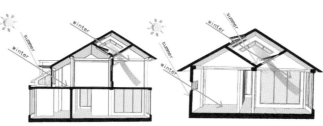

landscape Node

Choose plants with strong noise reduction ability

Meditate alone and immerse yourself in space

Communication space

Multi-level plant planting mode reduces noise from surrounding roads

Profile Map

光泽·栖居 Sunshine · Habitat 6

Solar PVT Heat Pump Thermoelectric Cooling Trigeneration

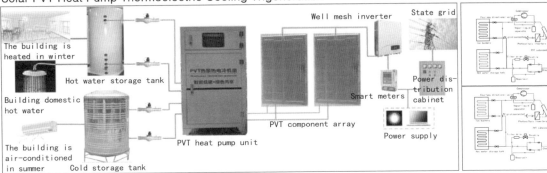

The use of solar photovoltaic solar thermal (PVT) integration technology and heat pump technology, increased heat pump circulation, so as to achieve the output of heat, electricity and cold energy on a set of systems, a multi-purpose, heat, electricity and cooling triple supply, while meeting the building's electricity demand, heat demand and summer air conditioning cooling demand, further improve the comprehensive utilization efficiency of solar energy.

Calculation of carbon emissions

Total energy consumption (KW/year)	Total carbon emissions (KW/year)	Carbon emission reduction by solar energy (KW/year)	Emission reduction rate (%)
827528	662022.4	460436	69.55

Building energy consumption table

	The measure of area (m²)	Energy consumption index	Company	Measured value	Company
Winter heating	2180	40	W/m²	87.2	KW
Cooling in summer	2180	60	W/m²	130.8	KW
Annual power consumption	2180	50	W/m²	109	KW
Domestic hot water	2180	50	L/人	5000	L

Solar system capacity calculation

	Laying area (m²)	Power generation (KW)	Heating area (m²)	Cooling area (m²)	Water quantity for preparing domestic heat (L)
PV	646	96.9			
PVT	216	32.4	1728	1080	5200
Ratio of solar energy capacity to meet building energy consumption (%)		118.62%	79.27%	49.54%	104%

Energy saving benefit calculation

Total energy consumption (KW/year)	Coal consumption (t/year)	Energy cost (yuan/year)	Solar energy capacity (KW/year)	Energy saving rate (%)	Energy saving cost (yuan/year)
827528	27.31	25671.4	575545	69.55%	17860

Indoor Nodes

East Elevation 1:200

有效作品参赛团队名单
Name List of All Participants Submitting Valid Works

作品编号	作者	单位名称	指导老师	单位名称
100639	李丽君、熊世英、徐华仁、冯腾飞、孟虎、赵佳佳、潘玥	昆明学院	左明星	昆明学院
100642	陈虎、杨紫涵	苏州科技大学	刘长春、金雨蒙	苏州科技大学
100650	刘璇、葛宵旗、李乐薇	华东交通大学	彭小云	华东交通大学
100654	曾天华、符盛、顾纪星、马佳薇	南京工业大学	刘强、薛洁	南京工业大学
100676	马志强、王凌豪、邹立君	浙江大学建筑设计研究院、新加坡国立大学、华东建筑设计研究院	—	—
100682	郑新杰、李润霖、陈怡凤、郑颖茵、张明斌	广州大学	庞玥、李丽	广州大学
100684	周可伊、陈言、郭奕岑、赵庆卓、过翔天、毛思异、陈思妍	重庆大学	黄海静	重庆大学
100690	雷咏润	河南工业大学	—	—
100699	刘杞铭、张军杰、吴辰懿、隋蕴仪	重庆大学	黄海静	重庆大学
100725	郑何山、何秀敏、李子昂、张景辉	西安建筑科技大学	何文芳、何泉、李帆	西安建筑科技大学
100727	孙滢洁、赵路轩、王敏、王怡心	南京工业大学	薛洁	南京工业大学
100729	周旭、俞悦、袁怡婷、徐欢	上海城建职业学院	张雪松	上海城建职业学院
100730	白玮佳、张佳欣、何华洋、张卓、张若寒	河北农业大学	王崇宇	河北农业大学
100748	吴雨蝶、马思远、杨晓雨、冼子竣	西安科技大学	孙倩倩	西安科技大学
100757	徐飞、黄雨晴、刘芳瑜	北京建筑大学	郝石盟	北京建筑大学
100759	张皓麒、李咏真、王子颖、林奕鑫	广州大学	庞玥、李丽	广州大学
100760	王兴元、王苗苗、国家璇、王磊	天津大学	杨崴	天津大学
100768	田昊伦	上海应用技术大学	—	—
100777	何丽诗、黄李、贺为皓、秦漓、肖峰、黄荟瑜、胡诗雨	四川农业大学	陈川、侯超平	四川农业大学
100783	陈曙琪、段于瑄、张满	河南工业大学	张华、马静	河南工业大学
100784	张培俊、金圣煜、薛莹莹、郑峥	中国美术学院风景建筑设计研究总院有限公司、浙江汇创设计集团有限公司	张军	中国美术学院风景建筑设计研究总院有限公司
100790	李恩正、谭彩燕、段羽琳	昆明学院	张楠、金禾	昆明学院

续表

作品编号	作者	单位名称	指导老师	单位名称
100796	赵才奇、曾杨昊、黄敏、陈沼宇	广州大学	席明波、万丰登	广州大学
100807	樊子铃、彭佳丽、蒋佩芸、文晓岚、俞楚葳	四川农业大学	陈川	四川农业大学
100812	孟宪涛、牟雨昕、韩晓雪、刘梦军	西安科技大学	孙倩倩	西安科技大学
100813	焦旋、赵彩霞、郭煜妍	西安科技大学	孙倩倩	西安科技大学
100821	郝泽厚、潘静涵、郑子豪、姚黄城	福州大学	王炜	福州大学
100835	王家伟、孙明辉	山东建筑大学	薛一冰、房涛	山东建筑大学
100836	樊泽宇、蒲丹妮、单司辰	西安科技大学	李雪平、孙倩倩	西安科技大学
100843	路鹏程、寇羽嘉、谢皓莹	北京理工大学	姚健、赵玫	北京理工大学
100845	曹裕龙、阮聪、刘伊琪、沈书睿	南京工业大学	薛春霖	南京工业大学
100846	付嫣然、李玲秀、甘杰文、林逸清	福州大学	林志森	福州大学
100848	黄梦悦、崔屿菲、潘森森	上海应用技术大学	王莉莉、周卓艳、程光	上海应用技术大学
100851	蒋帅、李涛、李雨蒙、苏小雪、毛雨晴、范凯琳	西南交通大学	张樱子	西南建筑大学
100854	孙锐、徐双雨、袁世博、孙艺丹、宁佳铭、朱南西、张涵清、卢佳宁、谷京昀、张吉祥、李鹏飞	哈尔滨工业大学	韩昀松、王艳敏	哈尔滨工业大学
100855	皇亚楠、王帅	上海应用技术大学	王莉莉	上海应用技术大学
100864	徐文宁、潘炜、黄琰、刘硕	南京工业大学	郭兰、彭克伟	南京工业大学
100867	汪昊尧、高梦娇、梁翰菲	沈阳建筑大学	高畅	沈阳建筑大学
100868	曹旭、黄宇、李上顾、阮灿健、仲琪	苏州科技大学	刘长春、金雨蒙	苏州科技大学
100871	许荣玮、陈雅菁、甘玉婷、张传杰	南京工业大学	刘杰文、刘强	南京工业大学
100873	岳国威、刘达、王瑞、余浩然、高溶	西安建筑科技大学	孔黎明	西安建筑科技大学
100874	祁茹钰、唐一平、杨雁翔、王晓娇、蔚诗轩	大连理工大学	郎亮、杨光	大连理工大学
100877	黎亚柏、黄文龙、杜鑫、刘晓晓	河南工业大学	马静、张华	河南工业大学
100878	潘静怡、陈雨薇、覃倚嘉、文柳伊、彭里江	湖南城市学院	刘培芳	湖南城市学院
100885	王钦、许港、姚歌、陈波波	华南理工大、哈尔滨工业大学、南京长江都市建筑设计股份有限公司、长安大学建筑学院	吴中平、金虹	华南理工大学建筑设计有限公司何镜堂建筑创作研究院、哈尔滨工业大学

续表

作品编号	作者	单位名称	指导老师	单位名称
100892	李果、徐永红、林裳、刘摇、白一帆、杨明倩	内蒙古工业大学	王婷、许国强	内蒙古工业大学
100893	杨婷婷、夏子惠、甘锦、高招奎、刘晨悦、陶志毅、赵悦然	内蒙古工业大学	伊若勒泰、许国强	内蒙古工业大学
100897	宋振宇、储一波、孙伟、李峥、张愉欣	南京工业大学	罗靖、郭兰、彭克伟	南京工业大学
100899	王艺霖、许乐萱、刘家锴、赵方	南京工业大学	杨亦陵	南京工业大学
100910	周允洁、张惠子、肖雅、夏禹晴	南京工业大学	罗靖	南京工业大学
100919	毕林昊、刘芸霏、石鸣春涧	西安建筑科技大学	成辉、李欣	西安建筑科技大学
100922	陈先鹏、秦佳楠	河南工业大学	马静、张华	河南工业大学
100924	刘忻杰、王文宇、章子豪	上海应用技术大学	程光	上海应用技术大学
100929	何勇杰、徐欣然、魏钰丰、刘孙卓然	北京工业大学	陈喆	北京工业大学
100931	宁馨儿、孙晴波、倪云杰、王彤	南京工业大学	罗靖、刘晓光	南京工业大学
100932	孙令涛、徐云展	天津城建大学	张平	珠海市设计院
100935	郑潇莹、王球锋、牛宇飞、高喆	西安交通大学	王宇鹏	西安交通大学
100936	杜天慧、武云杰、刘芳鸣、袁梓飞	华北理工大学、厦门大学、浙大宁波理工学院、华北理工大学	檀文迪	华北理工大学
100938	姚成琴、朱棋	西安建筑科技大学	成辉、李欣	西安建筑科技大学
100940	陈鹏程、黄卓然、陈宇开	上海应用技术大学	王莉莉	上海应用技术大学
100941	李昕、刘凌云、林昕、吕姚霏、陈晓莹	广州大学	席明波、万丰登	广州大学
100942	鲁俊逸、王少潜、韩旭	中国矿业大学	马全明、邵泽彪	中国矿业大学
100946	上官玉麒、赵雨婷、曹雯欣、覃文欢、陆澳晨	昆明理工大学	陆莹、毛志睿	昆明理工大学
100947	李锗淳、王慕娴、夏雨、刘至晴、翟佳棣、赵灵犀	昆明理工大学	陆莹、毛志睿	昆明理工大学
100948	李松桦、张多、包蕾	沈阳建筑大学	汝军红	沈阳建筑大学
100950	高千禧、刘洁	河南工业大学	马静、张华	河南工业大学
100951	全子威、杨欢	中建凯德电子工程设计有限公司、成都易美互动科技有限公司	—	—
100952	沙扬皓、王子倩、阮志鹏、李啸晗、虞静雯	合肥学院	丁蕾、司大雄	合肥学院

续表

作品编号	作者	单位名称	指导老师	单位名称
100955	王宝玉、吴斌、林溢轩、肖明星、黄崇劲、蔡滨鸿	北京建筑大学、北海艺术设计学院	晁军、孙克真、杨红、李英、庄乾阔	北京建筑大学、北海艺术设计学院
100961	王占阳、张志达、李阳、许人天	米兰理工大学	Angelo Lorenzi	米兰理工大学
100965	陈轲、欧阳秀妍、郭曦爻、冯思雨、赵伟翔、周恬如、贲小杰	西南科技大学	王大川、蒋琳、周铁军	西南科技大学、重庆大学
100967	瞿琳茜、邱潇仪、房璐杰、陈烁杨	重庆大学	张海滨	重庆大学
100978	杨旖文、王宜杉、曾建木、陈启祯	南京工业大学	徐善彬、罗靖、舒欣	南京工业大学
100982	卓金明、李舒祺、许浩川、周超超	西安建筑科技大学	李涛、孔黎明	西安建筑科技大学
100987	张颖、朱雅萱	河南工业大学	张华、马静	河南工业大学
100995	李春颖、贺川、颜廷旭	多伦多大学		
100996	冯雪珂、柴雨欣、李文瀚、秦志宁	兰州交通大学	高发文	兰州交通大学
100999	张恩源、谢继明、李清超、李昊嘉	河北建筑工程学院	王金奎、史建平	河北建筑工程学院
101004	刘思彤、郭志成、陈雨清、刘贺丹、何若萌、王子翱、伍君奇	北京交通大学	蒙小英、杜晓辉	北京交通大学
101005	靳宝	河北工艺美术职业学院		
101007	杨天心、张雨佳、梁骏嘉、程卓然、李圣哲	北京交通大学	王鑫、张文	北京交通大学
101009	李嘉梅、李凤莉	河南工业大学	马静、张华	河南工业大学
101012	支叶青、赵春燕、郑雅娴	北京建筑大学	孟璠磊、刘烨	北京建筑大学
101014	王成君、董丽英、刘青彪、朱巧凤、彭连飞、毛兴媛	潍坊科技学院	周超、武海泉	潍坊科技学院
101015	陈世建、陈倩、李波	艾杰国际建筑设计有限公司	陈世建	艾杰国际建筑设计有限公司
101018	洪未名、王曦曼	吉林建筑大学	周春艳、赫双龄	吉林建筑大学
101019	吕宇蓓、吴爽、陈茜、陈烁遂、陈迪菲	南京工业大学、同济大学	杨亦陵	南京工业大学
101020	刘晓晓、杜鑫、王淉宇、黎亚柏、黄文龙	河南工业大学	马静、张华	河南工业大学
101021	吕宇蓓、吴爽、陈茜、张佳宁	南京工业大学	郭兰、彭克伟	南京工业大学
101022	毕睿航、罗志远、任帅	南京工业大学	林杰文	南京工业大学
101024	胡一鸣、陈奕宏、黄志毅、刘宇翔、郑渊正、刘日尧	福州大学	邱文明	福州大学

续表

作品编号	作者	单位名称	指导老师	单位名称
101027	张瑞英、祝淑芬、闫海、张象龙、周禹辰	湖南大学、聊城大学、安徽建筑大学、山东建筑大学、大连理工大学城市学院	郑斐、王月涛	山东建筑大学
101028	陈漪臻、郑雨婕、陈雨昕、徐雨杭	南京工业大学	姜雷、郭兰	南京工业大学
101029	宋立、杨宝奎、张廷辉、祝双双、许浴泊	昆明学院	李江奇、陈虹羽、左明星	昆明学院
101030	殷欣睿、李智轩、蔡佳如	南京工业大学	刘强	南京工业大学
101031	郑龙纪、肖瑞、李霁玮、季乐宇	南京工业大学	舒欣、薛春霖	南京工业大学
101032	刘思懿、付雯雯、汪珈亦、车玉姝、赵鑫慧	西北工业大学	刘煜	西北工业大学
101034	艾里牙包拉提	西安科技大学		
101047	胡喆涵、叶雨菲、毛锐杰、杨政	南京工业大学	董凌、薛春霖	南京工业大学
101050	吴淑清、曲纯瑞、杨智胜	福州大学	崔育新	福州大学
101053	杨怡琳、祝浩艺、章子玥、李晨曦	南京工业大学	董凌	南京工业大学
101058	施语林、杨光、万沐霖、蒋旭亮、胡家皓	南京工业大学、东华大学	薛洁、刘强、周紫昱	南京工业大学
101059	李鸿宇、刘承宇、宋莉、丁雨、程琛、赵辉强	山东科技大学	冯巍、王雅坤	山东科技大学
101064	李轶群、刘银露、邱妤菲菲、朱浩博、高祚、陈雪柯、陈安娴	重庆大学	何宝杰	重庆大学
101065	曾广钊、曹时语、于威	青岛理工大学	薛凯	青岛理工大学
101068	张锟、韦海璐、王睿轩、王晟浩、杨莫、王悌娟、戴轩、角元昊	西北工业大学、东南大学	邵腾、王晋	西北工业大学
101075	杨婉瑜、黄易、张羽洁、李思晨	昆明理工大学	陆莹、毛志睿	昆明理工大学
101078	蔡昊哲、李季恒、王渮蝶、欧子怡、朱子跃	昆明理工大学	陆莹	昆明理工大学
101084	王珏、周子靖、宗顾涵	中国石油大学华东	李佐龙、王凌旭	中国石油大学华东
101086	高中岭、张汉昭、刘鲁、陈绪燕、刘心昊、范凯	山东建筑大学设计集团有限公司	赵学义	山东建筑大学设计集团有限公司
101091	王海力、赵磊、冯昌春、李晓琳、薛雅文、辛泓雨	潍坊科技学院	周超、武海泉	潍坊科技学院
101092	陈妍、何盈、孟倩影、张嘉伟	南京工业大学	薛春霖	南京工业大学
101093	凌宇、陆玉凤、索浩越	燕山大学、北京工业大学	徐振华、孙嘉男	北京城市学院、北京优优星球教育科技有限公司

续表

作品编号	作者	单位名称	指导老师	单位名称
101099	林凯、蒋敬亦、李吕雨阳、刘幸宇	嘉兴学院、昆明理工大学	陆莹、毛志睿	昆明理工大学
101104	崔创国、陈均基、王波峰	西安科技大学	石嘉怡	西安科技大学
101105	杨汶瑾、陈俊豪、古津铭、林涵晔、叶家瑞、赵晓婷	重庆大学	周铁军、张海滨、李骏	重庆大学
101106	郭思洁、李若楠	河南工业大学	张华、马静	河南工业大学
101107	张一鸣、王晓燕、宋婕、杨敏敏、邹宇飞、杨云辉	西北工业大学	刘煜	西北工业大学
101112	逄丽影、姜彩霞、刘起岳	山东建筑大学	薛一冰	山东建筑大学
101113	李上顾、阮灿健、曹旭、黄宇、仲琪	苏州科技大学	刘长春、金雨蒙	苏州科技大学
101114	梁丁鹏	木加建筑设计工作室		
101119	利紫晴、陈欣莹、陈阳、石慧	广州大学	庞玥、李丽	广州大学
101123	李若娴、束子玥、周宇航、王璇	南京工业大学	刘强、薛洁	南京工业大学
101128	卢秋羽、潘明慧、徐曾娜、俞嘉敏	南京工业大学	郭兰、彭克伟	南京工业大学
101132	殷嘉蔓、陈忠耀、王思懿、杨雄、尹家雪	重庆大学	黄海静	重庆大学
101135	肖维中、邱一天、柳影影、曹静、于丹丹	华中科技大学、北海艺术设计学院、北京工业大学	刘晖、杨红、李紫微	华中科技大学、北海艺术设计学院、北京工业大学
101138	许荣玮、陈雅菁、甘玉婷、张传杰	南京工业大学	林杰文	南京工业大学
101146	霍彦如、陈思洁、孙彤	南京工程学院、江西科技师范大学	周静	南京工程学院
101150	王路、王菡纭、倪赛博、王君宜	南京工业大学	张海燕、胡振宇	南京工业大学
101151	崔倍萱、曹予婕、傅妍妍、罗德倩	西安交通大学	张樱子、李百毅	西安交通大学
101153	陈佳怡、罗程、刘昕彤、白馨怡	重庆大学	周铁军	重庆大学
101157	陈思安、孙上词、钟锐、吴俊豪	沈阳建筑大学	侯静、武威	沈阳建筑大学
101158	田泽轩、吴月、唐梓榕、赵一凡、孙翊宸	西南民族大学	熊健吾、王晓亮、毛刚	西南民族大学
101161	王艺萌、唐中博、李家兴	北京交通大学	张文	北京交通大学
101165	代伦、陈聪、张超、张森	青岛理工大学	商选平、郝占鹏	陕西省建筑设计研究院西宁办事处、青岛理工大学
101168	张肖同、王欣、包雯依、籍研溪、季青云、葛潇然、黄琬涔、黄思宇、白鹏、张琦	浙江理工大学	文强	浙江理工大学

续表

作品编号	作者	单位名称	指导老师	单位名称
101178	王兰、李佳、于洽卿、孙剑、张文馨、马炳蔚	山东建筑大学	薛一冰、杨倩苗	山东建筑大学
101180	赖汐玥、袁榕齐、申思文、龙科洁、王佳、韩雨琪	成都理工大学	冯桢懿	成都理工大学
101181	戴滢婕、关怡平、肖霄	南京工业大学	刘强、薛洁	南京工业大学
101182	霍然、张小曼、尹航、邓晓颖	山东建筑大学	侯世荣、杨倩苗	山东建筑大学
101184	李秀赟、黄真贤、陈玥	福州大学	崔育新	福州大学
101185	王学程、罗恺蓓、张佳文、叶超	长安大学	刘凌	长安大学
101187	康孟琪、王君宜、王路、吕铭洁、姚望	南京工业大学	张海燕、胡振宇	南京工业大学
101188	朱雅夫、周天娇、严帅、刘媛卉	华南理工大学	王静	华南理工大学
101191	刘月、熊羽珂、程星、刘柠熙、林儒韬、王文瑄	昆明理工大学	陆莹	昆明理工大学
101201	马子雯、侯靖轩、曹高源	天津大学	朱丽	天津大学
101205	陈晓歌、陈千	黛北建筑设计张家口有限公司、中国邮政储蓄天津分公司	陈晓歌	黛北建筑设计张家口有限公司
101209	姚望、王君宜、吕铭洁、康孟琪	南京工业大学	胡振宇、张海燕	南京工业大学
101211	刘萱熙、戴姝瑶、张书馨	上海应用技术大学	周卓艳	上海应用技术大学
101216	费凡、罗明宇、叶清、程子瀚	合肥工业大学	李早、王旭	合肥工业大学
101219	杨艳	长安大学	张磊、刘凌	长安大学
101220	杨雁翔、王晓娇、蔚诗轩、祁茹钰、唐一平	大连理工大学	刘九菊	大连理工大学
101222	杨献巧、邓雅骏、夏乐芯	福州大学	吴木生	福州大学
101223	樊秀君、刘滢、董海音	南京工业大学	薛春霖	南京工业大学
101224	路易、梁伟豪、白盛鸿、李景秀、孙天伊、侯亚玲	天津大学、北京林业大学	郭娟利、李伟	天津大学
101225	薛思琪、武僖晴、吴梓会	西安建筑科技大学、北方工业大学、华侨大学		
101226	吕铭洁、王君宜、姚望、康孟琪、王路	南京工业大学	胡振宇、张海燕	南京工业大学
101227	詹科、汤洛行、王加俊	贵州一道创意景观设计有限公司、贵州工商职业学院		
101230	李嘉仪、叶方铭	汕头大学、福州工商学院	徐振华、孙嘉男	北京城市学院、北京优优星球教育科技有限公司

续表

作品编号	作者	单位名称	指导老师	单位名称
101231	毛越、孙艺萌、贺晓婷	南京工业大学	姜雷	南京工业大学
101234	赖彦豪、蔡志峰、张涛	广州大学	席明波、万丰登	广州大学
101237	陈璐倩、王可、韦寒雨、杨宇	南京工业大学	罗靖、刘晓光	南京工业大学
101248	林颖、马嘉、李文博、董晓晗	西安建筑科技大学	孔黎明、李涛	西安建筑科技大学
101253	何沁	山东建筑大学		
101258	信蔚林、孟雨欣、石宇豪、耿晨晓	长安大学	夏博	长安大学
101261	周凌波、丛觅雪、吕玉露	山东建筑大学	仝晖、金文研	山东建筑大学
101263	顾笑言	武汉大学	黄凌江、兰兵、舒阳	武汉大学
101274	陈巨鸿、周柯汛、张丙南	长安大学	任娟	长安大学
101288	张文聪、郭雨欣、张立飞、鲁彬	合肥工业大学	李早、贾丽丽	合肥工业大学
101289	范蕊、程玉红、张滨坚、胡彤	合肥工业大学	贾莉莉、何伟	合肥工业大学

2022台达杯国际太阳能建筑设计竞赛办法
Competition Brief on International Solar Building Design Competition 2022

竞赛宗旨：

生态文明建设是关系中华民族永续发展的根本大计。促进人与自然和谐共生，坚持生态优先、科学确立绿色发展理念是重中之重。本次竞赛以建设保护区科学考察站为赛题，探索人与自然的和谐共处方式，为维持生态系统稳定探索新的途径。

项目背景：

项目地位于四川省绵阳市平武县藏族木皮乡关坝村中。平武县位于四川盆地西北部、涪江上游，不仅处于大熊猫国家公园岷山片区大熊猫栖息地自然保护区中，并被四川省森林自然教育基地批准为关坝沟流域自然保护小区。全村林地总面积6499.3亩，森林覆盖率为96.3%。区内有大熊猫、金丝猴、红豆杉和珙桐等国家保护珍稀动植物70余种，生物资源丰富，生态环境质量高，是绵阳市重要的生态腹地和水源涵养地。

该项目计划建设关坝沟流域自然保护小区科学考察站，为村中建设以自然保护为主体的体验培训、研学、旅游等方面，提供来访人员居住、餐饮、集中活动的设施。一期计划建设包括休息空间、住宿、教育空间及相应的服务配套设置等。

竞赛主题：阳光·山水驿
竞赛题目：四川省平武县关坝沟流域自然保护小区科学考察站
主办单位：国际太阳能学会
中国建设科技集团中央研究院
中国建筑设计研究院有限公司

GOAL OF COMPETITION:

Ecological conservation is a fundamental task for the sustainable development of the Chinese nation. To promote the harmonious coexistence of man and nature, it is of utmost importance to give priority to the ecosystem and come up with a sound idea of green development. With the subject of a research station in the reserve, the competition aims to learn the harmonious coexistence of man and nature and explore new ways to maintain the stability of the ecosystem.

PROJECT CONTEXT:

The project is located in Guanba Village, Mupi Tibetan Township, Pingwu County, Mianyang City, Sichuan Province. Situated in the northwest of Sichuan Basin and the upper reaches of the Fujiang River, Pingwu County is in the Minshan Mountain Giant Panda Habitat Nature Reserve of Giant Panda National Park and has been approved as the Guanbagou Basin Nature Mini-Reserve by Sichuan Forest Nature Education Base. The village, with a forest coverage rate of 96.3%, has 6499.3 mu of woodland. There are more than 70 species of rare plants and animals under state protection, such as giant pandas, golden snub-nosed monkeys, yews, and dove trees. Blessed with abundant biological resources as well as high-quality ecosystem and environment, it is an important ecological hinterland and water conservation area in Mianyang.

The project is going to build Guanba Village Nature Reserve Research Station to provide facilities for visitors' living, eating, and activities in the village during their nature conservation-oriented experience, training, research, study, travels, and other activities. The first stage of construction includes space for rest, accommodation, space for education, and corresponding services and supporting facilities.

THEME OF COMPETITION:

Sunshine & Station in the Scenery

承办单位：国家住宅与居住环境工程技术研究中心
冠名单位：台达集团
支持单位：山水自然保护中心
　　　　　青年应对气候变化行动网路
评委会专家：**Deo Prasad**：澳大利亚科技与工程院院士、澳大利亚勋章获得者、澳大利亚新南威尔士大学教授。
杨经文：马来西亚汉沙杨建筑师事务所创始人、2016 年梁思成建筑奖获得者。
Peter Luscuere：荷兰代尔伏特大学建筑系教授。
崔愷：中国工程院院士、全国工程勘察设计大师、中国建筑设计研究院有限公司总建筑师。
王建国：中国工程院院士、教育部高等学校建筑类专业教学指导委员会主任委员、东南大学建筑学院教授。
庄惟敏：中国工程院院士、全国工程勘察设计大师、2019 梁思成建筑奖获得者、清华大学建筑学院院长。
林宪德：中国台湾绿色建筑委员会主席、中国台湾成功大学建筑系教授。
宋晔皓：清华大学建筑学院建筑与技术研究所所长、教授、博士生导师，清华大学建筑设计研究院副总建筑师。
钱锋：全国工程勘察设计大师，同济大学建筑与城市规划学院教授、博士生导师，高密度人居环境生态与节能教育部重点实验室主任。
仲继寿：中国建筑设计研究院副总建筑师、国家住宅科技产业技术创新战略联盟秘书长。
黄秋平：华东建筑设计研究总院总建筑师。
冯雅：中国建筑西南设计研究院顾问总工程师、中国建筑学会建筑热工与节能专业委员会副主任。
组委会成员：由主办单位、承办单位及冠名单位相关人员组成。办事机构设在国家住宅与居住环境工程技术研究中心。

设计任务书及专业术语等附件：

附件 1：四川省关坝沟流域自然保护小区科学考察站任务书
附件 2：四川省绵阳市平武县木皮藏族乡关坝村地形图
附件 3：四川省绵阳市平武县木皮藏族乡关坝村项目场地内现状建筑平面图
附件 4：专业术语
附件 5：参赛者信息表

SUBJECT OF COMPETITION：

Research Station in Guanbagou Basin Nature Mini-Reserve, Pingwu County

HOSTS：

International Solar Energy Society（ISES）
The Central Research Institute of China Construction Technology Group
China Architecture Design & Research Group（CADG）

ORGANIZER：

China National Engineering Research Center for Human Settlements（CNERCHS）

TITLE SPONSOR：

Delta Electronics

SUPPORTING UNITS：

Shanshui Conservation Center
China Youth Climate Action Network

EXPERTS OF JUDGING PANEL：

Mr. Deo Prasad：Academician of the Australian Academy of Technological Sciences and Engineering, Winner of the Order of Australia, and Professor of University of New South Wales, Sydney, Australia

Mr. King Mun YEANG：Founder of T.R.Hamzah&Yeang Sdn.Bhd（Malaysia）, 2016 Liang Sicheng Architecture Prize Winner

Mr. Peter Luscuere：Professor of Department of Archetecture, Delft University of Technology

Mr. Cui Kai：Academician of China Academy of Engineering, National Engineering Survey and Design Master and Chief Architect of China Architecture Design & Research Group（CADG）

Mr. Wang Jianguo：Academician of China Academy of Engineering, Chairman of the National Administry Committee on Teaching Architecture to Majors in Higher Education under The Ministry of Education, and Professor of School of Architecture, Southeast University

Mr. Zhuang Weimin：Academician of China Academy of Engineering, National Engineering Survey and Design Master, Winner of 2019 Liang Sicheng Architecture Prize, and Dean of School of Architecture, Tsinghua University

Mr. Lin Xiande：President of Taiwan Green Building Committee, China and Professor of Department of Architecture of National Cheng Kung University, Taiwan, China

Mr. Song Yehao：Director, Professor and Doctoral Supervisor of Institute of

奖项设置及奖励形式：

综合奖：

一等奖作品：1名，颁发奖杯、证书及人民币 100000 元奖金（税前）；

二等奖作品：3名，颁发奖杯、证书及人民币 40000 元奖金（税前）；

三等奖作品：6名，颁发奖杯、证书及人民币 10000 元奖金（税前）；

优秀奖作品：20名，颁发奖杯、证书及人民币 2000 元奖金（税前）；

入围奖作品：30名，颁发证书；

技术专项奖：名额不限，颁发证书；

设计创意奖：名额不限，颁发证书。

参赛要求：

1. 欢迎建筑设计院、高等院校、研究机构、绿色建筑部品研发生产企业等单位，组织专业人员组成竞赛小组参加竞赛。

2. 请参赛者访问 www.isbdc.cn，按照规定步骤填写注册表，提交后会得到唯一的注册号，即为作品编号，一个作品对应一个注册号。提交作品时把注册号标注在每副作品的左上角，字高6mm。注册时间2022年3月26日～2022年8月15日。

3. 参赛者同意组委会公开刊登、出版、展览、应用其作品。

4. 被编入获奖作品集的作者，应配合组委会，按照出版要求对作品进行相应调整。

注意事项：

1. 参赛作品电子文件须在 2022 年 9 月 15 日前提交组委会，请参赛者访问 www.isbdc.cn，并上传文件，不接受其他递交方式。

2. 作品中不能出现任何与作者信息有关的标记内容，否则将视其为无效作品。

3. 组委会将及时在网上公布入选结果及评比情况，将获奖作品整理出版，并对获奖者予以表彰和奖励。

4. 获奖作品集首次出版后 30 日内，组委会向获奖作品的创作团队赠获奖作品集 1 册（盖官方纪念章）。

5. 竞赛活动消息发布、竞赛问题解答均可登陆竞赛网站查询。

Architecture and Technology, School of Architecture, Tsinghua University, and Deputy Chief Architect of Architectural Design and Research Institute of Tsinghua University

Mr. Qian Feng: National Engineering Survey and Design Master, Professor and Doctoral Supervisor of College of Architecture and Urban Planning Tongji University (CAUP), Director of Key Laboratory of Ecology and Energy-saving Study of Dense Habitat (Tongji University), Ministry of Education

Mr. Zhong Jishou: Deputy Chief Architect of China Architecture Design & Research Group (CADG) and Chief Commissioner of Solar Energy Building Committee of China Renewable Energy Society

Mr. Huang Qiuping: Chief Architect of East China Architectural Design & Research Institute (ECADI)

Mr. Feng Ya: Chief Engineer of China Southwest Architectural Design and Research Institute Corp. Ltd. and Deputy Director of Special Committee of Building Thermal and Energy Efficiency, Architectural Society of China (ASC)

MEMBERS OF THE ORGANIZING COMMITTEE:

It consists of members of competition hosts, the organizer, and the title sponsor. The administration office is in China National Engineering Research Center for Human Settlements (CNERCHS).

THE DESIGN SPECIFICATIONS AND PROFESSIONAL GLOSSARY:

Annex 1: Design Specification of the Research Station of Guanba Village, Mupi Tibetan Township, Pingwu County, Mianyang City, Sichuan Province

Annex 2: Topographic Map of Guanba Village, Mupi Tibetan Township, Pingwu County, Mianyang City, Sichuan Province

Annex 3: Floor Plan of the Site of Guanba Village Project in Mupi Tibetan Township, Pingwu County, Mianyang City, Sichuan Province

Annex 4: Professional Glossary

Annex 5: Information Table

AWARD SETTING AND AWARD FORM:

General Prize:

First Prize: 1 winner

The Trophy Cup, Certificate and Bonus RMB 100,000 (before tax) will be awarded.

Second Prize: 3 winners

The Trophy Cup, Certificate and Bonus RMB 40,000 (before tax) will be awarded.

Third Prize: 6 winners

The Trophy Cup, Certificate and Bonus RMB 10,000 (before tax) will be awarded.

Honorable Mention Prize: 20 winners

所有权及版权声明：

参赛者提交作品之前，请详细阅读以下条款，充分理解并表示同意。

依据中国有关法律法规，凡主动提交作品的"参赛者"或"作者"，主办方认为其已经对所提交的作品版权归属作如下不可撤销声明：

1. 原创声明

参赛作品是参赛者原创作品，未侵犯任何他人的任何专利、著作权、商标权及其他知识产权；该作品未在报纸、杂志、网站及其他媒体公开发表，未申请专利或进行版权登记，未参加过其他比赛，未以任何形式进入商业渠道。参赛者保证参赛作品终身不以同一作品形式参加其他的设计比赛或转让给他方。否则，主办单位将取消其参赛、入围与获奖资格，收回奖金、奖品及并保留追究法律责任的权利。

2. 参赛作品知识产权归属

为了更广泛推广竞赛成果，所有参赛作品除作者署名权以外的全部著作权归竞赛承办单位及冠名单位所有，包括但不限于以下方式行使著作权：享有对所属竞赛作品方案进行再设计、生产、销售、展示、出版和宣传的权利；享有自行使用、授权他人使用参赛作品用于实地建设的权利。竞赛主办方对所有参赛作品拥有展示和宣传等权利。其他任何单位和个人（包括参赛者本人）未经授权不得以任何形式对作品转让、复制、转载、传播、摘编、出版、发行、许可使用等。参赛者同意竞赛承办单位及冠名单位在使用参赛作品时将对其作者予以署名，同时对作品将按出版或建设的要求作技术性处理。参赛作品均不退还。

3. 参赛者应对所提交作品的著作权承担责任，凡由于参赛作品而引发的著作权属纠纷均应由作者本人负责。

声明：

1. 参与本次竞赛的活动各方(包括参赛者、评委和组委)，即表明已接受上述要求。

2. 本次竞赛的参赛者，须接受评委会的评审决定作为最终竞赛结果。

3. 组委会对竞赛活动具有最终的解释权。

4. 为维护参赛者的合法权益，主办方特提请参赛者对本办法的全部条款、特别是"所有权及版权"声明部分予以充分注意。

国际太阳能建筑设计竞赛组委会

网　址：www.isbdc.cn

The Trophy Cup, Certificate, and Bonus RMB 2,000 (before tax) will be awarded.

Finalist Award: 30 winners

The Certificate will be awarded.

Prize for Technical Excellence Works:

The quota is open-ended. The Certificate will be awarded.

Prize for Innovative Designs:

The quota is open-ended. The Certificate will be awarded.

PARTICIPATION REQUIREMENTS:

1. Institutes of architectural design, colleges and universities, research institutions, green building parts R&D and production enterprises, and other units are welcome to organize professional staff to team up for the competition.

2. Please visit www.isbdc.cn and follow the instruction to complete the registration form. After submitting the registry, you will get a registration number, which is also the number of your entry. One registration number for one entry. The registration number must be marked on the top left corner of each submitted work with a word height of 6mm. Registration time: March 26, 2022 - August 15, 2022.

3. Participants must agree that the organizing committee may publish, print, exhibit, and apply their works.

4. Authors whose works are compiled into the publication should cooperate with the organizing committee to adjust their works according to publication requirements.

IMPORTANT CONSIDERATION:

1. Electronic documents of all entries must be uploaded to the official website www.isbdc.cn before September 15, 2022. Other manners for presenting entries are invalid.

2. Entries shall not have any marking which can identify the participant. Otherwise, the entry shall be considered invalid.

3. The organizing committee will publicize the process and result of the competition online in time, compile and publish the winning entries. Winners will be honored and awarded.

4. Within 30 days after the first publication of the collection, the organizing committee will present the team of the winning entry with one winning portfolio (with an official commemorative seal).

5. Check the competition website for competition activities, news, and FAQs

ANNOUNCEMENT ABOUT OWNERSHIP AND COPYRIGHT:

Before submitting entries, participants must check the following terms with full understanding and consent.

By relevant laws and regulations, hosts deem that any "participant" or "author" who has submitted works has made the following irrevocable statement regarding the ownership and copyright of the submitted work.

1. Announcement of Originality

The entry must be original without infringement of any patent, copyright,

组委会联系地址：北京市西城区车公庄大街 19 号（100044）
国家住宅与居住环境工程技术研究中心

联系人：鞠晓磊、张星儿、郑晶茹

联系电话：86-010-88983377、86-010-88983383

E-mail：isbdc2021@126.com　　QQ 交流群：49266054

微信公众号：国际太阳能建筑设计竞赛

trademark, and other intellectual property. Ensure the entry has not been published in any other media, including newspapers, periodicals, magazines, and webs. The entry has not been patented or registered for copyright. It has not been involved in any other competition and has not been put in any commercial channels. The participant must assure that the work shall not be put in any other competition or transferred to others in the same form. Otherwise, sponsors will disqualify you from participating, being shortlisted, and being awarded, with all resulting consequences including forfeiture of prizes and awards. Hosts reserve the right of legal liability.

2. Ownership of Intellectual Property of Entries

To better promote competition results, all copyrights of entries except the right of authorship shall be enjoyed by the organizers and the title sponsor. Copyright owners may exercise their copyrights by ways including but not limited to the followings: exploiting works by redesigning, producing, selling, exhibiting, publishing, and publicizing; using works on construction for self-use or authorizing others to use. Hosts enjoy rights to display and publicize all entries. Without authorization, no unit or individual (including authors themselves) may transfer, copy, reprint, disseminate, excerpt, publish, distribute and license in any form. Organizers and the title sponsor are allowed to sign the author's name for using entries and do technical processing according to the requirements of publication and construction. All entries shall not be returned to the author.

3. Participants shall be held liable for the copyrights of their entries. Authors shall be liable for all disputes over copyrights

ANNOUNCEMENT：

1. By participating in this competition, all parties involved (including participants, judges, and the organizing committee) indicate that they have accepted the requirements above.

2. All participants must accept the judgment of the judging panel as the final result of the competition.

3. The organizing committee reserves the final explanation right of the competition.

4. In order to safeguard the legitimate rights and interests of the participants, hosts call attention to all clauses in this document, especially the part of "Announcement about Ownership and Copyright".

Organizing Committee of International Solar Building Design Competition
Website：www.isbdc.cn
Address of Organizing Committee：
No.19, Chegongzhuang Street, Xicheng District, Beijing, 100044
China National Engineering Research Center for Human Settlements
Contacts：Ju Xiaolei, Zhang Xing'er, Zheng Jingru
Tel.：86-010-88983377, 86-010-88983383
E-mail：isbdc2021@126.com　　QQ Group：49266054
WeChat Official Account：国际太阳能建筑设计竞赛（Guo Ji Tai Yang Neng Jian Zhu She Ji Jing Sai; International Solar Building Design Competition）

附件1：四川省平武县关坝沟流域自然保护小区科学考察站
Annex 1：Research Station in Guanbagou Basin Nature Mini-Reserve, Pingwu County, Sichuan Province

1. 项目背景

项目地位于四川省绵阳市平武县木皮藏族乡关坝村，北纬39°33′18″，东经104°33′58″紧邻G247国道，九绵高速出口位于G247国道与村口相接点的北侧4公里处。关坝村位于大熊猫国家公园的岷山片区大熊猫栖息地自然保护区，并为四川省森林自然教育基地和关坝沟流域自然保护小区。全村林地总面积6499.3亩，森林覆盖率为96.3%，区内有大熊猫、金丝猴、红豆杉、珙桐等国家保护珍稀动植物70余种，生物资源丰富，生态环境质量高，是绵阳市重要的生态腹地和水源涵养地。

2. 项目要求

该项目计划建设为四川省平武县关坝沟流域自然保护小区科学考察站，为来访客人、科考人员、研学学生提供工作和生活的场所。目前已建成一栋2层221.88m²的关坝村村委会办公楼和一栋3层572.25m²的带有5间住宿功能和公共卫生间的展览展示楼。一期计划结合现有建筑建设具有科考研学展示、接待住宿驿站、管护办公等功能的建筑群，二期利用既有农房用地，建设300m²的数字体验中心。

3. 气候条件

项目所在地位于木皮藏族乡中部，夏季凉爽多雨，冬季寒冷多雪，1月平均气温4℃，年平均气温13.9℃。年平均降雨量800mm。

4. 基础设施

项目所在地有基本给水、电力设施，无排水、天然气管道等，无集中供暖。

5. 竞赛场地

项目位于四川省绵阳市平武县木皮藏族乡关坝村村口附近，项目用地位于东偏南（高）向西偏北（低）走向的山谷地带，谷底小溪常年有水由东向西流入，项目建设用地位于小溪南侧，山谷南侧山势较为陡峭，北侧山势略微平缓。规划

1. Project Context

The project is located in Guanba Village, Mupi Tibetan Township, Pingwu County, Mianyang City, Sichuan Province. It's at longitude 104°33′58″ East and latitude 39°33′18″ North and is next to G247 National Highway. The Jiuzhaigou-Mianyang Highway exit is located 4 km north of the point where G247 National Highway meets the village entrance. Situated in the Minshan Mountain Giant Panda Habitat Nature Reserve of Giant Panda National Park, Guanba Village has been approved as the Guanbagou Basin Nature Mini-Reserve by Sichuan Forest Nature Education Base. The village has 6499.3 mu of woodland which accounts for a forest coverage rate of 96.3%. There are more than 70 species of rare plants and animals under state protection, such as giant pandas, golden snub-nosed monkeys, yews, and dove trees. Blessed with abundant biological resources as well as a high-quality ecosystem and environment, it is an important ecological hinterland and water conservation area in Mianyang.

2. Project Requirements

The project is scheduled to be built into Guanba Village Nature Reserve Research Station to provide working and living places for visitors, researchers, and field-trip students. At present, there are a 2-story 221.88m² office building of Guanba Village Committee and a 3-story 572.25m² exhibition and display building with 5 accommodations and a public restroom. The first phase of the project is planned to leverage the existing building to build a complex with functions such as research, field-trip display, reception, accommodation, management, protection, and office. The second phase will leverage the existing farmhouse to build a 300m² digital experience center.

3. Climate

The project is located in the central part of Mupi Tibetan Township. It is cool and rainy in summer, and it is cold and snowy in winter. Its yearly average temperature is 13.9℃, and the average temperature in January is 4℃. Rainfall averages 800mm a year.

4. Infrastructure

There are basic water supply and electricity supply facilities but no drainage, natural gas pipeline, or central heating system.

5. Site

The project is near the village entrance of Guanba Village, Mupi Tibetan

用地 10709m², 现有管护及村委办公楼、展示接待楼和民居等三栋建筑物，一期占地面积 2293.96m², 地势为周边高中间低的地形，综合考虑用地外既有的管护及村委办公楼和展示接待楼两栋建筑，建设自然保护小区科学考察站。既有民居占地 567.89m², 待房屋拆迁后用于二期建设用地。

图1 项目所在区位图
Figure 1　Project Location Map

图2 项目所在地俯视图
Figure 2　Vertical View of the Project Site

图3 规划用地范围
Figure 3　Scope of Planning Land

图4 项目建设用地平面图
Figure 4　Floor Map of the Land for Project Construction

图5 项目建设分期图
Figure 5　Map of Phased Construction

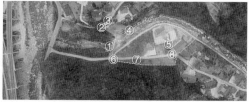

图6 用地及周边相机方位图
Figure 6　The Site and the Position and Direction of Surrounding Cameras

Township, Pingwu County, Mianyang City, Sichuan Province. This site is in a valley extending from the southeast (high) to the northwest (low). At the bottom of the valley, a stream flows from east to west all year round. The project site is in the south of the stream. The valley is relatively steep on the south side and slightly gentle on the north side. The planning land is 10709m². There are 3 existing buildings—an office building for forest management, protection, and the village committee; an exhibition and reception building; and a residential building. The first phase covers an area of 2293.96m², which is high all around and low in the middle. The nature reserve research station is to be built with overall consideration of the existing office building for forest management, protection, and village committee and the exhibition and reception building outside the planning land. The existing residential building is 567.89m², which will be used for the second phase of construction after the demolition and resettlement.

6. Design Requirements

1) Consider local geographical conditions and existing buildings such as the village committee and the exhibition and reception center. Design projects including the research station; the office building for forest management, protection, and the village committee; the reception accommodation; and the cultural square within the competition site. Optimize existing buildings according to the requirements of utilization functions, which can be found in the annex. The total floor area is 2100m² (including existing buildings). Design a parking lot for 1 bus and 5 small cars at the same time.

2) As the site for the second phase construction, the existing residential building and its site will be converted into a digital experience center. The digital experience center to be built in the second phase should be easily accessible to the reception center of the first phase. It should facilitate phased construction.

3) New buildings must be suitable for local terrains and the surrounding environment. Principally, buildings should be one floor only. Two floors are acceptable if necessary.

4) Material selection, construction methods, operation, and maintenance should be low-carbon and reduce emissions.

5) Take into account the local climate and nature. Combine characteristics of local construction materials and buildings. As the building needs residential heating in winter, while needs ventilation and insulation instead of cooling in summer, participants have to consider using active or passive solar energy technology and a low-carbon construction system.

图7 ①号相机图像
Figure 7 ① Camera Image

图8 ②号相机图像
Figure 8 ② Camera Image

图9 ③号相机图像
Figure 9 ③ Camera Image

图10 ④号相机图像
Figure 10 ④ Camera Image

图11 ⑥号相机图像
Figure 11 ⑥ Camera Image

图12 ⑦号相机图像
Figure 12 ⑦ Camera Image

图13 ⑧号相机图像（航拍）
Figure 13 ⑧ Camera Image（Aerial Picture）

图14 用地现状图
Figure 14 Status Map of the Site and its Surroundings

6. 设计要求

1）在给定的竞赛用地范围内，结合考虑地形地貌及现状的村委会和展览接待中心等既有建筑，设计自然保护小区科学考察站、管护办公及村委会、接待客栈及文化广场等项目，并结合使用功能要求对现状建筑的功能进行优化设计。具体要求见附表，总建筑面积 2100m²（含既有建筑）。设置可同时停放一辆大巴和五辆小车的停车场。

2）建设用地内的既有民居及所占用地，作为二期用地，用于数字体验中心的建设，二期建设的数字体验中心应与一期建设的接待中心联系便利，并利于分期建设。

3）新建建筑应与现状地形和周边环境有机结合，建筑层数以 1 层为主，必要时可局部 2 层。

4）材料选型、建造方法、运行维护全过程应考虑低碳减排的设计方法和建造措施。

5）考虑当地的气候特点和自然环境，结合当地的建筑材料和建筑特点，解决建筑的冬季需要采暖，夏天不需要制冷，强调通风隔热等特点，合理选用主、被动太阳能技术及低碳建造技术。

6）在规划用地之外，G247 国道与进村路连接处，因山体原因在南向北方向存在视线盲区，建议结合保护区标识设置的同时做相应处理。

6) Outside the planning site, the juncture of G247 National Highway and the road into the village has a blind spot in the south to north direction due to the mountain, so it is suggested to do the corresponding treatment while placing a nature reserve sign.

7) Set up ecological measures such as garbage disposal and sewage harmless treatment according to the supporting facilities in Gunaba Village.

8) Consider the economy and popularization of the project.

9) Function settings of the land in the community are in the table below:

Table of Function Settings and Land Use

Functional Space	Function	Quantity	Usable Area (m²)	Note
Visitors' Reception	Reception hall	1	80	The visitor reception buffer space such as the hall, which is adjacent to and interconnected with the visitors' rest area
Visitors' Rest Area	Rest area	1	80	Seating and resting areas for visitors to take a break. An extension of the buffer space next to the reception hall
	Break room	1	20	An area for providing boiled water and selling beverages, tea, or other drinks

7）建议结合关坝村市政配套设施，合理选择相应的垃圾处理和污水无害化处理等生态环保的应对措施。

8）考虑项目的经济性和可推广性。

9）竞赛用地及功能设置表如下表所示：

用地及功能设置表

功能空间	内容要求	数量	使用面积（m²）	备注
来访接待	接待大厅	1	80	门厅等来访人员接待缓冲空间，与访客小憩空间相邻互通
访客小憩	休息区	1	80	为到访人员小憩提供座椅休息区，与接待大厅相邻、扩展集散缓冲空间
	水吧	1	20	可提供开水并销售饮料、茶水等
	卫生间	1	30	综合考虑访客集中到达、平时人员较少的使用特点
	特产展售	1	40	展示及销售本地区特色产品
	储藏室	1	30	特产库房
展览展示	实物、展板、影像	1	300	通过实物、展板、影像等方式宣传展示自然保护区的自然和生态人文地理
研学教室	教学互动	2	120	为来访、科学考察、研学等人员、提供工作与交流的空间，空间要求可合并为一个大空间，共可最多容纳60人
	储藏室	1	30	为研学教室的多功能使用提供一定的收纳储藏空间
办公接待	办公室	2	40	两间行政管理办公空间
	洽谈室	1	30	12人小型会议室，用于内部接待洽谈
其他	走道等辅助空间		100	用于室内交通等功能
管护办公	办公室	4	140	保护区及村办公用空间
	值班室	1	15	24小时值班，与监控室相邻
	监控室	1	15	接待中心及村域范围视频监控
	会议室	1	30	办公行政会议等
	走道等辅助空间		25	门廊、外廊楼梯等

续表

Functional Space	Function	Quantity	Usable Area (m²)	Note
Visitors' Rest Area	Restroom	1	30	Sometimes, crowds of visitors come here at the same time, while usually there are few people
	Exhibition and selling of featured products	1	40	Exhibition and selling of local featured products
	Storeroom	1	30	Stores of featured products
Exhibition and Display	Objects, boards, and images	1	300	Publicizing and showing the nature and eco-human geography in the nature reserve through objects, boards, and images
Research Classroom	Teaching and interaction	2	120	Rooms of working and communicating for visitors, researchers, and students. Rooms can be put into a big one with a capacity of 60 people
	Storeroom	1	30	Certain storage space for the multiple functions of research classrooms
Office and Reception	Office	2	40	Two rooms for the administrative office
	Negotiating room	1	30	12-person small meeting room for internal reception and negotiation
Others	Auxiliary space such as corridors		100	For indoor movement and other functions
Management, Protection, and Office	Office	4	140	Offices for the reserve and the village
	Watchroom	1	15	24 hours on duty. Next to the monitoring room
	Monitoring room	1	15	Video surveillance covering the reception center and the village
	Meeting room	1	30	A room for office, administration, meetings, etc
	Auxiliary space such as corridors		25	Porches, stairs of verandas, etc
Accommodation and Reception	Guest Rooms	10	240	10 rooms for the reception and accommodation of visitors, researchers, students, and other people
	Reception	1	10	Hall, counter distributing space
	Auxiliary rooms	2	10	Supporting service rooms, storage rooms, etc

续表

功能空间	内容要求	数量	使用面积（m²）	备注
住宿接待	客房	10	240	来访、科考、研学等人员的住宿接待，提供客房10间
	接待	1	10	门厅、柜台集散空间
	辅助	2	10	配套服务用房、库房等
食堂厨房	餐厅	1	90	最多容纳50人同时用餐餐厅
	厨房	1	35	最多提供50人同时用餐厨房
其他	走道等辅助空间		160	楼梯走廊、库房、洗衣房（含自助）等空间
公共厕所		1	30	服务于室外公共区域
数字互动	VR数字体验、辅助用房	1	300	二期用地
合计			2000	

7. 评比办法

1）由组委会审查参赛资格，并确定入围作品；

2）由评委会评选出竞赛获奖作品。

8. 评比标准

1）参赛作品须符合本竞赛"作品要求"的内容；

2）作品应具有原创性，鼓励创新；

3）作品应满足使用功能、绿色低碳、安全健康的要求，建筑技术与太阳能利用技术具有适配性；

4）作品应充分体现太阳能利用技术对降低建筑使用能耗的作用，在经济、技术层面具有可实施性。

评比指标	指标说明	分值
规划与建筑设计	规划布局、建筑空间组合、功能流线组织、建筑艺术	50
被动太阳能采暖技术	利用建筑设计与建筑构造实现建筑采暖与蓄热	25
主动太阳能利用技术	太阳能光伏、光热等主动太阳能技术的利用	10

续表

Functional Space	Function	Quantity	Usable Area (m²)	Note
Restaurant and Kitchen	Restaurant	1	90	Restaurant for up to 50 people at the same time
	Kitchen	1	35	Kitchen for up to 50 people at the same time
Others	Auxiliary space such as corridors		160	Staircase corridor, storage room, laundry room (including laundromat), and other spaces
Public Restroom		1	30	In outdoor public area
Digital Interaction	VR digital experience, auxiliary room	1	300	The site for the second phase of construction
Total			2000	

7. Appraisal Methods

1) The organizing committee will check up all entries and select shortlist entries.

2) The judging panel will select out winning entries.

8. Appraisal Standards

1) Entries must meet the demands of the "Requirements for Works".

2) Entries should be original and innovative.

3) Outcomes should be available, green, low-carbon, safe, and healthy. Construction technology and solar energy technology must be compatible.

4) Entries should reflect the solar energy technology's role in reducing building energy consumption. They must be economically and technologically feasible.

Appraisal Indicator	Explanation	Score
Planning and architecture design	Planning design; architectural space combination; functional division and traffic flow; and architectural art	50
Passive solar heating technology	Ensure heating and thermal storage of buildings through design and structure	25
Active solar utilization technology	Utilize and transmit solar energy through solar photovoltaic systems and collectors	10
Other technologies	Build and operate the community with green, low-carbon, safe, and healthy technologies	15
Operability	Feasibility, economy, and popularity of works	10

9. Drawing Requirements

1) Entries should meet the project requirement of design depth. Main technologies should have relevant technical drawings and indicators. Drawings and text should be clear and readable with accurate data.

续表

评比指标	指标说明	分值
采用的其他技术	建造与运行过程中的绿色、低碳、安全、健康技术	15
可操作性	作品的可实施性，技术的经济性和普适性	10

9. 图纸要求

1）设计深度达到方案设计深度要求，主要技术应有相关的技术图纸和指标。作品图面、文字表达清楚，数据准确。

2）需提交方案设计说明，应包括方案构思、太阳能技术、低碳技术与设计创新（限200字以内）、技术经济指标表。

3）提交作品需进行竞赛用地范围内的规划设计，提供总平面图（含活动场地及环境设计）。

4）提交作品需体现规划用地范围内的新建建筑、既有建筑、景观布局及交通流线组织，与项目周边自然环境和村庄的关系，进村道路与G247国道连接点等部位的标志与交通组织的处理。

5）充分表达建筑与室内外环境关系的平面图、立面图、剖面图，比例不小于1:200。

6）能表现出技术与建筑结合的重点部位、局部详图及节点大样，比例自定；低碳及其他相关的技术图、分析图。

7）整体社区、地块、建筑单体效果表现图。

10. 文字要求

1）"建筑方案设计说明"采用中英文双语，其他为英文（建议使用附件2中提供的专业术语）。

2）排版要求：A1展板（594mm×841mm）区域内，统一采用竖向构图，作品张数应为4或6张（偶数）；中文字体不小于6mm，英文字体不小于4mm。

3）文件分辨率300 dpi，格式为JPG或PDF文件。

4）提交参赛者信息表，格式为JPG或PDF文件。

5）上传方式：参赛者通过竞赛网页上传功能将作品递交竞赛组委会，入围作品由组委会统一编辑板眉、出图、制作展板。

2) Submit a schematic design description containing design concepts, solar energy technology, low-carbon technology, innovative design (less than 200 words), and technical and economic indicators.

3) Provide a planning design within the outline of the competition site and a floor plan (including the venue/environment design).

4) Entries should reflect the relationship of new buildings, existing buildings, landscape layout, and traffic flow organization within the planning site, with the natural environment and the village around the project. Entries should also show the signs and traffic organization of the juncture of the road into the village and the G247 National Highway.

5) Provide floor plans, elevations, and sections with a scale not less than 1:200, which can fully express the relationship between the architecture and the indoor and outdoor environment.

6) Show the key parts of the combination of technology and architecture, details, and magnifying detail drawings with self-defined scale; low carbon and other related technical drawings and analysis drawings.

7) Render perspective drawing of community, land, and single building

10. Text Requirements

1) The submission should be in English (technical terms in Annex 3 are recommended), in addition to "architectural schematic design description" in both English and Chinese.

2) Typesetting Requirements: Entries should be put into 4 or 6 (even numbers) exhibition panels, each 594mm × 841mm (A1 format) in size (arranged vertically). Word height of Chinese is not less than 6mm and that of English is not less than 4mm.

3) File resolution: 300 dpi in JPG or PDF format.

4) Information tables of participants should also be submitted in JPG or PDF format.

5) Uploading: Entries should be submitted to the organizing committee through the competition's official website. Shortlist works will be compiled, printed, and made into exhibition panels by the organizing committee.

村委会、管护站照片（道路方向拍摄）　　村委会、管护站南侧照片（展示接待中心方向拍摄）　　展示接待中心西北侧照片（村委会方向拍摄）　　展示接待中心东北侧照片（小溪方向拍摄）

村委会、管护站南侧照片（门前空地方向拍摄）　　村委会、管护站北侧照片（小溪方向拍摄）　　展示接待中心西南侧照片（道路方向拍摄）　　用地鸟瞰图

附件 2：专业术语
Annex 2: Professional Glossary

百叶通风	— shutter ventilation
保温	— thermal insulation
被动太阳能利用	— passive solar energy utilization
敞开系统	— open system
除湿系统	— dehumidification system
储热器	— thermal storage
储水量	— water storage capacity
穿堂风	— through-draught
窗墙面积比	— area ratio of window to wall
次入口	— secondary entrance
导热系数	— thermal conductivity
低能耗	— lower energy consumption
低温热水地板辐射供暖	— low temperature hot water floor radiant heating
地板辐射采暖	— floor panel heating
地面层	— ground layer
额定工作压力	— nominal working pressure
防潮层	— wetproof layer
防冻	— freeze protection
防水层	— waterproof layer
分户热计量	— household-based heat metering
分离式系统	— remote storage system
风速分布	— wind speed distribution
封闭系统	— closed system
辅助热源	— auxiliary thermal source
辅助入口	— accessory entrance
隔热层	— heat insulating layer
隔热窗户	— heat insulation window
跟踪集热器	— tracking collector
光伏发电系统	— photovoltaic system
光伏幕墙	— PV façade
回流系统	— drainback system
回收年限	— payback time
集热器瞬时效率	— instantaneous collector efficiency
集热器阵列	— collector array
集中供暖	— central heating
间接系统	— indirect system
建筑节能率	— building energy saving rate
建筑密度	— building density
建筑面积	— building area
建筑物耗热量指标	— index of building heat loss
节能措施	— energy saving method
节能量	— quantity of energy saving
紧凑式太阳热水器	— close-coupled solar water heater
经济分析	— economic analysis
卷帘外遮阳系统	— roller shutter sun shading system
空气集热器	— air collector
空气质量检测	— air quality test (AQT)
立体绿化	— tridimensional virescence
绿地率	— greening rate
毛细管辐射	— capillary radiation
木工修理室	— repairing room for woodworker
耐用指标	— permanent index
能量储存和回收系统	— energy storage & heat recovery system
平屋面	— plane roof
坡屋面	— sloping roof
强制循环系统	— forced circulation system
热泵供暖	— heat pump heat supply
热量计量装置	— heat metering device
热稳定性	— thermal stability
热效率曲线	— thermal efficiency curve
热压	— thermal pressure
人工湿地效应	— artificial marsh effect
日照标准	— insolation standard
容积率	— floor area ratio
三联供	— triple co-generation
设计使用年限	— design working life
使用面积	— usable area
室内舒适度	— indoor comfort level
双层幕墙	— double facade building

中文	English	中文	English
太阳方位角	— solar azimuth	屋面隔热系统	— roof insulation system
太阳房	— solar house	相变材料	— phase change material (PCM)
太阳辐射热	— solar radiant heat	相变太阳能系统	— phase change solar system
太阳辐射热吸收系数	— absorptance for solar radiation	相变蓄热	— phase change thermal storage
太阳高度角	— solar altitude	蓄热特性	— thermal storage characteristic
太阳能保证率	— solar fraction	雨水收集	— rain water collection
太阳能带辅助热源系统	— solar plus supplementary system	运动场地	— schoolyard
太阳能电池	— solar cell	遮阳系数	— sunshading coefficient
太阳能集热器	— solar collector	直接系统	— direct system
太阳能驱动吸附式制冷	— solar driven desiccant evaporative cooling	值班室	— duty room
太阳能驱动吸收式制冷	— solar driven absorption cooling	智能建筑控制系统	— building intelligent control system
太阳能热水器	— solar water heating	中庭采光	— atrium lighting
太阳能烟囱	— solar chimney	主入口	— main entrance
太阳能预热系统	— solar preheat system	贮热水箱	— heat storage tank
太阳墙	— solar wall	准备室	— preparation room
填充层	— fill up layer	准稳态	— quasi-steady state
通风模拟	— ventilation simulation	自然通风	— natural ventilation
外窗隔热系统	— external windows insulation system	自然循环系统	— natural circulation system
温差控制器	— differential temperature controller	自行车棚	— bike parking
屋顶植被	— roof planting		